Shaping the Future of Electric Machines

Alexey Matveev

Shaping the Future of Electric Machines

Springer

Alexey Matveev
Trondheim, Norway

ISBN 978-3-031-95247-0 ISBN 978-3-031-95248-7 (eBook)
https://doi.org/10.1007/978-3-031-95248-7

This Springer imprint is published by the registered company Springer Nature Switzerland AG
The registered company address is: Gewerbestrasse 11, 6330 Cham, Switzerland

Introduction

The book proposed to the reader investigates a relatively underexplored topic—the management of the evolution of electric machines. Typically, the word "evolution" is associated with history, but this book is not about history; it is about shaping the future. While classical books on electric machines and Electromechanics usually provide a predominantly "static" picture of theory and practice in the field of electric machines, showing what has been and what is, this book offers a "dynamic" perspective—indicating possible changes and development potential.

To some extent, this book is an attempt to reassess the material known to professionals, providing a new perspective and introducing new tools for analysis. Some proposals may even seem radical, as they challenge dominant narratives and practices in modern Electromechanics—particularly in areas such as machine classification and comparison.

One of the main goals of this book is to encourage sophisticated and free thinking in the field of Electromechanics. This is achievable through a comprehension of reality with the help of concepts from the philosophy of technology, independent synthesis of knowledge about trends and patterns of evolution, and inventive practices.

The book is structured as follows:

In Chap. 1, electric machines are presented as objects of the technosphere, their functionality, purpose, dimensions, materials from which electric machines are produced, and the fundamental laws governing their operation are briefly described. The mechanism of *informational selection*, acting in the process of techno-evolution, along with its criteria, is outlined. The specifics of some popular applications of electric machines, such as renewable power generation, propulsion systems of marine vessels, and aerial vehicles, are discussed and the selection process in these applications is demonstrated, utilizing the corresponding key performance indicators (criteria). Given that weight and size are among the most important criteria, attention is paid to the empirically derived law describing the "pre-determination" of machine mass through its power and rotational speed. Universal expressions for estimating the mass and dimensions of electric machines are derived based on this law. Based on these expressions, a method for comparing machines is proposed as one of the tools for informational selection. Attention is also paid to the cost of the final product, as one of the most important criteria, and examples of cost models are provided.

One of the key ideas underpinning this book is that the evolution of electric machines follows certain trends and patterns, and to some extent, the evolutionary process can be managed. In Chap. 2, the driving forces and directions of electric machines evolution are discussed. The role of major inventions and minor improvements is highlighted, and examples of evolutionary trends and patterns are provided. The evolution of electric machines is described in the context of the "supersystem–system–subsystem" axes. Examples of the evolution of materials used in electric machines are examined, and trends in the development of design and manufacturing methods are outlined. Specific examples of the directions of electric machines evolution are presented, along with discussions on opposing forces and limitations.

In Chap. 3, the central question of managing the evolutionary process is posed. The main actors in the evolutionary process and their roles are examined. The role of universities, companies, investors, and governmental and supranational bodies in the evolution of electric machines is discussed. Technological platforms and management of key performance indicators within the same platform are considered. Inventors are separately introduced as the key players in the evolutionary process, and their psychology is explored. Historical examples of inventions are discussed. The use of intellectual property (IP) in industrial competition is demonstrated, as well as the role of IP in the development of electric machines.

Rebuilding Electromechanics for the coming technological era may require seemingly radical steps, and certain things will have to be reconsidered. Chapter 4 proposes the new classification of electric machines, based on a set of features different from the conventional one. The new approach appears more useful for inventors working in the field of electric machines and for the application of artificial intelligence aimed at creating new types and configurations of electric machines. Also, in Chap. 4, the concept of organizing and transmitting knowledge about electric machines is presented as a set of logical blocks, graphically represented by a pentagram—a well-known symbol and geometric figure. The pentagram demonstrates the connectivity and interdependence of processes in electric machines and can serve as a navigator when creating new electric machines.

In the process of managing evolution, specialized thinking methodologies will undoubtedly be necessary. Chapter 5 demonstrates the application of the theory of inventive problem solving (TRIZ) in electromechanical problems. The minimum necessary toolkit for independent work in this direction is provided. As supplementary information, system resources are described (using a specific example), and the utilization of physical and chemical effects occurring in electric machines is showcased. Techniques for predicting the development of electric machines are briefly reviewed.

To discover one's own vision of the results of the evolutionary process and to define one's place in the process, it is necessary to go beyond the boundaries of technical sciences and turn to the humanities, primarily to philosophy. In Chap. 6, philosophical questions are posed regarding the sources of knowledge, the philosophical essence of electric machines, the uniqueness of Electromechanics, and several other questions. Ideas of various philosophers of technology related to these questions are examined. An attempt is made to comprehend the modern situation and "integrate" Electromechanics into the context of ideas and concepts of the philosophy of technology. The synthesis of ideas from various philosophers is undertaken, and an attempt is made to propose elements of a worldview into which the professional activities of electrical engineers can fit. Several hypotheses are put forward. Based on the proposed worldview and the formulated hypotheses, answers to the posed questions are provided, and tasks for electrical engineers are outlined. Theses for discussion, proposed for inclusion in a special section of the philosophy of technology—*the philosophy of Electromechanics*—are presented.

Author believes that participation in the management of the evolution of electric machines is possible based on the three blocks of acquired knowledge and concepts detailed in this book:

- Knowledge of evolutionary trends and patterns for electric machines and the mechanisms of informational selection will help to build a picture of objective reality with regard to the development of electric machines.
- Engineering creativity methodologies, such as TRIZ, operating within the space of a new classification of electric machines, will assist in creating new solutions in the field of electric machines and related systems.
- The philosophy of technology will provide ethical reference points, teach paradoxical and complex thinking, help recognize one's role in the present and future, develop a personal strategy for participation in the evolutionary process, and create a new reality for Electromechanics.

In conclusion to the preface, it is worth noting that the material in this book is a kind of "Lego set" that each reader will assemble in their own way.

Competing Interests The author has no competing interests to declare that are relevant to the content of this manuscript.

Contents

List of Symbols and Abbreviations

Symbols

A	Availability
a	Transportation cost
C_{tm}	Time and financial costs to meet demands
C_{ey}	Annual cost of consumed energy
C_{Σ}	Total costs
C_e	Cost of consumed energy
C_M	Machine constant
C_p	Power coefficient of the machine constant relative to power
C_n	Power coefficient of the machine constant relative to rotational speed
CO_2	Annual carbon dioxide emissions
CoE	Normalized cost of electricity
D	Diameter
D_{δ}	Rotor diameter
D_{tr}	Distance over which passengers are transported
E_{Σ}	Total energy generated
E_t	Annual energy production in year t
H	Harmful actions ("repayment factors")
I	Degree of ideality of demands
IC	Investment costs
IC_t	Investment costs in year t
i, j, k, l	Indexes of variables
$k, k_1, k_2, k_3, k_4, k_5$	Coefficients
K_t	Machine type factor
K_e	Efficiency level factor
K_c	Cooling factor
K_u	Voltage factor
K_{fp}	Frequency and pole factor
K_s	Shape factor
K_i	Structural components factor
K_o	Overload factor

K_x	Other (miscellaneous) factors
L	Length
l	Base dimension
l_δ	Magnetic core (stack) length
T	Torque
T'	Normalized torque or specific torque
T_{max}	Maximum torque
m	Mass
N	Number of trips per year
N_p	Number of transported passengers
lt	System lifecycle
n	Rotational speed
n_{max}	Maximum rotational speed
n'	Normalized rotational speed
$OpEx$	Operating expenses over the pump lifecycle
$OpEx_t$	Operating and maintenance costs in year t
P	Power
P'	Normalized power or specific power
Qn	Quantity of demands
Ql	Quality of demands
r	Discount rate
T_w	Time during which the system fully functions as intended
T_s	Time during which the system is unavailable or nonfunctional
U	Voltage
V	Volume
α, β	Adjustment coefficients
ϵ	Price
$\propto$	Coefficient

Abbreviations

DFIM	Doubly-fed induction machine
EM	Electric machine
HF PMSM	High-frequency permanent magnet synchronous machine
HTSC	High-temperature superconductors
SCIM	Squirrel cage induction machine
TS	Technical system

1 Criteria for Evaluation and Selection Mechanisms in the Process of Evolution

Electric Machines as Objects of Technosphere[1] and Universe

This book is for those who know what electrical machines are and are familiar with classic books on this subject, for example, [1]. We will not repeat the material that is well worked out in classical literature. Instead, in the first chapter we will try to look at electrical machines not from the perspective of the development of specific technology, applications, or competition of industrial players for some market niches, but as the part of Technosphere of our planet. For our research, in the following chapters, it will be suggested to sometimes look at the research subject from different perspectives and imagining ourselves not only in the role of an engineer or a scientist, but also, for example, as philosophers, inventors, etc. So, let's assume that we have decided to explain what electrical machines are to someone who has never thought about them, doesn't know what they are, or, at best, has only a vague idea—let's imagine a child or, even better, a resident of an extraterrestrial civilization. We will use an approach known since ancient times—almost any object can be characterized by the following attributes: appearance, functions, and operating principles, size, the materials from which the object is made, its behavior, impact on humans, relationship with the surrounding world, where the object comes from, and where it disappears to. We will go through this list and at the same time test the strength of some well-known truths in the world of Electromechanics. In the end, we will obtain one of many possible descriptions of electrical machines as objects of the Technosphere.

1. **Appearance and Shape.** Rotating machines[2] can have a shape close to that of a cylinder or a disk, sometimes a rectangular parallelepiped. For example, Figure 1.1a shows a motor (in cross-section) designed to drive a pump in an oil

[1] Technosphere—part of the biosphere transformed using technical means for socio-economic purposes.

[2] Sometimes instead of "electrical machine" we write just "machine" or use abbreviation "EM" to make the sentences more compact.

A. Matveev, *Shaping the Future of Electric Machines*,
https://doi.org/10.1007/978-3-031-95248-7_1

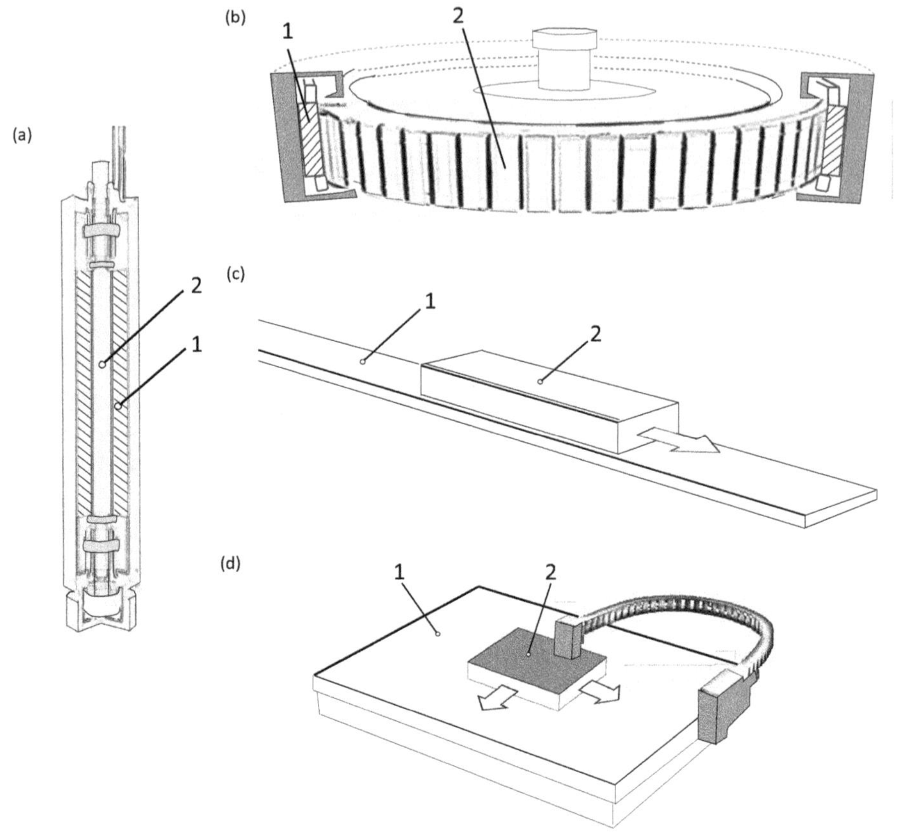

Fig. 1.1 Examples of electrical machines of various shapes (1 - stationary part, 2 - moving part)

well. The elongated shape of the motor is due to the limited space within the well. Figure 1.1b shows a generator to be driven by a hydro turbine. Its shape, with a large diameter and short axial length, is coming from the need for high torque at low rotational speeds. Linear and planar machines typically have a rectangular parallelepiped shape. Figure 1.1c shows one variant of a linear machine, while Figure 1.1d shows a planar one. When a machine is built into another object, its shape can sometimes be difficult to determine. Overall, electric machines can have a wide variety of structural designs.

2. **Function**. Electric machines are energy converters, and this is their primary human-assigned function. Based on the purpose for which the machines are created, they are typically viewed as converters of electrical energy into mechanical energy (motors) or, conversely, mechanical energy into electrical (generators). According to the Second Law of Electromechanics, all electrical machines are reversible [2], meaning that, in principle, any machine can function both as a motor and as a generator.

Let's imagine a group of alien researchers who have advanced means for measuring physical quantities, decided to test this! Indeed, if one connects an electrical load (a group of resistors) to the stator terminals of a synchronous machine with permanent magnets and a cage winding on the rotor, and then begins to spin the shaft, current will be generated in the stator circuit—demonstrating generator mode. If one connects the stator circuit of this same machine to an AC power source, the machine's shaft will begin to rotate—demonstrating motor mode.

However, if one connects the load to the stator terminals of an induction machine with a squirrel-cage rotor and start spinning the shaft, no current will appear in the stator circuit! The same is true, for example, for a synchronous reluctance machine. We know that in these last two cases, additional measures are needed for the operation in the generator mode to take place, but the alien researchers might draw incorrect conclusions about the reversibility of electric machines. This means that, in principle, the machines are reversible, but in practice, this is sometimes only realized under specific conditions.

Let's return to the types of converted energies. Electrical machines are otherwise called electromechanical converters, reflecting in the name only two types of energy—electrical and mechanical. As is well known, according to the First Law of Electromechanics [2], all electric machines have an efficiency of less than 100%, meaning that part of the energy is lost, released as heat and mechanical vibrations. The energy flow diagrams for motor mode (a) and generator mode (b) are shown in Fig. 1.2. Thus, in any mode, an electric machine also acts as a generator of thermal energy and mechanical vibrations.

It's interesting that an electric motor operating with an efficiency below 50%, and thus producing more heat losses than mechanical energy, could be identified by an extraterrestrial researcher as an electrothermal converter rather than an electromechanical one! This line of reasoning will be useful to us in Chap. 5 when we discuss forecasting the electric machine evolution.

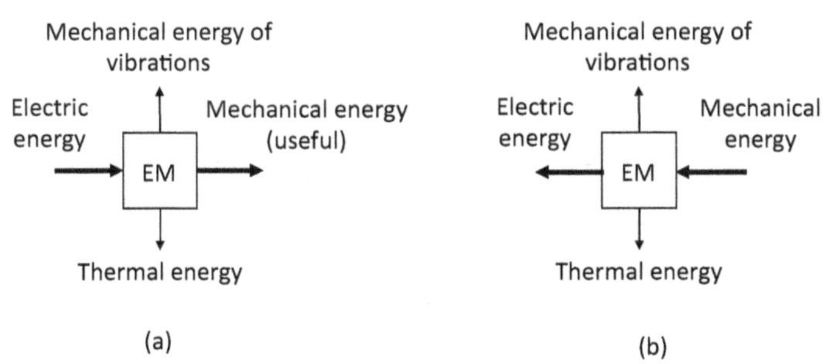

Fig. 1.2 Energy flow diagrams in electric machine: (**a**) motor mode, (**b**) generator mode

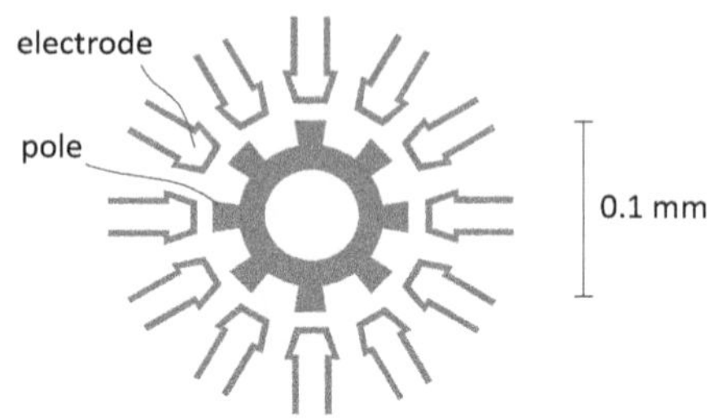

Fig. 1.3 MEMS motor concept and size

3. **Quantitative Characteristics.** As quantitative characteristics, one can take size and amount of energy being converted. Let's first note that electric machines can be either natural (occurring in nature) or "artificial"—created by humans. Traditional human-made machines, used for power generation, in industry, or in the infrastructure of modern daily life, can have power ratings ranging from fractions of a watt to about 2 gigawatts. Their sizes (the largest dimension of the machine—either diameter or length) range from roughly 1 mm to 20 meters. In other words, electric machines vary greatly in terms of both size and energy they convert. For comparison with other energy converters created by man, the following example can be given: the most powerful rocket engine has a power of about 30 GW, that is, an order of magnitude more than the most powerful electric machine.

   There is a class of devices, also created by humans, that are smaller than the smallest traditional electric machines—these are microelectromechanical systems (MEMS), which integrate interconnected mechanical and electrical components on a microscopic scale. MEMS consist of components sized from 1 to 100 micrometers (i.e., from 0.001 to 0.1 mm). The overall size of MEMS devices ranges from 0.02 to 1 mm. An example of a MEMS motor concept is shown in Fig. 1.3. Actuation of the motor happens by creation of electrostatic forces between electrodes of the stator and poles of the rotor by sequential excitation of the electrodes.

   Finally, the smallest man-made "molecular" electric motor measures just 200 nm.

   For comparison, let's consider the extreme sizes and power of natural electric machines:

   - The smallest natural "bacterial" electric motor has a diameter of about 45 nm. The power of a bacterial flagellum is about 10^{-17} watts.
   - On the other end of the spectrum is the planet Earth, which, according to I.P. Kopylov [3] and several other scientists, is a giant electric machine of the unipolar-type. The Earth's diameter is 12,600 km, and its power output is approximately 6.5×10^{29} watts.
   - The positioning of both natural and man-made electric machine on a logarithmic scale of sizes is shown in Fig. 1.4.

4. **Materials from Which the Object Is Made.** Primarily, metals are used for the production of electrical machines, although polymers, plastics, and composite materials are also used to a comparatively lesser extent. Perhaps the only abso-

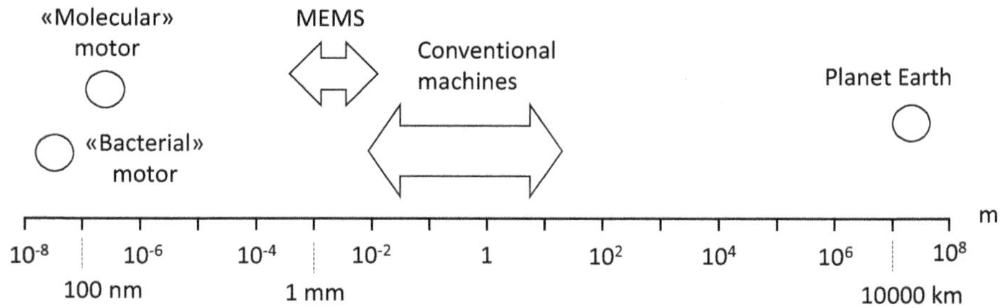

Fig. 1.4 Sizes of various types of electrical machines on a logarithmic scale

lutely necessary component of an electric machine is the conductors for carrying electric current, as energy conversion is impossible without electrical currents. The other components (magnetic cores, shafts, housings, etc.) are optional.

5. **Relationship with the Surrounding World, Impact on the Environment and Humans.** Electrical machines are "integrated" into the world through various systems that are powered by the machines—pumps, fans, conveyors, lifting mechanisms, etc.—or through systems that drive electric generators. Overall, electric machines are part of a hierarchy of "supersystems-systems-subsystems" (discussed in Chap. 4). Their impact on humans and the environment (beyond their primary function) can occur through the generated heat, electromagnetic radiation, vibrations, and acoustic noise. In emergency situations, such as short circuits, the machines can pose a danger to human life and health.

   Electric machines can operate in a vacuum, under high pressures, in aggressive environments, and at temperatures ranging from absolute zero to hundreds of degrees Celsius.
6. **Emergence and Disappearance.** This book deals mainly with the machines created by humans. It is obvious that it is people who invent, design, build, operate, and then decommission and dispose the machines after the end of their service life. But one may ask: who exactly decides *which* electric machines should appear or disappear? One of the possible answers to this question is in the next section.

Informational Selection

Techno-evolution is a directed gradual and natural change in the types of products over a series of generations. As one of the types of technical products, electric machines are part of the process of techno-evolution. One of the key mechanisms of techno-evolution is *informational selection* (a term used by B.I. Kudrin[3] in [4]). Informational selection operating in the Technosphere has common features with natural selection operating in the biosphere, based on variability, heredity, and

[3] The concept proposed in this book differs from the concept proposed by Kudrin.

selection. The mechanisms of informational selection will be described below, highlighting their similarities and differences with natural selection.

In the modern economy, there is a growing number and variety of manufactured products, including electrical machines. Even the product lines of a single major manufacturer exhibit significant diversity. One reason for the increase in the variety of types is the optimization of electric machines for specific, sometimes very narrow, applications.

The basic unit of techno-evolution is a group of products (individuals) of the same type, occupying a space with defined boundaries. We will refer to such a group as a "population." For each population of electric machines as a "species" (described as series, type rating), the limits of all physical and chemical factors (such as load ranges, environment—temperature, humidity, pressure, etc.) can be determined. As a result, an n-dimensional volume can be described—an ecological niche, fixed in space and time.

In biology, it is established that more organisms of each species are born than can find sustenance, survive, and reproduce. Similarly, in technology, more product types are developed than there are available ecological niches. If two machine types begin to compete for the same niche, the more competitive type will displace the other.

The variability of product types in the process of techno-evolution is obvious. Every product is created according to documentation. An electric machine fundamentally cannot be created without documentation (drawings, calculations, technical specifications, software code, etc.). Any document undergoes changes, and the evolution of a product type is a consequence of changes in its documentation.

The following definitions can be given, borrowing terms from biology:

- Genotype—the design of a product as described in its documentation; the totality of all documents that define the product.
- Phenotype—the realized set of characteristics of the product; a functional, fine-tuned product (the realized genotype).

The creation of a new product's genotype (through modification of the existing or development of new documentation), which leads to the emergence of a new product type, is analogous to the mutation of genetic information in biological systems.

The fundamental difference between natural selection and informational selection lies in the fact that the destruction of a biological organism simultaneously results in the destruction of its genetic information, while survival leads to the reproduction of a similar organism and identical genetic information. In contrast, the destruction or preservation of a product has no direct impact on the documentation—its "genetic information" [5]. Thus, we highlight an important distinction: in the biosphere, genotype and phenotype are unified, whereas in the Technosphere, they are separated.

Products that emerge as a result of attempts to implement new ideas compete to enter specific *communities* of technical products, where the number of ecological niches is limited. Populations of products that possess characteristics enabling them

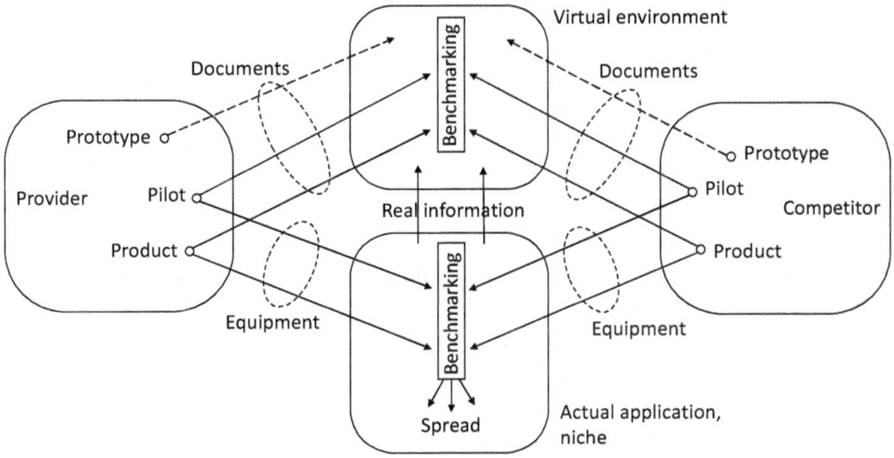

Fig. 1.5 Diagram of informational selection for electrical machines

to occupy new or redistribute existing ecological niches in their favor will survive. In the Technosphere, both genotypes (documentation and design) and phenotypes (realized products) engage in the "struggle for existence" having limited material and energy resources. The mechanism of informational selection is illustrated in Fig. 1.5.

On the path to its ecological niche, a product goes through several stages. The first step is prototyping, which represents the "internal loop" of development, confined within the company or research center. The finished prototype typically differs from what was originally specified in the documentation, necessitating further refinement, testing, and validation. Even at this stage, it is possible to discuss the process of transforming hereditary information into phenotypic characteristics, which is reflected in the manifestation of the product's individuality (e.g., assigned names or numbers).

If successful, the next step is to install the electric machine in a real environment, often referred to as a pilot project. Here, the product first enters the ecosystem and undergoes direct comparison with other types of machines. If experience of the operation is successful, serial, mass production begins. The new type of electric machine becomes a "product." At the stage of transitioning to mass production, all improvements are documented and incorporated into production, installation, commissioning, maintenance, repair, and disposal programs.

In the modern economy, the primary process of informational selection is the competition among marketing and sales departments of the electric machines providers, which often takes place within the procurement and engineering departments of integrators, original equipment manufacturers, or end-users. This process is usually not perfect and does not solely rely on technical and economic parameters; political factors at various levels can play a role, errors can occur, and even manipulations may take place. For instance, a test report or marketing information with inaccurate details can influence the selection outcomes.

In the process of informational selection, there is a repeated creation of provisional information—*opinions* (at various organizational levels) about whether a product is functional or needs refinement, whether it is "better" or "worse," and so on. Opinions are formed at all stages of development of a new population (prototypes, pilot projects/installations, products). Cycle after cycle, informational selection is implemented in the form of the documentation of a set of the "better-worse" opinions.

In the technical environment, as in biocenoses, competitive relationships are most strongly manifested near equilibrium: during the development of a new market or application or the creation of an industrial enterprise, the range of product combinations might be broader. In contrast, for established markets, the introduction of a new type of electric machines is more challenging, facing greater resistance and potentially negative outcomes even for the best machine options due to system conservatism and high entry barriers.

It is important to note that regardless of how significant or insignificant the changes made to prototypes or pilot installations in a real working environment may be, they will not be reproduced if they are not documented, as only genetic changes are inherited. *Heredity* is a mandatory feature of informational selection: every document contains *hereditary* information, materialized experience from previous generations. The selection of genotypes leads to the realization of phenotypes.

One of the elements of securing genetic information is achieving agreement throughout the entire supply chain, from sub-suppliers to consumers. Any population of electric machines without the corresponding documentation (e.g., lost documentation) becomes a source of unconfirmed information. If documentation is lost due to, for example, disasters, company bankruptcies, or internal conflicts (which is less likely for medium and large companies but possible for startups), the loss of the genotype may also mean the extinction of the population. Reverse engineering based on phenotypes is possible but rare and more likely an exception than the rule.

Overall, the displacement of existing populations occurs because their performance metrics are "worse" compared to those of newly emerging ones. This results in the cessation of production of old populations and, consequently, their discontinuation. With the "virtualization" of the world, the competition among genotypes is becoming increasingly important (one might say, decisive) compared to the competition among phenotypes.

In the future, it may become possible to use comprehensive mathematical modeling (AI-enhanced) to evaluate the impact of a new population on the techno-ecosystem and the ecosystem's impact on the new population, i.e., to assess the potential outcomes of the struggle for existence. This could speed up evolution in the future.

Selection Criteria

To answer the question of which machine of the two given candidates is "better" or "worse," and to form a well-founded opinion on the matter, a set of criteria can be selected to compare the chosen machines. For example, let's take two machines

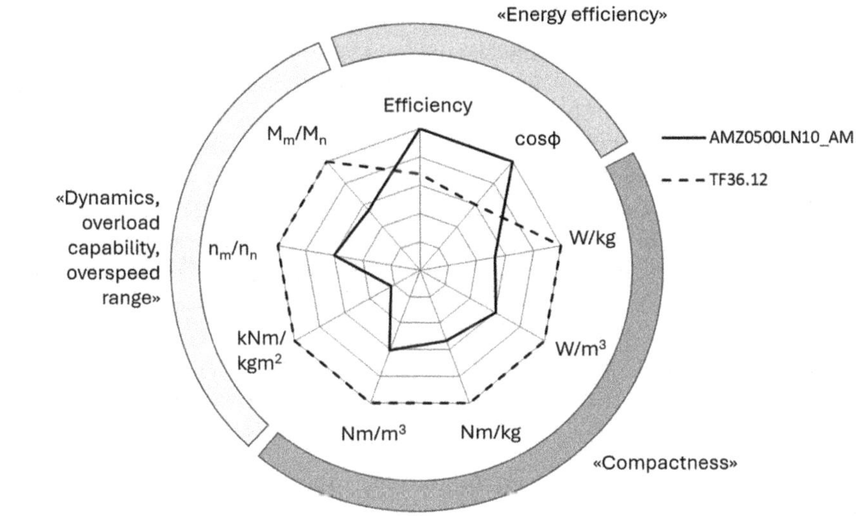

Fig. 1.6 Comparison visualization

produced by leading European manufacturers—Oswald's TF36.12 [6] and ABB's AMZ0500LN10_AM [7]. The machines are of permanent magnet synchronous (PMSM) type, have comparable power ratings (both are in the range of 300–500 kW) and the same rated rotational speed (250 rpm). The following criteria can be selected: specific power and torque, power and torque density per volume, the ratio of maximum possible speed to rated speed, the ratio of maximum overload torque to rated torque, the ratio of maximum torque to moment of inertia, efficiency, and cosφ. The parameters can be normalized for graphical representation (as in Fig. 1.6) in such a way that the closer a point corresponding to a certain characteristic is to the periphery of the diagram, the better that machine's characteristic is. For ease of visual perception, several criteria can be grouped into conditional categories: "energy efficiency", "compactness", and "dynamics".

In such a comparison, the ABB machine lags behind the Oswald machine in terms of compactness and dynamics but is more energy efficient.[4] It's impossible to conclude that one machine is unequivocally better than the other. It's important to understand that an electric machine can only be considered "better" or "worse" relative to a specific application. For example, in the metalworking industry, "dynamics" is high priority, while in pump and fan systems, dynamics may be less important, with efficiency being the key factor. Comparing machine characteristics outside the context of their applications usually does not make much sense. Finally, in practice,

[4]The results of the comparison between these specific ABB and Oswald machines should in no way lead to conclusions about the overall capabilities of these electric machine providers. In principle, they are capable of producing machines with different "focus"—efficiency, dynamics, compactness, and so on.

the machine's price will inevitably be a factor in any comparison. A considerably lower price for the machine might justify its lag in certain characteristics. It's also important not to overlook frequency converters and other system components (if they are used together with the machine), whose characteristics may be directly influenced by the choice of electric machine.

So, since a meaningful comparison of machines can only be made through the lens of the applications for which they are designed, let us consider several relevant application examples. In the first three examples, we will demonstrate the dependence of system efficiency on the type of electric machine and drive train structure, as well as the importance of efficiency in these example applications.

Example 1 Let's consider the so-called "motor-generator" (M-G)[5] of a hybrid propulsion system (Fig. 1.7). This electric machine can operate in the following modes:

- "Generator mode": when the vessel is powered by the main diesel engine (MDE), the M-G can feed consumers on the main bus, allowing the diesel-generators to be turned off (energy flow "2" in Fig. 1.7).
- "Booster mode": M-G working as a motor, adding extra torque to the main shaft, helping the MDE (energy flows "1" and "3").
- "Electric propulsion": working as the sole motor driving the main shaft, powered from the main bus. This mode is activated when the MDE fails or in situations where the operation of the MDE is not justified, and it is turned off and mechanically disconnected from the main shaft (energy flow "3").

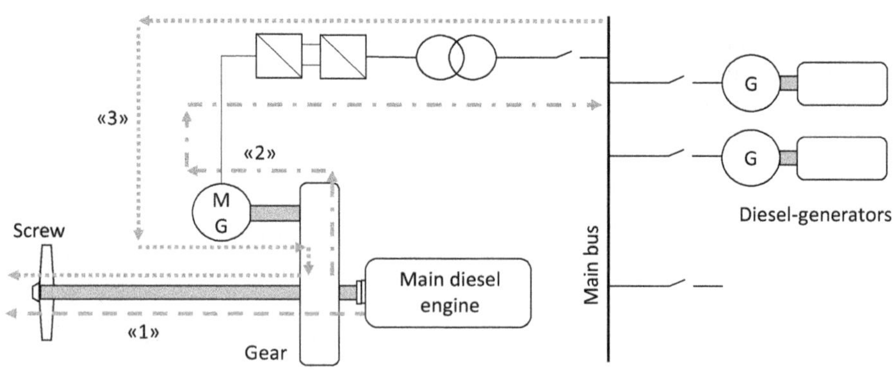

Fig. 1.7 Hybrid propulsion system

[5] This name is used to underline that in this application the electric machine is used in both motoring and generating modes.

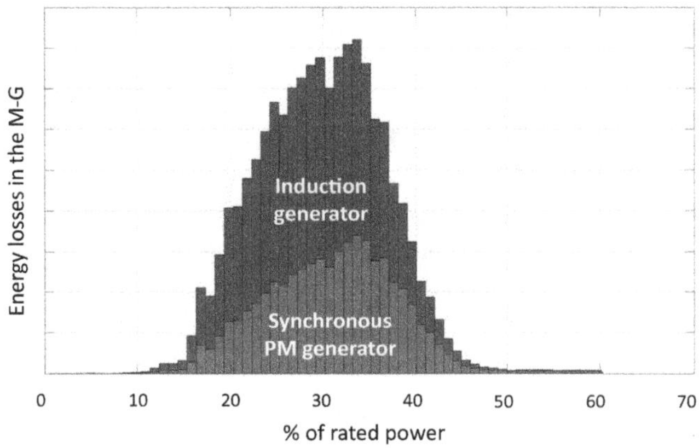

Fig. 1.8 Comparison of losses in the M-G

Let's consider the two alternatives for the M-G type for this application: squirrel cage induction machine (SCIM) and permanent magnet synchronous machine (PMSM). In the hybrid propulsion system, the size and weight of the M-G is not particularly critical. However, the impact of the machine type on energy losses and fuel consumption may be high—high efficiency of the machine is critical, both when it is run at rated speed and torque or close to them or when it is run across a wide range of speeds and loads.

In Fig. 1.8, the possible losses for the SCIM and PMSM variants of M-G are shown (qualitatively) for several months of operation of the M-G on a fishing trawler with the respective propulsion system. The primary mode in this example is generator mode, while the other modes are not utilized. Here, the generator mostly operated at 20–40% of its nominal power. The use of PMSM as the M-G (instead of SCIM) would significantly reduce losses, leading to energy savings and, consequently, fuel savings on the vessel. Whether the higher cost of PMSM-based system is justified over an SCIM-based system should be determined through a techno-economic analysis, where the input data would include the purchasing cost of both machines and the cost of fuel.

Note 1. In some applications, efficiency of the machine across the wide range of speeds and loads can be the key criterion in the informational selection. Cost, however, will always be the accompanying criterion.

Example 2 Now, let's demonstrate the dependence of system efficiency on drive train structure in the task of choosing the optimal drive system for a variable speed wind turbine. The selection of the electric generator for the system is influenced by both the structure of the entire power take-off system and the characteristics of the generator itself. The type and structure of the generator windings will determine the

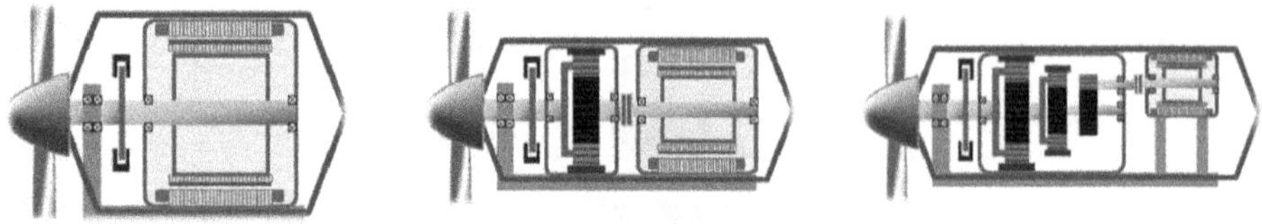

Fig. 1.9 Three of the possible drive train structural alternatives compared (the illustration does not reflect the actual layout and proportions of the components)

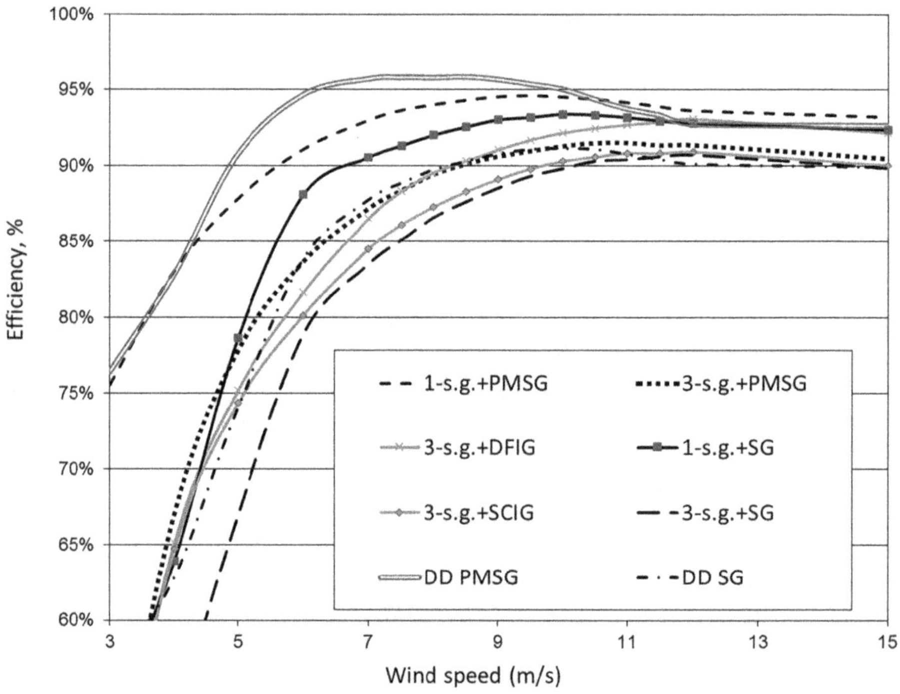

Fig. 1.10 System efficiency for various combinations of gearboxes and generators

type and structure of the frequency converter and transformer, which are the necessary components of the turbine. Typically, in such a task, the choice is made between the possible structural options for the drive train: direct drive (Fig. 1.9a) and systems with a mechanical gearbox. The gearboxes can be with one stage (Fig. 1.9b), two-stages, and three-stages (Fig. 1.9c). As in the first example, one of the most important criteria for the design of the drive train is the system efficiency. Since wind turbines must generate energy at varying wind speeds, it is essential to consider the system's efficiency across a wide range of rotational speeds.

Figure 1.10 presents the comparison of the efficiency of eight alternative system designs utilizing various structures and generator types (data from [8]). This comparison allows assessing which drive train provides better performance for different wind conditions, helping to choose the optimal balance between complexity, cost, and efficiency of the wind turbine. The system components included in the efficiency calculation in this example are the mechanical gearbox, generator, frequency converter, and transformer. Each of these components incurs losses, and the efficiency of each component usually decreases as the load decreases. The labels in Fig. 1.10 are as follows: *1-g.s.* single-stage gearbox, *3-g.s.* three-stage gearbox, *PMSG* permanent magnet synchronous generator, *DFIG* doubly-fed induction generator, *DD* direct drive, *SCIG* squirrel cage induction generator, *SG* synchronous generator with excitation windings. The diagram shows that the "1-g.s. + PMSG" system provides the best efficiency at high wind speeds (11–15 m/s), but at medium speeds (5–9 m/s), it is outperformed by the direct drive system with a PMSG.

The choice of the optimal system, and consequently the selection of the generator, will be influenced by the statistical distribution of wind speeds at the turbine's installation site. The weight of the drive train will also be of great importance, and based on this criterion, the systems with the best efficiency may not be the favored ones. The transition from system efficiency to a numerical assessment of the generated or saved energy will be demonstrated in the next example.

Note 2. Electric machines and other drive train components such as gears, converters, used together with the machines, can "weaken" or "reinforce" each other. In informational selection the machine is "seen" not only through the prism of the application but also through the lens of the system it is regarded as being part of.

Example 3 Now, we will consider the efficiency of the two competing system topologies for the azimuthing thruster of a marine vessel. In this example, two options shown in Fig. 1.11 are proposed for evaluation: the system with two bevel

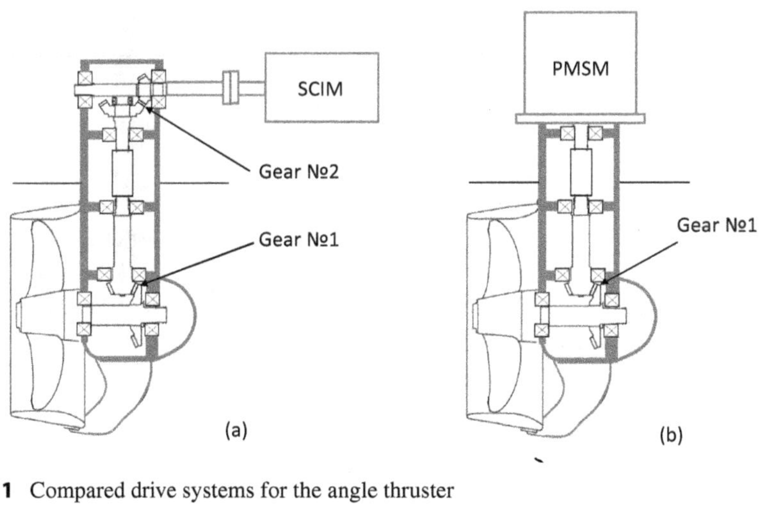

Fig. 1.11 Compared drive systems for the angle thruster

gearboxes and horizontally mounted SCIM (Fig. 1.11a), and the system with one bevel gearbox and a vertically mounted PMSM (Fig. 1.11b). The rotational speeds of the two machines are obviously different (speed of SCIM is higher by the ratio of the gear number 2). The machine in the system with one bevel gearbox operates at a lower speed and delivers higher torque. These two options can be compared based on the energy consumption measured over a specific representative period.

During navigation, the vessel operates at various speeds, as shown in the example statistical distribution of operating time at different speeds in Fig. 1.12a. High cruising speeds correspond to a high rotational speed of the thruster propeller and,

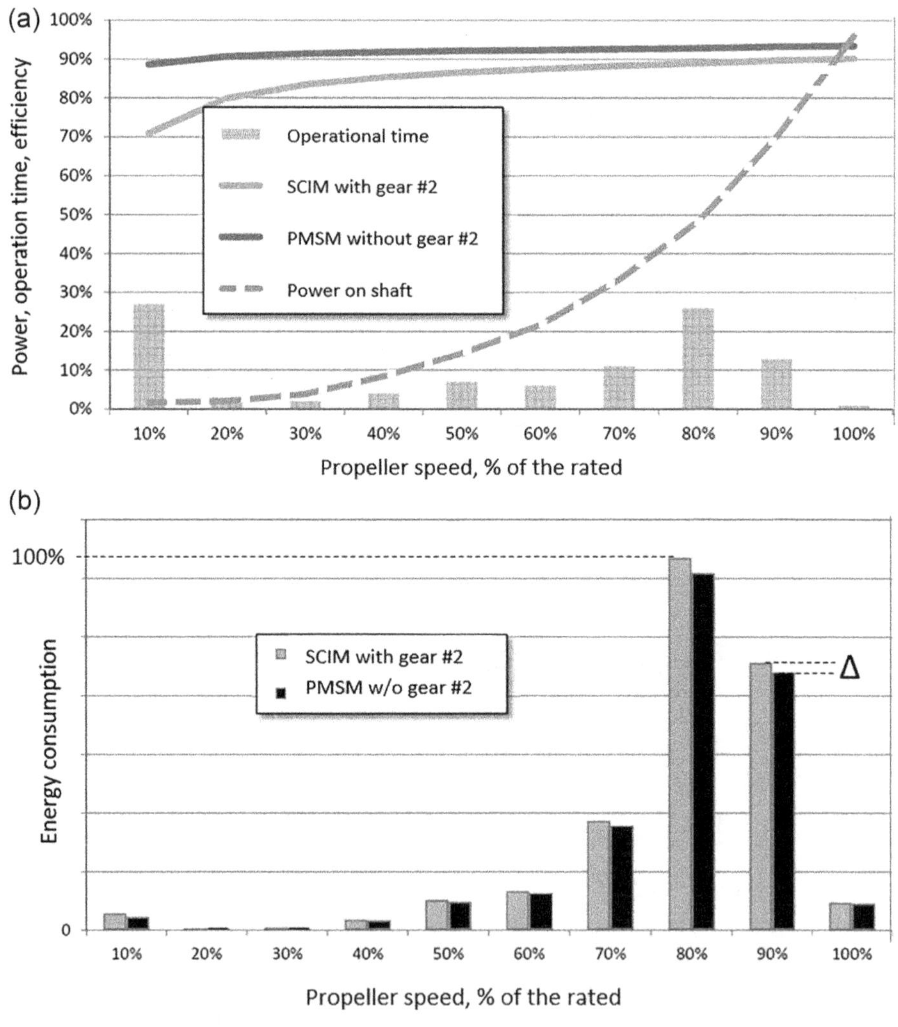

Fig. 1.12 Calculation of the difference in energy consumption

consequently, the electric machine. The energy consumed by the thruster at each load level directly depends on the efficiency of the system at that level. At higher rotational speeds, the load and power on the shaft increase, and even a relatively small difference in the efficiency of the systems at higher rotational speeds can significantly impact energy losses and overall energy consumption.

At low rotational speeds, the efficiency difference between the systems is significant, which makes the difference in energy consumption noticeable despite the relatively low load levels. The annual energy consumption at different speeds for the two systems being compared is shown in Fig. 1.12b. To calculate the total difference in the energy consumed, denoted as ΔE_c, the differences Δ at all speeds must be summed. The value of ΔE_c (in kWh per year) can then be used to assess the payback period of the investment if, for example, the existing system with SCIM (like in Fig. 1.11a) is to be replaced by a more expensive system with the PMSM (like in Fig. 1.11b). If the price of energy (possibly recalculated from the price of fuel) per kWh was C_{kWh}, then the payback period of the investment would be

$$t_{pb} = \frac{I_c}{C_e},$$

where $C_e = \Delta E_c C_{kWh}$, I_c—investment cost to replace the system with SCIM with the system with PMSM.

Note 3. In some applications, the choice of certain machine types can simplify the drive train and improve its characteristics.

Example 4 Let us consider the selection of an electric motor for a battery-powered UAV, often referred to as a drone. Issues related to the choice of optimal motor technology are usually an integral part of higher-level design tasks associated with the overall drone construction, including the design of the entire power system, which consists of the battery, multiple propeller consoles with electric motors, and electronic speed controllers.

Overall, the decision-making space for drone developers can be visualized as shown in Fig. 1.13. In each specific case, developers must seek a compromise

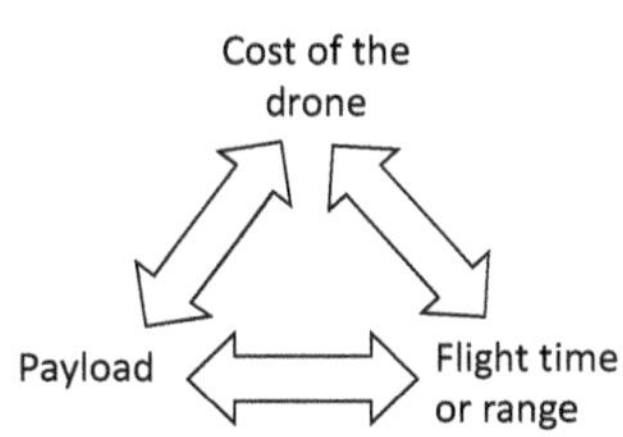

Fig. 1.13 The space of system-level trade-offs

between cost, payload capacity, and flight range of the drone. Improving one parameter often comes at the expense of another. To understand how an electric motor is selected, it is essential to understand the space of systemic trade-offs. Let's first consider the impact of battery capacity (and, accordingly, its size) on payload capacity and flight duration:

- With a fixed battery size, reducing the payload allows for longer flights. This is because the propulsion systems operate with less torque and at lower rotational speeds, consuming less energy.
- With a fixed payload weight, a larger battery size results in longer flight durations, provided that the propulsion systems can lift the heavier drone without critical overloads and maintain the same efficiency.
- It is also evident that a higher-capacity (and heavier) battery increases the cost of the drone.

Now let's take the electric motor into consideration. Currently (2020s), almost all drones use permanent magnet synchronous motors. The technology providers offer various PMSM products, characterized by different levels of efficiency and weight. In general, PMSM, which belongs to the same class, can be designed for higher efficiency by using more active materials, which consequently results in greater mass. Replacing the existing motor with a more efficient one will lead to less battery drain during operation, allowing for longer flights—this is certainly true when mass of the existing motor and the new one are the same. However, when higher efficiency is achieved at the cost of increased motor mass, the choice becomes less straightforward. Moreover, a heavier motor usually comes with a higher price tag, making the selection of the right motor even more complex. The designer's task includes finding the right balance between the weights of the motors and the battery.

But it is more important to select the optimal combinations of the propeller and the motor—this is even more critical than having high efficiency or low weight for an individual motor. Let's take an example. Figure 1.14 shows the loads on the propulsion system during the flight of a delivery drone. The diagram highlights three main load levels corresponding to the following modes: "1"—horizontal flight, "2"—hovering, and "3"—maneuvering and maintaining stability in wind gusts.

Let's consider the load characteristics of the "propeller-motor" system (Fig. 1.15). The rotational speed and torque are normalized:

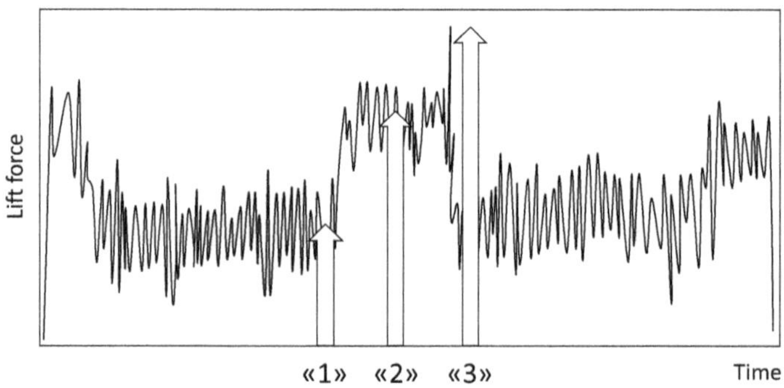

Fig. 1.14 Variations in lift force, three levels

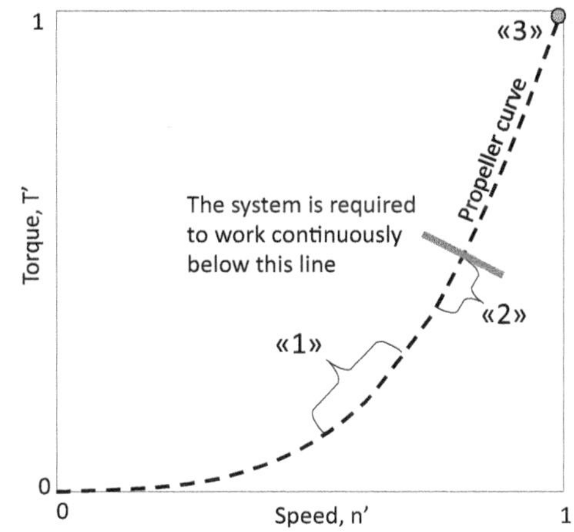

Fig. 1.15 Propeller characteristics and operating regions

$$n' = \frac{n}{n_{\max}} \tag{1.1}$$

$$T' = \frac{T}{T_{\max}} \tag{1.2}$$

Point "3" on the propeller curve in Fig. 1.15, at which the maximum torque and the corresponding maximum rotation speed are reached, has the coordinate [1;1]. The points and areas "1", "2", "3" in Fig. 1.15 refer to the levels "1", "2", and "3" in Fig. 1.14.

Typically, the curve part close to point "3" is the critical mode area in which the propulsion system only needs to work for a few seconds. Point "3" determines the overload capability of the motor. Modes "1" and "2" are the long-term operation modes. Since the system spends more time in mode "1", it is important to have high efficiency in this region of the load curve. Mode "2" sets the thermal load limit for the motor. Yet, there is another even more important aspect that needs to be considered. Figure 1.16 shows efficiency maps of the two motors considered as the candidates for the drone propulsion system. In the first case (Fig. 1.16a), the motor has an efficiency of up to 95%, but when operating on the load curve in long-term mode it will have an efficiency of about 80–85%. In the second case (Fig. 1.16b), although the motor has a maximum efficiency of only 90%, it will operate close to this efficiency level because the area of maximum efficiency coincides with the region of continuous operation on the propeller load curve. Thus, an optimal efficiency map of the motor (for the given propeller) may be more important than the maximum efficiency value of the motor.

To conclude this example, it's important to note that the primary requirement for civilian drones is flight safety, which is largely ensured by the reliable operation of the propulsion systems. Based on this, a cheaper system with high efficiency and low weight may lose out in the informational selection process to a more expensive and less energy-efficient one if the latter demonstrates higher reliability.

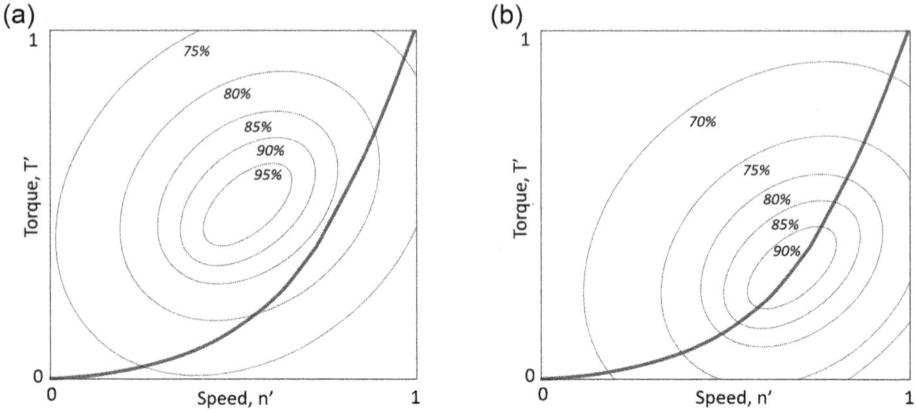

Fig. 1.16 Propeller characteristics on the motor efficiency map: (**a**) non-optimal map, (**b**) optimal map

Note 4. The optimal match between the characteristics of the driven equipment and the electric machine may be one of the most important factors.

To complete the description of possible assessment and selection criteria, let's consider an example with a even more expanded set of criteria.

Example 5 When designing tidal turbines, a multicomponent criterion is used—levelized cost of energy (LCE), graphically represented in Fig. 1.17. The principle behind this criterion, which serves as the main objective function, is as follows: the numerator represents all costs (over the lifespan of the turbine or the entire turbine farm), and the denominator represents all the energy produced (over the same period). The values are discounted over time, which is depicted by the prismatic shape of the graphic elements containing the text. Improvements (reductions) in the criterion's value can be achieved by reducing costs and/or increasing energy production, as indicated by the arrows in Fig. 1.17.

Reductions in LCE can be obtained, for example, by increasing power conversion efficiency, reducing costs both for the purchase of the equipment and for its repair. One of the important factors is *availability* of the turbine—the degree to which the turbine is in working order, which is calculated using the following formula: $A = \frac{T_w}{T_w + T_s}$, where T_w is the time during which the turbine is fully operational and working for its intended purpose, T_s is the time during which the turbine is unavailable or not performing its intended function ("downtime"). Higher availability usually indicates higher system reliability and ease of repair. Achieving high

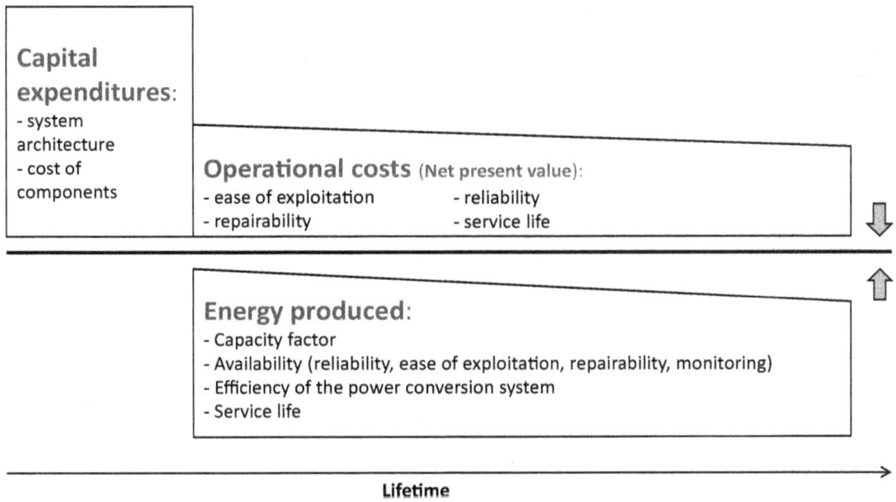

Fig. 1.17 Graphical illustration of LCE

availability often involves implementing redundancy, fault tolerance, and recovery strategies to minimize downtime.

The key question is how the choice of a particular electric generator will impact the system and its characteristics, and consequently, the system design criteria. In the process of selection, the generator is evaluated through the "filter" of criteria, which are interdependent and often contradictory. In our example, we can point out at least two contradictions ("C"), illustrated in Fig. 1.18:

- C1—high efficiency can be achieved by using more active materials in the generator, but this will increase its cost and, thus, capital costs of the project.
- C2—less thermally loaded generator, is more reliable, but at the same time larger in size and heavier, i.e., more expensive. This means that capital costs are in direct conflict with availability.

It is also possible to show the direct influence ("DI" in Fig. 1.18) of one of the criteria on another: low reliability, leading to the need for frequent repairs, or even replacement of electric machine (EM), leads to increased operational costs.

It is important to understand that in addition to the main criteria, many other criteria should be considered, and they can determine the results of informational selection. Let's list some of these criteria for the case of considering a radically new electric generator as an alternative to a conventional, proven one (for the same case of tidal turbine):

- *Sensitivity of LCE to changes in the system*: How will the LCE change if the novel generator proves unreliable and needs to be replaced with a proven generator of a different type, requiring system-level modifications? If the costs of the replacement are high, the project's LCE could increase to an unacceptable level.
- *Replacement feasibility*: If the mechanical interface of the generator is unique, replacing it could be problematic, as it might require redesigning the turbine nacelle.

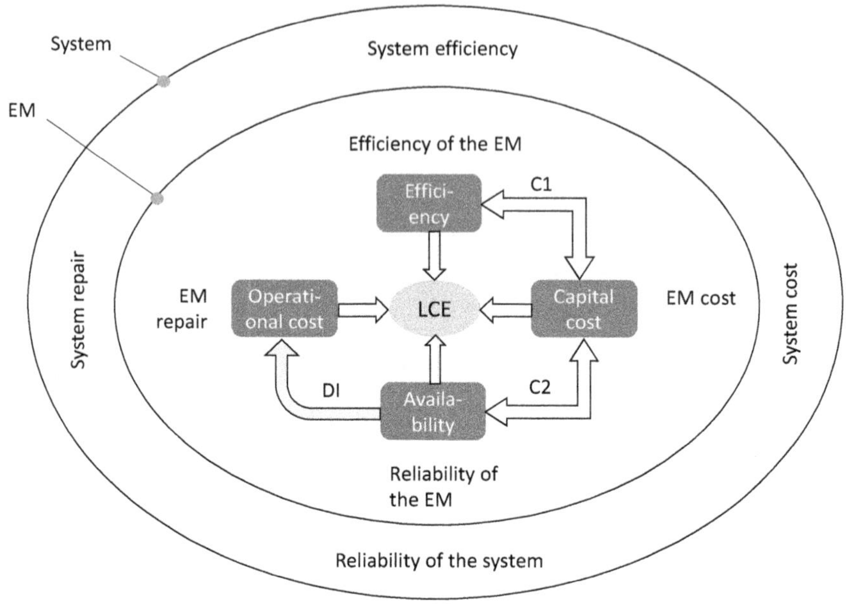

Fig. 1.18 Criteria interdependence in the system context

- *Freedom to operate in the context of intellectual property rights*: If the selected generator is of a special design, it likely involves specific solutions that, although the probability is usually low, may be patented by someone else. This could potentially create legal issues for the turbine supplier and the tidal park owner.
- *Component availability*: If the novel generator uses a highly specialized pressure compensator or a unique frequency converter topology, future repairs or replacement of faulty parts could become a problem for the turbine farm operator.
- *Ease of transportation*: If the diameter of the new generator is large (several meters), which can be the case with direct-drive solutions, transportation to the turbine assembly site may not be possible by road, requiring alternative transportation methods (such as by sea or helicopter).
- *Technology readiness level*: If the generator is still a prototype, selecting it for the project could be risky.
- *Technology readiness year*: If the prototype hasn't been sufficiently tested, certification of the generator could take considerable amount of time, which might be unacceptably long for the project.
- *Repairability*: The possibility of repairing the generator on-site or only at the manufacturer's facility can be the important time and cost driver.
- *Complexity of maintenance*: do the generator's auxiliary systems (cooling, pressure compensation, bearings) require maintenance during operation.
- *Testability*: Availability of a test bench for full-scale testing of the entire drive system before it is deployed on the underwater turbine.

- *Service life*: if the electric machine is totally new, there is no statistical data to predict its service life, whereas for a traditional, proven generator, there can be available comprehensive data showing a high likelihood of reliable operation for many years.

Note 5: In some applications, the candidate electric machine is evaluated through the prism of criteria which are interdependent and contradicting each other.

Following the examples discussed above, the concluding summary can be given:

- It makes little sense to evaluate electric machines in isolation from their applications; the real set of evaluation criteria is determined by the applications.
- Usually in each application there can be identified the most "visible" criterion (like efficiency in the first three examples), but on a more detailed level it will be found that the total number of criteria is large.
- Individual criteria can be combined into multicomponent criteria (like LCE).
- The criteria may be contradicting each other.

In the world of electric machine applications, there are many other criteria and objective functions not mentioned in the five examples above, for instance positioning accuracy, low torque ripple, low vibrations and noise, high dynamics, etc.—the list may be long.

Price and Cost

Price is one of the most important criteria when selecting an electric machine for any application, so it must be included in the set of key performance indicators (KPIs), alongside weight, size, efficiency, etc. When developing a new machine, it's essential to consider the market situation and the prices of existing machine series that are expected to be the competitors. The question of the machine's potential price should be raised as early as possible during the development process. Unlike other KPIs, cost is a "floating" and "flexible" parameter that can change over time.

When developing a new machine, it's important to try to find answers to the following questions:

- How can the target price be identified, and how reliable is this value?
- How might the prices of competing machines change over time, particularly in situations of competitive pressure?

It is important to understand the structure of the price, including components such as the built-in profit margin, adjustments for expected inflation, various risks, warranties, taxes, material costs, labor expenses, and the operational costs of the enterprise. This understanding will help, among other things, to plan measures for reducing the cost of the machine being developed, such as selecting the optimal manufacturing process or determining the division of labor between subcontractors.

The planned price of the product can influence other KPIs. For instance, requirements for a low price may force the designer to compromise on other characteristics such as, for example, efficiency level.

The cost structure can change with the rapid development of one of the active materials used in the machine's design. For example, as shown in [9], the cost of HTS tape could decrease by an order of magnitude with improvements in the tape itself and the transition to mass production. In this case, the share of HTS tape in the cost structure of a superconducting machine would significantly decrease.

Examples of Simplified Price Models

Price models can be built based on an analysis of available catalogues and price lists. A model describing a series of machines can be more reliable than price information about one individual machine. In general, machine mass, determined by the amount of used materials, mainly metals, can be taken as the main parameter of the model, since cost of the materials is usually the main cost driver.

Example 1 In the online course *DriveConstructor*, described in [10], the price ϵ for general-purpose industrial machines is approximated based on the machine's mass m, type, voltage level, protection rating, housing material, and mounting type:

$$\epsilon = k_1 k_2 k_3 k_4 k_5 m^{\propto}, \tag{1.3}$$

where $\propto$ and k_1 are coefficients dependent on voltage, k_2 is coefficient for machine type, k_3 is coefficient for degree of protection, k_4 is coefficient for housing material, and k_5 is coefficient for the mounting type.

Example 2 The price model for PMSM developed for UAV propulsion, with voltage level of 10–40 V, designed open for cooling with external air, can also be based on the machine's mass [11]:

$$\epsilon = 400 m^{0.4}, \tag{1.4}$$

where ϵ is price in Euro, m is mass in kg.

It's challenging, and perhaps not practical, to build a universal price model applicable to any type of machine. However, it is essential to develop price and cost models for specific types of machines and individual series.

Factors Determining the Weight and Size of Electric Machines

It's important to understand the factors that determine the mass and size of an electric machine. To have a meaningful discussion about the objectives and outcomes of machine design, reliable reference points are necessary. Comparing a new machine with the current state of the art can serve as a tool for evaluating the quality of the chosen technical solutions and technologies and allows for the selection of strategies for further improvement. The main concept we'll base our discussion on is the *hypothesis about the mass of an electric machine*, formulated in [14] as follows: "*The mass of any rotating electric machine is proportional to its power raised to the power of 0.75 and inversely proportional to its rotational speed raised to the power of 0.5*". The corresponding expression is:

$$m \equiv \frac{P^{0.75}}{n^{0.5}} \tag{1.5a}$$

In a more general form, the hypothesis can be presented as:

$$m \equiv P^{C_p} n^{C_n} \tag{1.5б}$$

where C_p and C_n are coefficients that reflect the general trend in the variation of the mass of electric machines as a function of power and rotational speed. In [14] it is suggested that C_p and C_n are determined by the "nature" of electric machines as a class of technical devices. The values $C_p = 0.75$ and $C_n = -0.5$ were derived and tested in [12–14]. To explain the value $C_p = 0.75$, one can refer to [13], which shows that for geometrically similar machines, with constant current density and flux density, the power P is proportional to the base dimension l to the fourth power: $P{\sim}l^4$. Then, by making an assumption that the mass of the machine is proportional to its volume and the basic dimension to the third power $m{\sim}V{\sim}l^3$, we can derive the expression $m{\sim}P^{3/4}$. Thus, the mass is proportional to the power raised to the power of 0.75, leading to the conclusion that $C_p = 0.75$. No physical explanation for $C_n = -0.5$ has been proposed so far.[6]

> *The mass of any rotating electric machine is proportional to its power raised to the power of 0.75 and inversely proportional to its rotational speed raised to the power of 0.5"*

Another key element necessary for understanding the factors that determine the mass and size of electric machines is a logarithmic map (Fig. 1.19), which describes the distribution of specific power and torque of electric machines (without division into types) depending on power levels and rotation speeds. In [14] it is stated that *electric machines are a class of devices, largely follow logarithmic laws when it comes to changes in mass as a function of power and rotational speed.*

From the *hypothesis about the mass*, there stems the possibility of introducing machine constant C_M defined as:

$$C_M = \frac{m}{P^{C_p} n^{C_n}}, \tag{1.6}$$

[6]Until the explanation is found for why the square root of speed works so well for estimation of mass of rotational electric machine, we can refer to James Jeans' quote from *The Mysterious Universe:* "We have already considered with disfavor the possibility of the universe having been planned by a biologist or an engineer; from the intrinsic evidence of his creation, the Great Architect of the Universe now begins to appear as a pure mathematician".

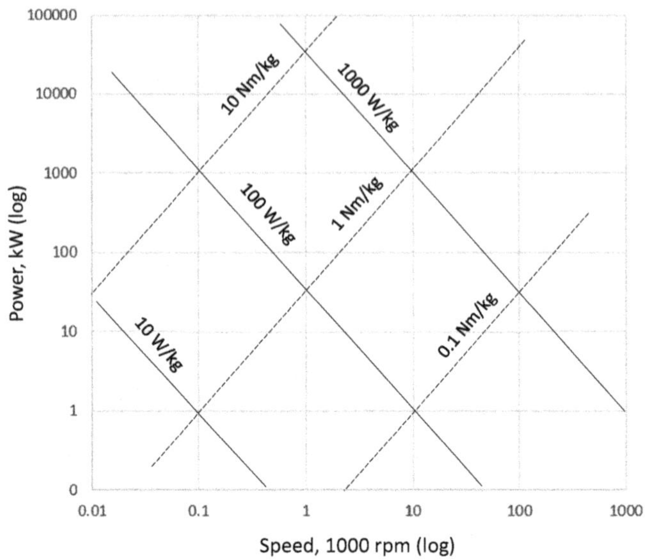

Fig. 1.19 Lines of equal specific power and torque (approximation)

Accordingly, the mass can be expressed through C_M as:

$$m = C_M P^{C_p} n^{C_n}, \tag{1.7a}$$

or, considering the proposed values of the coefficients C_p and C_n:

$$m = C_M P^{0.75} n^{-0.5}, \tag{1.7b}$$

where m is in kg, P—in kW, n—in rpm.

To some extent, the expressions (1.7a, 1.7b) describe something analogous to the law of gravity, where the "gravity" is represented by the ratio $\frac{P^{0.75}}{n^{0.5}}$, and which designers attempt to overcome by working on reducing the value of the machine constant C_M.

The machine constant C_M depends on many factors. In [12], eight key factors were proposed, and the constant was represented as follows:

$$C_M = K_t K_e K_c K_u K_{fp} K_s K_i K_o K_x. \tag{1.8}$$

where K_t is the machine type factor, K_e is the efficiency level factor, K_c is the cooling factor, K_u is the voltage factor, K_{fp} is the frequency and poles factor, K_s is the shape factor, K_i is structural components factor, K_o is the overload factor, and K_x represents other factors.

It should be noted that the factors are constructed so that the most standard machine designs correspond to factor values equal to one. This applies to all factors except K_t. Let's go through the aforementioned factors that influence the machine constant C_M and thus define the machine's mass.

Machine Type Factor (K_t): The lower the value of C_M, the more compact the machine. Compactness is a common term related to the volume, and in practice, the mass and volume of a machine are usually directly correlated. "High compactness" is often correctly understood as "low mass." This is the sense in which the term "compactness" will be used below. Compactness is described by the machine type factor K_t. For example, it is well known that machines with permanent magnets or superconducting windings will be more compact than conventional machines. In [13], the suggested value is $K_t = 1100$ for a doubly-fed induction machine (DFIM), $K_t = 1000$ for a squirrel-cage induction motor (SCIM), and $K_t = 700$ for an induction machine with a solid rotor. In [14], a value of $K_t = 500$ is proposed for permanent magnet machines. Since only a few of machine types are mentioned, this can be regarded as an illustration of the principle and not the final and complete set of data. In the future it would be interesting to build a hierarchy of compactness for different types of electric machines based on the values of the machine type factor K_t.

Voltage Factor (K_u): The voltage level determines the thickness of the slot and inter-phase insulation in the winding and, consequently, the intensity and efficiency of heat dissipation in the active parts of the machine, which in turn affects the machine's size. In [13], a linear expression for the voltage factor was derived, which can be used for voltages above 400 V: $K_u = 1 + 0.035\,(U - 400)/1000$, where U is the voltage in volts. To cover electric machines with voltages below 400 V, [14] proposes a more universal expression that should work for the entire possible voltage range—both below and above 400 V:

$$K_u = 0.2\ln\left(U^{0.8}\right). \tag{1.9}$$

Cooling Factor (K_c): In [13], based on the analysis of SCIM series, values of K_c for selected cooling types and protection degrees were proposed, as shown in Table 1.1. This table contains only a handful of variants, so it can serve just as an example and needs to be expanded in the future. For instance, in the case of UAV motors and some all-electric aircraft motors with air flowing through the machine gap, cooling could be even more effective than the most efficient cooling methods listed in Table 1.1, with value of K_c going down to 0.4 for such machines. For naturally cooled machines with a closed design and no through ventilation (e.g., IP55, IC400), a value of K_c between 1.1 and 1.2 can be suggested. It's important to note that the type of coolant, its temperature, and the intensity of its flow through the machine can vary significantly, making all proposed values approximate. The cooling oncepts from Table 1.1 are illustrated in Fig. 1.20.

Efficiency Level Factor (K_e). The size of the machine designed for maximum efficiency can be significantly larger than the size of the machine optimized for low weight and having some low but still acceptable level of efficiency. In [14], a Pareto-optimal "frontier" of efficiencies for 1 MW PMSM was presented, obtained through series of automated electromagnetic calculations for various designs of the active

Table 1.1 Examples of values of the cooling factor

Cooling and degree of protection	Factor K_c
IC411, IP55	1
IC71W, IP55	0.80
IC611, IP55	0.66
IC81W, IP55	0.55
IC01, IP23	0.53

parts (the data reproduced in Fig. 1.21). Beneath the "frontier" curve there are many points representing non-optimal design options. An acceptable compromise for the designer may be point "1" on the "knee" of the curve. In Fig. 1.21, the point "2" represents the option acceptable in the case of mass optimization with an acceptable

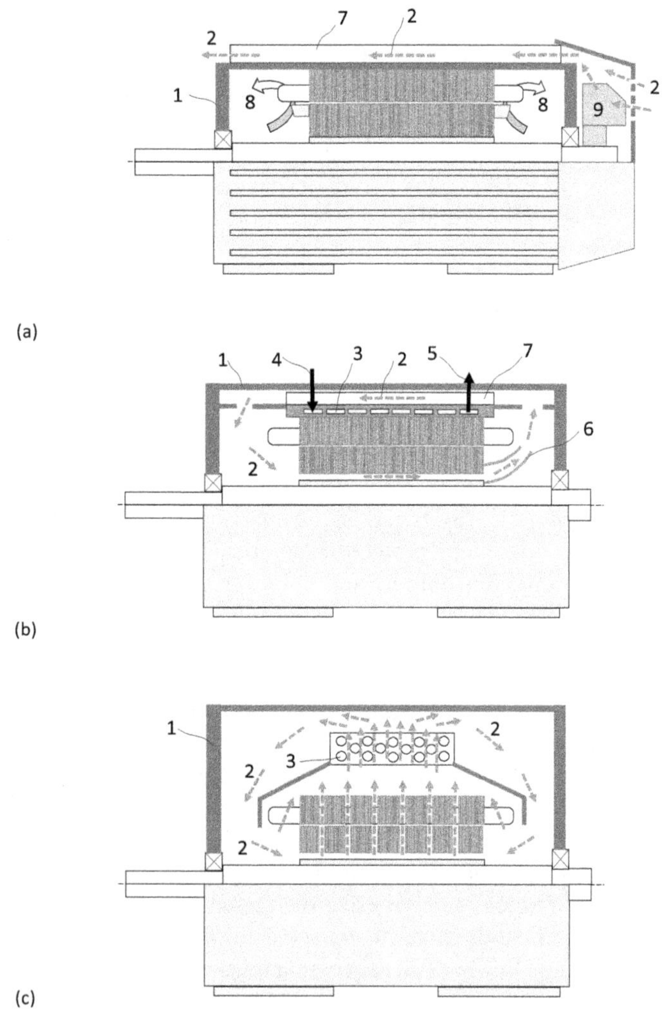

Fig. 1.20 Considered types of cooling: IC411 (**a**), IC81W (**b**), IC71W (**c**), IC01 (**d**), IC611 (**e**). (1—housing/frame, 2—air flow, 2a—cold air, 2b—heated air, 3—channels for liquid coolant, 4—liquid coolant inlet, 5—liquid coolant outlet, 6—guide for air circulation, 7—cooling ribs, 8—internal air circulation, 9—on-shaft fan, 10—screen, 11—external cooling air, 11a—cold external air, heated air)

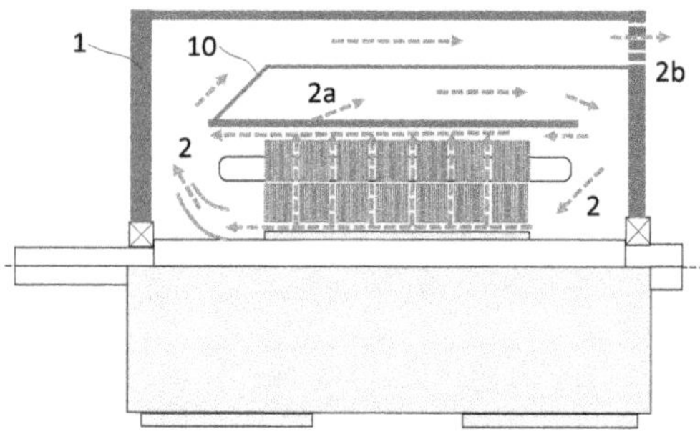

(d)

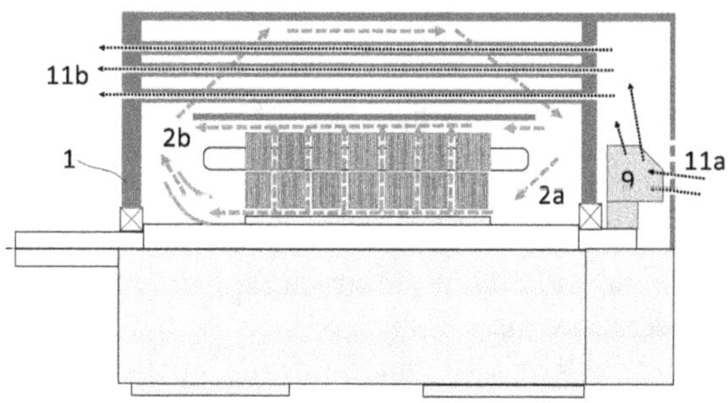

(e)

Fig. 1.20 (continued)

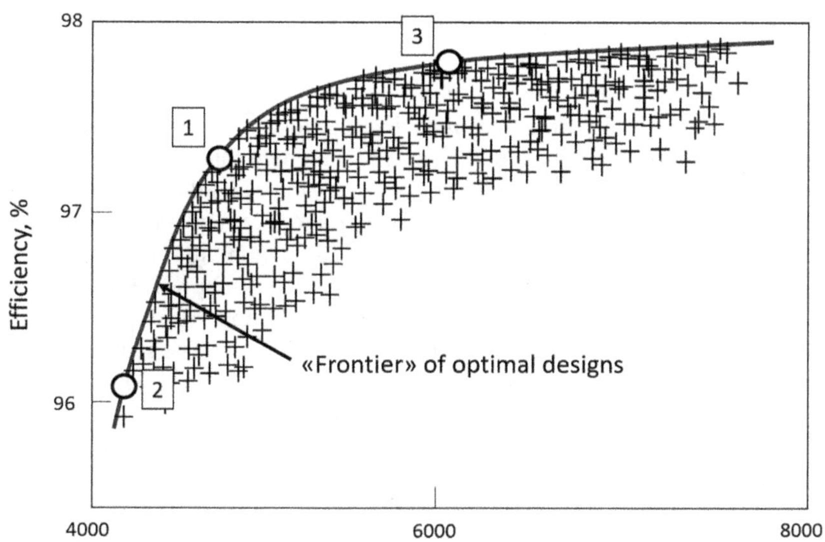

Fig. 1.21 "Frontier" of Pareto optimal design solutions

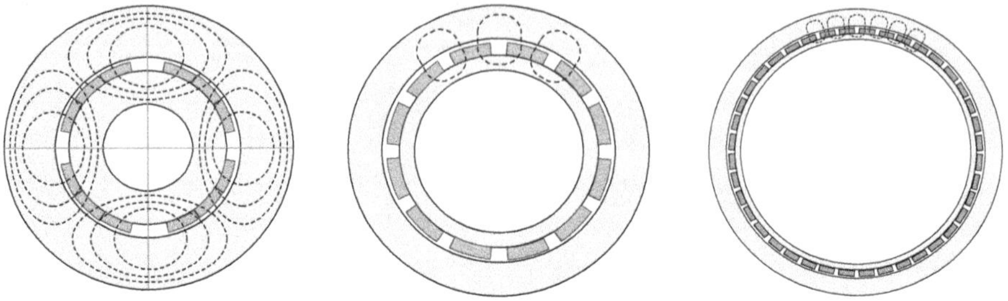

Fig. 1.22 Permanent magnet machines with different number of poles: 4 (**a**), 16 (**b**), and 80 (**c**)

low level of efficiency, while point "3" represents an option suitable when maximum efficiency is required.

For practical use, the following values of efficiency level factors were proposed for the case when only active parts are considered (in [14]): $K_e = 1$ for an average level of efficiency (near point "1" in Fig. 1.21), $K_e = 0.8 \ldots 0.9$ for a low level of efficiency (near point "2"), and $K_e = 1.2 \ldots 1.3$ for a high level of efficiency (near point "3"). With the addition of the structural parts of the machine (housing, shaft, etc.) into consideration, the influence of the selected efficiency level on the size of the entire machine will be less—the following values of K_e are proposed (for a cast iron or steel housing): $K_e = 1$ for an average efficiency level (the same as in the case of only active parts), $K_e = 0.9 \ldots 0.95$ for a low efficiency level, and $K_e = 1.1 \ldots 1.15$ for a high efficiency level. It should be noted that if the housing is lightweight (made of aluminum), the values may be as follows: $K_e = 0.9$ for a low efficiency level and $K_e = 1.15$ for a high efficiency level. Again, like with the previous factors, these values need refinement and can be used just as rough guidance.

Frequency and Pole Factor (K_{fp})**.** The combination of high supply frequency and high number of poles can result in a significant reduction in machine weight. Figure 1.22 schematically illustrates machine topologies with 4, 16, and 80 poles, with the corresponding magnetic field lines. It is evident that machines with fewer poles require a greater yoke thickness for both the stator and the rotor. As the number of poles increases, the core becomes thinner and, consequently, the machine's weight decreases. It should be noted that for the same nominal rotational speed, an increase in the number of poles requires an increase in supply frequency. As the supply frequency increases, losses in both steel and copper go up, so there is a certain limit to the use of this approach for weight reduction. The successful implementation of this method is not feasible for all types of machines. Machines with a large number of poles are usually synchronous machines.

Traditionally, when designing electric machine, designers based their work on supply frequency of 50 Hz. Then, to achieve the rotational speed of 3000 rpm, a 2-pole machine must be used. To get a double rotational speed of 6000 rpm, a doubled supply frequency (f = 100 Hz) is required, which can be provided by a frequency converter. For even higher rotational speeds, a correspondingly higher supply frequency can be used (while maintaining the same number of poles). To achieve a lower rotational speed such as, for example, 600 rpm, when using supply frequency of 50 Hz, a multi-pole machine can be employed.

This approach to selecting the supply frequency and number of poles is illustrated by the lower row of machines in Fig. 1.23 (##1–5). An alternative is to use a significantly higher frequency (e.g., 10 times higher!) illustrated by the upper row of machines in Fig. 1.23 (##6–8). Machines in the upper row typically have lower mass compared to those in the lower row. To account for the impact of this design approach (higher frequency and large number of poles), the following approximate values are suggested for the corresponding Frequency and Pole Factor: $K_{fp} = 1$ for $2 < p < 20$, $K_{fp} = 0.8$ for $20 < p < 30$, $K_{fp} = 0.6$ for $30 < p < 40$, and $K_{fp} = 0.5$ for $p > 40$. The presence of structural components significantly affects the value of this factor, as will be shown below.

Overload Factor (K_o). Requirements for overload capability can influence the mass of the machine. Different types of machines have different overload capabilities. For example, the permissible short-term overload for PMSM is typically 150–200% of the rated capability, for induction motor it is 200–300%, and for slotless PMSMs, it can be up to 500–700%. If the overload, for example, during start-up or during operation exceeds 200%, then when designing a PMSM, it will be necessary to assume a greater size and mass of the machine than for the case without overloads. At the same time, when designing a SCIM for short-term overloads of 200%, the machine size and mass can be the same as for the case without overloads. The proposed methodology suggests the following approximate values for the overload factor: $K_o = 1$ when overloads do not impact the mass of the machine, $K_o = 1.2$ for some impact of overloads on the mass, and $K_o = 1.4\ldots2$ for significant impact of

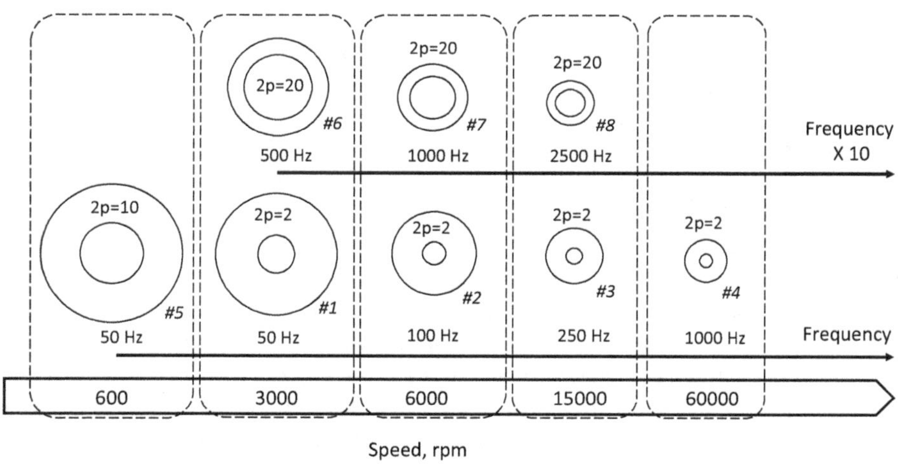

Fig. 1.23 Various options for implementing electric machine for some rotation speeds

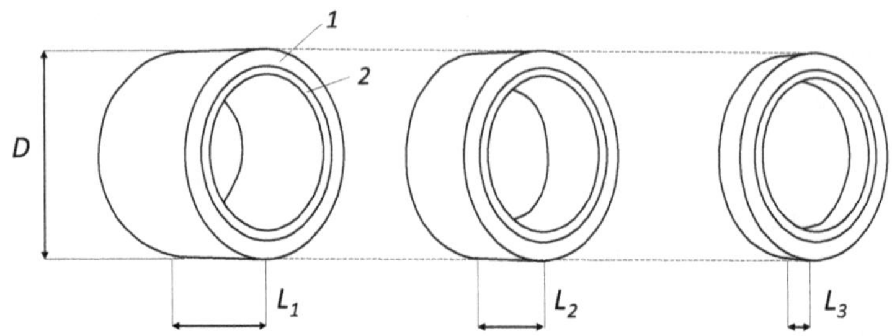

Fig. 1.24 Examples of different ratios of diameter and length of active parts (1—stator, 2—rotor)

overloads on the mass. Also, the factor K_o can reflect the operation of the machine in intermittent mode without overloads (then $K_o < 1$).

Shape Factor (K_s). The mass of the active parts of the machine depends on their shape, i.e., on the ratio of diameter (D) and length (L). Figure 1.24 shows three variants of machine shapes, where the same diameter is maintained, but the axial length varies. In (1.14) it is claimed that if we take $\frac{D}{L} = 2$ as the standard ratio corresponding to the factor value $K_s = 1$, then options with $\frac{D}{L} > 2$ will correspond to $K_s < 1$, i.e., greater compactness, and options with $\frac{D}{L} < 2$ will correspond to $K_s > 1$, i.e., less compact (and higher mass). The following expression is proposed in [14]:

$$K_s = 1.1 - 0.052\frac{D}{L} \tag{1.10}$$

Structural Components Factor (K_i). There are two extreme cases for how a machine can be represented in a catalog or specification: (1) only the active parts of the machine are presented (frameless motor), and (2) the "complete" machine is presented, including the housing, bearings, end shields, shaft, terminal box, etc. Between these extremes, there are many intermediate cases. For example, the motor shown in Fig. 1.25, designed for integration into the frame of an all-electric aircraft (AEA), lacks a shaft and bearing shields. In the case of the wheel-motor shown in Fig. 1.26, the load-bearing mechanical structures of the wheel also serve as the load-bearing structures of the rotor with permanent magnets, meaning that only the stator has the supporting structure, and the rest of the machine consists almost exclusively of the active parts. The mass of all structural elements of a complete machine is typically approximately equal to the mass of the active parts (if the housing is made of cast iron or steel). This means that the mass of the complete machine is roughly twice the mass of the active parts. However, if the casing is made of aluminum, which is 2.5 times lighter than steel or cast iron, the difference between the mass of the complete machine and the mass of its active parts will be less.

The proposed methodology suggests the following approximate values: $K_i = 1$ for the complete machine, $K_i = 0.5$ for only the active parts. If the housing is lightweight (aluminum), $K_i = 0.8$ can be used for the complete machine.

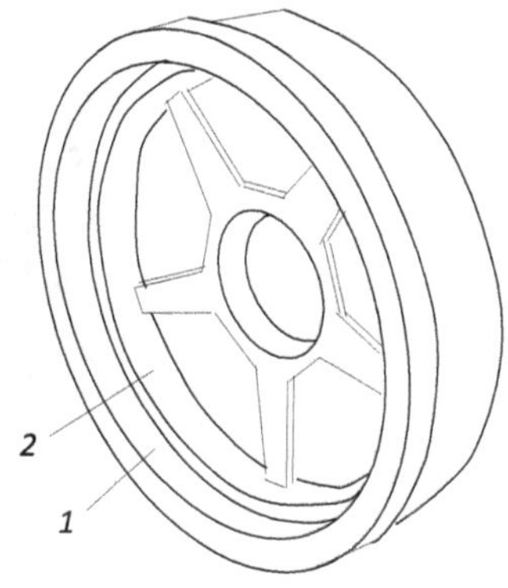

Fig. 1.25 Motor for integration into electric aircraft propulsion system (1—stator, 2—rotor)

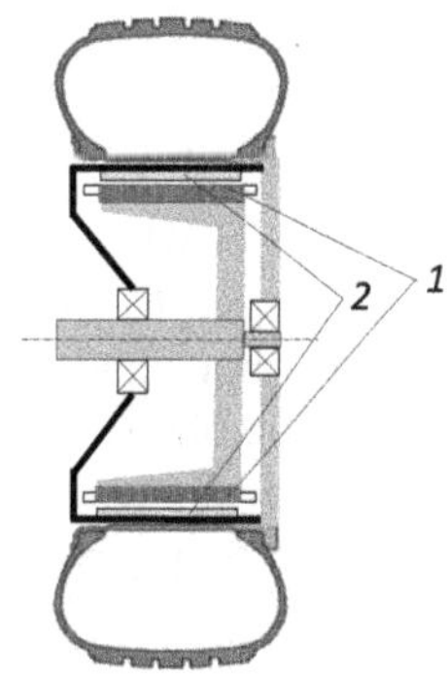

Fig. 1.26 Wheel–motor (1—stator, 2—rotor)

Other Factors (K_x). The configurations of electric machines are so diverse that the factors described above cannot cover all possible variations. Therefore, the methodology includes a factor K_x, which encompasses all other influencing factors.

The following examples of solutions can be given when the mass of the machine is slightly lower than usual ($K_x < 1$): (1) design with an external rotor, (2) the use of expensive materials with improved characteristics, and (3) a machine designed to operate in relatively low ambient temperatures.

One can also give examples of situations where the mass of the machine will be slightly higher than usual ($K_x > 1$): (1) operation at high altitudes above sea level, for example in an aircraft; (2) high ambient temperature, for example a machine in a deep oil well; and (3) liquid-filled machines having high friction losses in the gap between stator and rotor, such as in a fluid-filled "flooded" machine.

Interdependence of the Factors. It is important to note that the factors are not independent of each other and these interdependencies should be studied. Here are some examples:

- Interdependence of K_i and K_{fp}: When the structural components make up half of the machine's mass, reducing the thickness of the "ring" of active parts by choosing higher pole numbers will have less impact on the overall mass compared to a machine consisting only of active parts. This effect, illustrated in Fig. 1.27, should be considered when selecting K_{fp}: when $K_i = 1$, K_{fp} is closer to 1 than when $K_i = 0.5$.

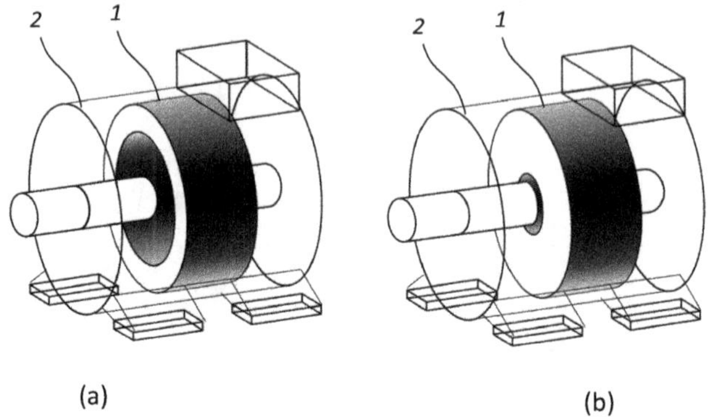

Fig. 1.27 Machines with a small number of poles (**a**) and with a large number of poles (**b**) (1—active parts, 2—carrying structures)

- Interdependence of K_c and K_e. High efficiency can make cooling easier. For example when losses are low, the transition from air cooling to water cooling does not provide a significant benefit, and the value of K_c can be above 0.8. On the other hand, intensive cooling allows for having lower efficiency if this is acceptable based on the design requirements.
- Interdependence of K_i and K_s: as the diameter D increases, the mass of the structural components can begin to grow disproportionately. Therefore, with $K_i = 1$, an excessively high $\frac{D}{L}$ ratio can lead to a lower deviation of K_s from 1.

It is also important to note that the minimum achievable values of the machine constant for the same machine technology vary in the space of characteristics and parameters, such as supply frequency, rotation speed, power, etc. Different technologies may demonstrate their advantages in different areas of the performance and parameter space—this idea is graphically represented in Fig. 1.28. For example, synchronous machines with a field winding on the rotor may be competitive when designed for low and medium speeds, but due to mechanical strength limitations, they may have reduced performance when designed for high rotational speeds. On the other hand, induction machines are not well-suited for continuous operation at low rotational speeds, where, at frequencies of 50 Hz, a large number of poles is typically used this leads to increased leakage fluxes and a reduction in the performance of induction machine.

To conclude the presentation of the method based on the hypothesis about the electric machine mass, a general unifying illustration is given in Fig. 1.29 indicating some values of the factors as examples.

In formula (1.7b), the right-hand side consists of the two components: the first is C_M and the second is the product $P^{0.75}n^{-0.5}$. These represent the two levels of "tuning" in the mass estimation process: the "rough" level is the product $P^{0.75}n^{-0.5}$ (with variability around 8–9 orders of magnitude), and the "precise" level is the machine

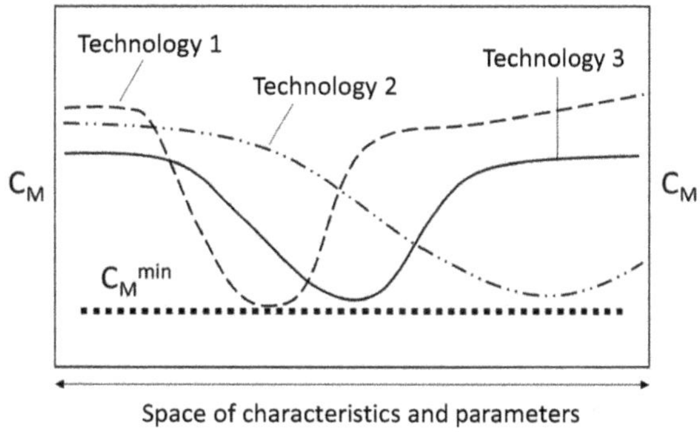

Fig. 1.28 Qualitative description of the optimal regions of characteristics and parameters for various technologies

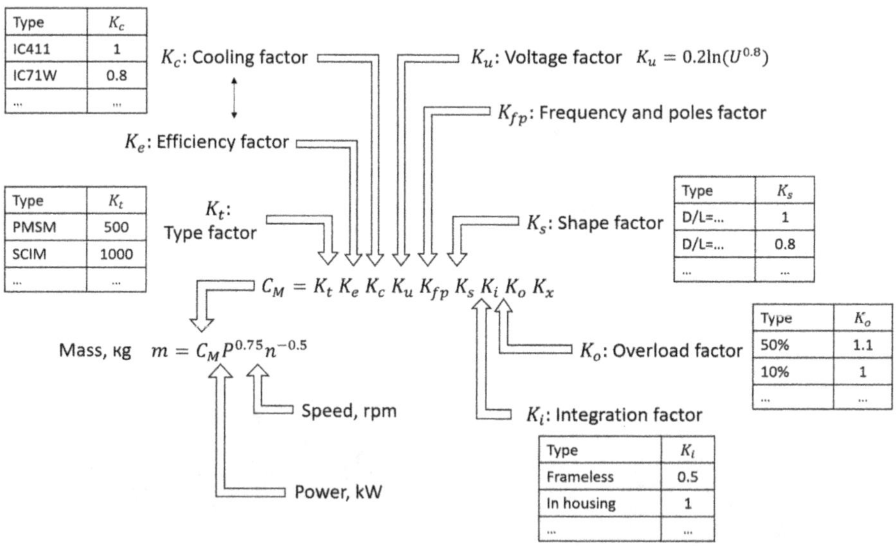

Fig. 1.29 Graphical presentation of the key factors determining mass of an electric machine

constant C_M (with variability around two orders of magnitude). The value of the ratio $\frac{P^{0.75}}{n^{0.5}}$ is decisive for the mass of the machine. The various parameters combined in the constant C_M have significantly less influence. This is described graphically in Fig. 1.30 where the value of $\frac{P^{0.75}}{n^{0.5}}$ sets the value on the mass scale higher ("1") or lower ("2") and the value of C_M adjusts the position a bit up ("A") or down ("B). The value of $\frac{P^{0.75}}{n^{0.5}}$ is maximum for high powers and low speeds and minimum for low powers and high speeds (Fig. 1.31).

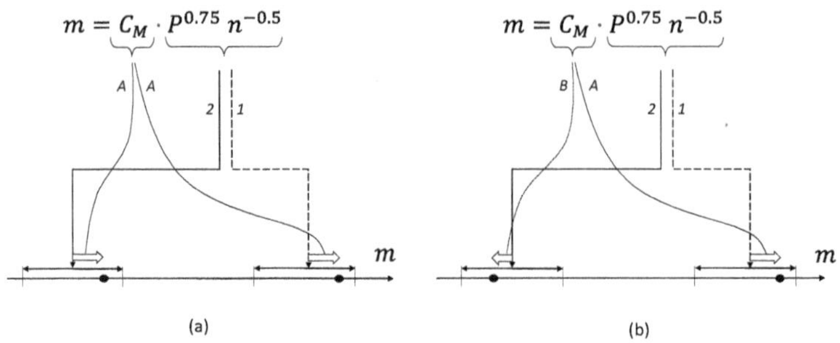

Fig. 1.30 Illustration of the influence of power and speed

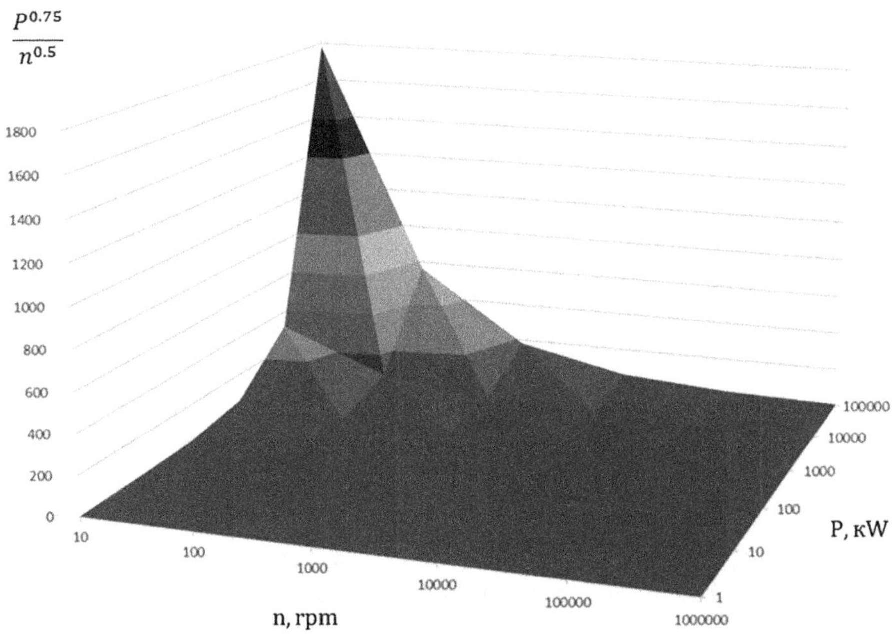

Fig. 1.31 Graphical representation of the ratio $\dfrac{P^{0.75}}{n^{0.5}}$

C_M is the result of a recent research, being the alternative to the generally accepted machine constants. A comparison of the C_M constant and the Arnold constant is given in Appendix 1. Further details can be found in [12–14].

Table 1.2 Specific power and torque of ABB's induction machines from [15]

rpm	3000	1500	1000	3000	1500	1000	3000	1500	1000
кW	Mass, kg			Specific torque, Nm/kg			Specific power, W/кg		
1.1	18	26	28	0.19	0.27	0.38	61	42	39
11	127	135	172	0.28	0.52	0.61	87	81	64
110	880	900	1000	0.40	0.78	1.05	125	122	110
1000	4400	4700	–	0.72	1.35	–	227	213	–

Comparison of Machines

When comparing electric machines, it is a common practice to use specific power P' or specific torque T' as the criteria. However, as we saw in Fig. 1.19, the specific indicators P' and T' vary significantly depending on the level of power and rotational speed, which naturally limits the situations where their use won't lead to incorrect conclusions. To illustrate this statement, let's select several machines of different power ratings from the catalogue of one of the leading motor manufacturers [15], ensuring that the difference between the rated powers of the selected electric machines is roughly one order of magnitude (10 times): 1.1, 11, 110, and 1000 kW. The results of the specific power and torque calculations are shown in Table 1.2. We can see that the values of both P' and T' within the same series of machines from the same provider can differ by several times. Additionally, it is worth noting that a machine with a higher specific power may have lower specific torque compared to some other machine. Therefore, relying on only one of these two criteria for comparison is unsafe in terms of drawing valid conclusions, except for the cases where the powers and rotational speeds of the machines being compared are identical. It is easy to check that for the same machines (from Table 1.2), the constant C_M will vary within a range of about 50%, making C_M a significantly more reliable criterion than P' and T'. Moreover, C_M serves as an integral metric, in a way combining the two criteria into one, which can simplify the comparison process.

Even within the same product series of the same manufacturer, machines can be designed in such a way that their compactness[7] expressed even through C_M, which as we saw is more stable than specific torque or power, may differ by tens of percent. As an illustration of this fact, let's compare the series of low-voltage machines with cast-iron frames from ABB (data from [15]) and Siemens (data from [16]). Both series have the same efficiency class (IE2), cooling type (IC411), and protection class (IP55). As the compactness criterion, we will use C_M, since it, as we know, is proportional to mass (as per eq. (1.6)) but in general is independent of power and rotational speed. The results of the C_M calculations for the 4-pole and 6-pole machines from both manufacturers are presented in Fig. 1.32. We can see that for the 4-pole machines in Fig. 1.32a, Siemens machines are more compact (C_M is lower). At power ratings of 4 kW or 7.5 kW, the machines of ABB and Siemens are

[7] Compactness in this context means relatively low mass and small size.

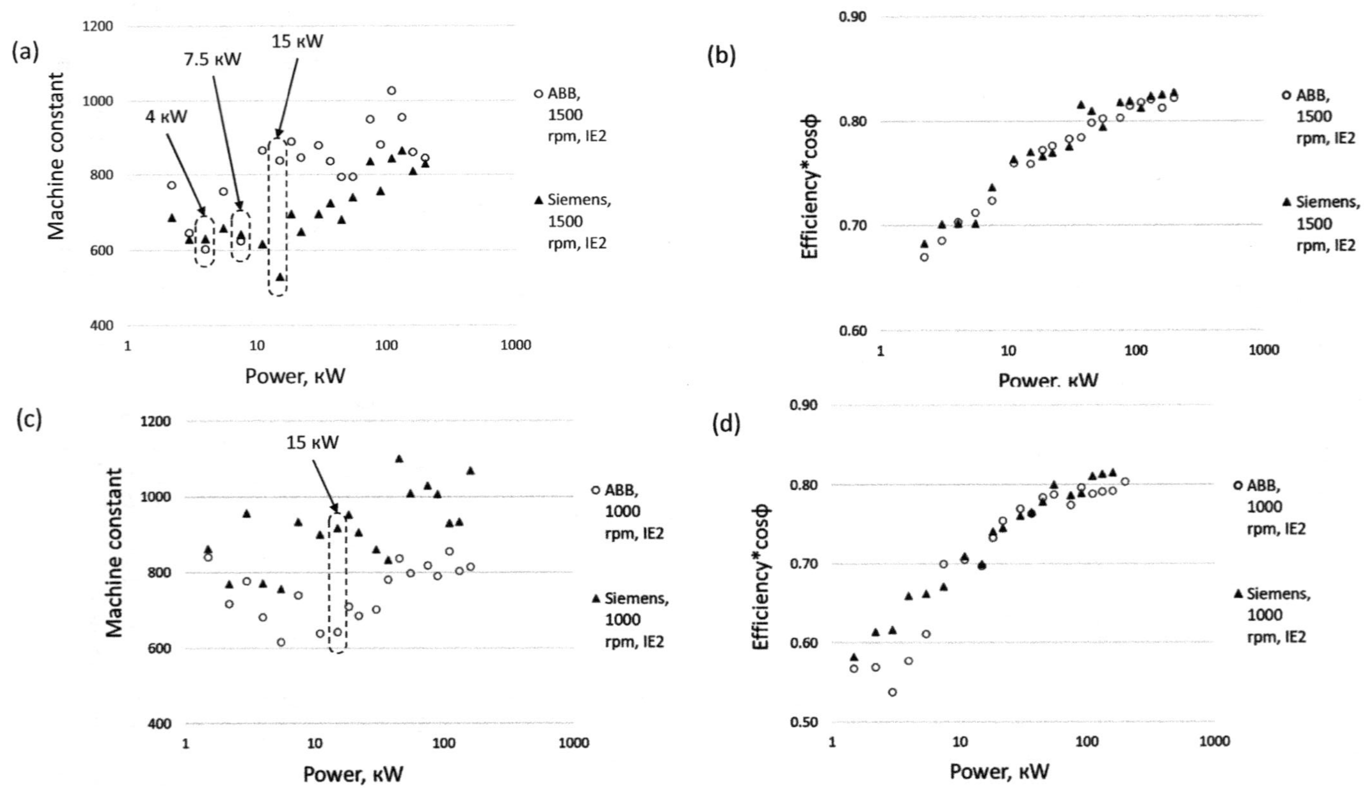

Fig. 1.32 Comparison of SCIM series of ABB and Siemens

comparable, but, for example, at 15 kW, Siemens has up to a 40% advantage in compactness. This difference in the compactness of 4-pole machines between the two series cannot be explained by differences in energy efficiency, as we see in Fig. 1.32b that the product of efficiency and cosφ is at the same level for both manufacturers.

Having carried out a similar analysis of 6-pole machines, we, however, come to completely opposite results. This time, ABB machines demonstrate greater compactness—see Fig. 1.32c, and at a power of 15 kW the advantage of ABB is the same 30–40%. In general, based on the results of comparing the 4 and 6 pole machines, it can be said that the ABB and Siemens machines in general look equal. Their product series are simply designed slightly differently.

The results of this analysis make it clear that comparing some new machine versus just a single machine from a catalogue or a research paper is not meaningful, as it can lead to completely incorrect conclusions. Additionally, we see that comparing series of machines or capabilities of manufacturers based on a few machines for some selected synchronous rotational speed (one selected pole number) is also misleading. Instead, all available rotational speeds should be considered. A more appropriate comparison would involve estimating the average value of the machine constant C_M for the entire series of machines.

To summarize, the parameter C_M allows not only comparison of individual machines but also comparison of entire series (product families), machine types, and, indirectly, the technologies used by the manufacturers. It is important to note that when comparing two individual machines, to draw adequate conclusions, the following conditions should be met:

- The machines (or series of machines) should have identical cooling systems and the same degree of protection.
- They should have the same rated voltage.
- They should be designed for the same application (with similar overload conditions, duty cycles, environment, etc.).

Additionally, it is important that the compared machines have the same composition of elements, such as housing, terminal box, and so on. The comparison should also take into account the efficiency level and power factor. It is also possible to compare only the active parts of the machines (without the housing, shaft, and other supporting structures), but then both machines being compared should not have a housing, shaft, etc. Obviously, if the choice between two electric machines leads to different system solutions, it becomes necessary to compare the entire systems.

Besides machine comparison, the constant C_M can also be used for scaling—estimating the mass and dimensions of a new electric machine based on machines with known mass and dimensions (see Appendix 2).

Final Comments

When introducing innovations involving electric machines, it is really important to recognize that each novel machine is evaluated in the context of informational selection through the prism of applications and system solutions, of which the machine is going to be a part of.

The task of estimating mass and size is the part of both the development of electric machines and the design of systems where electric machines serve as components. Comparing different machine options serves as the selection tool, which is why special attention was given to this topic and the key principle of how mass and size depend on certain factors was demonstrated. It should be noted that, firstly, the details of the proposed approach still need to be refined through practical application, and secondly, the methodology is suitable only for rotating electric machines whose primary function is efficient energy conversion, rather than, for example, precise positioning. In the future, the proposed approach may be adapted for linear electric machines as well.

References

1. A. Ivanov-Smolensky, *Electrical Machines* (Mir Publishers, 1988)
2. I. Kopylov, *Electrical Machines* (Energoatomizdat, 1986)
3. I. Kopylov, *Electromagnetic Universe*, 2nd edn. (Publishing House of MEI, 1999)
4. B. Kudrin, *Technetics: New Paradigm of Philosophy of Technology (the Third Scientific Worldview)* (Publishing house of Tomsk State University, 1998), 40 p.
5. B. Kudrin, Studies of technical systems as communities of products—Technocenoses, in *Systemic Research. Methodological Problems. 1980*, (Nauka, Moscow, 1981), pp. 236–254. http://www.kudrinbi.ru/public/10091/index.htm
6. Catalogue «Torque motors PM synchronous motors. Series TF water-cooled» Oswald Elektromotoren GmbH, www.oswald.de
7. Catalogue "ABB BU/Permanent magnet motors". GB 05–2004, www.abb.com
8. Z. Zhang, A. Matveev, S. Øvrebø, R. Nilssen, A. Nysveen, State of the art on generator technology for offshore wind energy conversion system, in *International Electric Machines and Drives Conference (IEMDC 2011), Niagara Falls, USA*, (IEEE International, 2011), pp. 1131–1136
9. R. Schiferl, A. Flory, W. Livoti, S. Umans, High-temperature superconducting synchronous motors: Economic issues for industrial applications. IEEE Trans. Ind. Appl. **44**, 1376–1384 (2008)
10. A.V. Matveev, K.A. Oladipo, V.V. Plytkevich, V.A. Kuzmichev, S.V. Shirinskii, Implementation of the Online Course "DriveConstructor", in *IV International Conference on Information Technologies in Engineering Education (Inforino-2018), Moscow, 23–26*, (2018)
11. P. Ruhland, K. Kvinnesland, A. Matveev, S. Coutandin, K. Nielsen, J. Fleischer, New production techniques for electric motors in high performance lightweight applications, in *Electric Drives Production Conference (E|DPC), 08–09 Dec*, (Ludwigsburg, 2020)
12. A.V. Matveev, The new machine constant and estimation of weight and dimensions of electric machines. Electrichestvo **11**, 45–53 (2019)
13. A.V. Matveev, Analyzing the types and series of induction machines on the basis of an alternative machine constant. Electrichestvo **7**, 41–56 (2020)
14. A.V. Matveev, Testing the hypothesis about the dependence of the mass of electric machines on the power and speed. Electrichestvo **7**, 4–18 (2021)
15. Catalogue "Low voltage process performance motors according to EU MEPS", 9AKK105944 EN, October 2013, ABB Motors and Generators. www.abb.com
16. Catalogue "SIMOTICS low-voltage motors type, series 1LE1, 1MB1 and 1PC1", D 81.1, 2013, Siemens. www.siemens.com

2 Driving Forces and Directions of Electric Machines Evolution

The Role of Major Inventions and Minor Improvements

In the long evolution of any kind of technical systems (such as electric machines), from time to time there take place interesting events—designers manage to modify and improve system characteristics that were initially considered fundamentally unchangeable and impossible to enhance. Often, these changes happen in completely unexpected ways, previously unknown or outright rejected. The idea of such a radical improvement is called an invention. The inventor has to break through real or perceived limitations, psychological inertia, and professional stereotypes. Naturally, not all inventions are of equal complexity and value. In the Theory of Inventive Problem Solving (TRIZ), inventions are categorized into five levels based on complexity. Below are the levels (from highest to lowest) with examples from the field of electrical machines[1]:

- **Level 5** (the highest). Creation of a completely new system with a new function. An invention of the 5th level is usually preceded by a great discovery. Examples: (1) the "voltaic pile" (Volta, 1800)—the first galvanic cell, preceded by the discovery of electricity; (2) the electromagnet (Sturgeon, 1820), preceded by the invention of the solenoid.
- **Level 4.** Creation of a new generation of a system that performs the same main function but operates on a new principle. The solution often lies in science rather than technology, and problems are frequently solved by integrating other fields of science; for instance, "mechanical" problems may be solved using chemical methods. Examples: Jacobi's direct current motor (1834), predecessors of induction machines[2] (works by Bailey, Ferraris, Tesla in the 1870–1880s).

[1] The inventions in the examples are classified into specific levels based on the author's subjective opinion.

[2] These inventions could also be classified as level three.

A. Matveev, *Shaping the Future of Electric Machines*,
https://doi.org/10.1007/978-3-031-95248-7_2

- **Level 3.** A radical improvement of an existing system using methods not known in the given field but known in other fields. These methods remain within the same branch of science, for example, mechanical problems are solved by mechanical methods, and chemical problems by chemical methods. On this level, at least one element of the system is completely changed. Examples: machine types emerging as developments of the early DC current machines (switched reluctance motor with mechanical commutator, Davidson, 1842) or the early induction machines (squirrel-cage induction motor, wound-rotor motor—works by Tesla and Dolivo-Dobrovolsky, 1887–1890).
- **Level 2.** The problem is solved using methods known within the given field. A skilled engineer is usually able to find a solution, but it may require considering 10–100 known options. Examples: many modern patents in the field of electrical machines.
- **Level 1.** The problem is solved using methods well known in the field. Almost any engineer can find a solution by reviewing 1–10 known options. Examples: most modern patents in the field of electrical machines, though the solutions presented in such patents can hardly be called inventions; they are more like "engineering improvements" to existing systems.

Many of the electric machine types widely used today were invented in the nineteenth century. The history of inventions in the field of electrical machines in that period is analyzed for example in [1]. The electric machine development history can be summarized in the simplified diagram shown in Fig. 2.1, where inventions of the 4th and 5th levels are labeled as "outstanding," and those of the 3rd level as "other significant" inventions.

Overall, the level of inventions decreases as the discipline progresses through history. In the twentieth century, new types of machines corresponding to the 3rd level were also invented, such as the stepper motor (1918, Woods) and the rolling rotor motor (1944, Moskvitin), but these were either "direct descendants" of the nineteenth-century machines or did not gain widespread use (and, accordingly, importance).

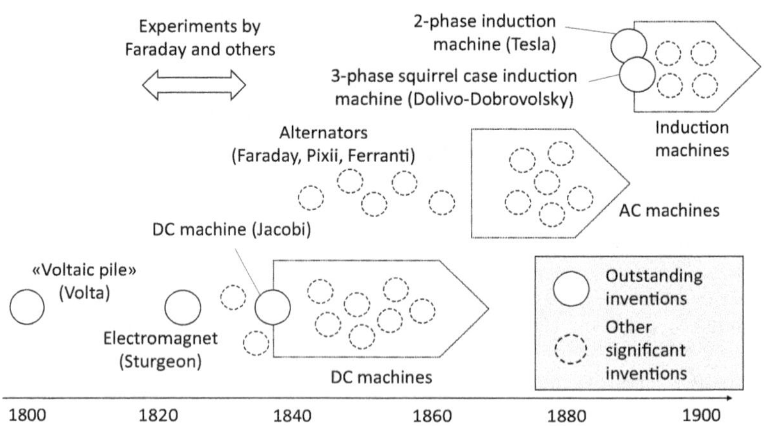

Fig. 2.1 Major inventions in the field of electrical machines in the nineteenth century

There is a stereotype that an invention is the result of a "sudden insight," meaning that invention and associated innovations are somewhat "random" processes. The randomness of discoveries is simply a characteristic of the very early stages of development of a particular science or engineering discipline, when there are few scientists, and they are exploring more or less random topics (in Electromechanics, this period was in the early nineteenth century). Thus, various interesting things are uncovered in a random order. But once some initial "excavations" succeed, the situation changes fairly quickly. The field attracts many researchers (in Electromechanics, already in the 1820s), and companies (in Electromechanics, from the mid-nineteenth century), and the next level of discoveries becomes almost inevitable—if one person does not make the discovery, someone else will, as it becomes a race to innovate.

Major inventions give rise to secondary tasks of various types—eliminating shortcomings, improving the sub-systems, solving secondary issues, adapting to new niches, and so on. All of these secondary problems cluster around the central invention and its subsequent implementation. While acknowledging the importance of inventions, we should not underestimate the role of minor improvements. In the effort to enhance a system, developers explore various options for improvements, taking certain steps, each simple and logical, testing them, and using each as a foundation for further steps. Numerous small improvements, though insignificant on their own, have a cumulative effect. In [2], technological progress is presented as a chain reaction of accumulating knowledge and skills, a process of learning[3]—an accumulation of experience that works very well within the framework of specific innovative activities and is primarily practical rather than theoretical. This accumulation of experience leads to faster changes. The chain reaction of development proceeds at a speed proportional to the achieved level (the more experience accumulated, the faster the development). At a certain point, when enough experience has been accumulated, a "breakthrough" occurs, leading to a new level—a kind of "quantum leap." Thus, the process of technological development is nonlinear, progressing through a series of "explosive" changes.

Usually, the "quantum leap" is represented by a single invention or a cluster of inventions, typically of the 2nd level, which are often formalized as patent applications. Thus, we can say that inventions give rise to secondary tasks and minor improvements. It can also work the other way round: accumulation of improvements can lead to significant inventions. In other words, one "major" invention can trigger a series of minor ones, and vice versa (Fig. 2.2). Here we can note that even the outstanding inventions of Tesla and Dolivo-Dobrovolsky were based on the experience accumulated before them, e.g., by Arago and Bradley.

Today, we are unlikely to witness revolutionary inventions in the field of Electromechanics. It can be said that the current development in electrical machines

[3]An important part of the aforementioned learning process is mathematical modeling and prototype testing, which allows for a significant reduction in the risks of implementing an idea directly on a full scale. Naturally, the success of modeling does not always guarantee success in the real-world operation of the machine, as models are limited and cannot capture all the system-wide interactions and effects.

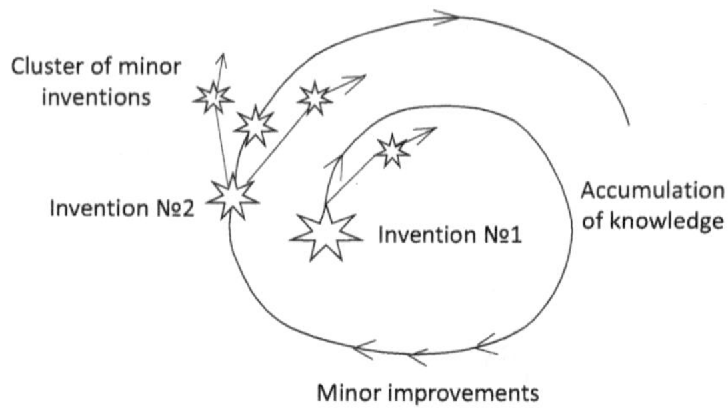

Fig. 2.2 Illustration of knowledge accumulation following a major invention and the emergence of new inventions

is largely driven by small, useful improvements (inventions of the 1st and 2nd levels). This means that the evolution of electric machines represents gradual, slow, but continuous progress. It is important to note that this kind of "normal" development follows certain strong patterns, which should be studied to create more competitive technologies and products. These matters are most comprehensively explored in TRIZ [3]. Having evolved from a methodology for solving technical problems into a science of the evolution of technical systems, TRIZ has provided the corresponding conceptual framework and analytical tools [4].

The emergence and development of technical systems, such as electric machines, is just one of many elements of technological progress, which in turn is one element of the evolution of human civilization. Technological progress can be viewed as the interweaving of numerous "streams," each of which includes the creation, implementation, operation, and eventual disappearance of some technical system. Technological progress is achieved through the practical realization of numerous and diverse improvements. Any technical system develops through the implementation of various ideas, which either modify the system or lead to the creation of a new one.

Evolutionary Trends, Patterns, and Lines

The background for the evolution of electric machines, as well as its driving forces, includes various modern *Trends*, such as, for example, increased attention to the environment and the growing use of polymer and composite materials. Evolution can be regarded as being the result of the interaction of numerous Trends with different weights and directions.

The evolution of technical systems is diverse and multidimensional, following a set of *Patterns* that coexist over time. Examples of such Patterns formulated in TRIZ [4] are the "Reduction of human involvement in the life of the technical system" and "Increased dynamism and controllability of the system". The same Patterns can apply to different technical systems. These Patterns illustrate and predict the changes occurring in a technical system during its evolution. Patterns represent stable, historically recurring tendencies, while Trends tend to be more changeable. Let us consider the two Patterns given as examples in more detail. The realization of the Pattern "Reduction of human involvement in the life of the technical system" in the case of an electric machine may lead in the near future to machine design done with the help of artificial intelligence (AI), manufacturing by robots without human involvement, diagnostics by AI using data from sensors integrated into the machines, and maintenance and repair carried out, again, by robots. The implementation of the Pattern "Increased dynamism and controllability of the systems" in the context of electric machines may imply a greater role for power and control electronics, as well as the emergence of more mechanical degrees of freedom due to the presence of more active parts or more complex and flexible options for mechanical integration. Here are a few more Patterns from [4] relevant to the evolution of electrical machines:

- Increasing the "intelligence" of systems created by humans
- Replacing human labor with machines in situations poorly suited for people
- Increasing the ideality[4] of systems
- Using resources untypical for biological evolution, such as very high pressure and temperature, powerful energy sources, dangerous substances, etc.

Evolutionary Trends and Patterns of higher-level systems ("supersystems") are the driving forces for systems of lower levels.

In addition to evolutionary Trends and Patterns, there exist also so-called *Lines of Evolution*, which show the sequence of changes a technical system undergoes during its evolution. Lines of evolution do not show the end result of evolution but rather how the system arrives at that result—this distinguishes Lines of Evolution from Evolutionary Patterns. One Evolutionary Pattern may include several Evolution Lines. An example of a Line of Evolution for a technical system is the sequence of two transitions (this concept is also coming from TRIZ): the first transition from a constant field to the use of a pulsating field, and the second transition from a pulsating field to a field pulsating at a synchronized frequency. Electric machines went through these transitions in the nineteenth century: from the Faraday's unipolar machine (constant field) to Henry's reciprocating motor (pulsating field), and then to Tesla's two-phase AC machine (synchronized pulsating fields, creating rotating field).

The evolution of electric machines can also be discussed in the context and terminology of technological paradigms[5] (we are currently transitioning to the 6th paradigm) or industrial revolutions (we are now in the 4th revolution). Electric machines emerged during the third technological paradigm and continue to exist in subsequent paradigms (the 4th and 5th). There is increasing mutual influence between electric machines and technologies from newer paradigms. For example, electronics, sensors, and artificial intelligence are playing an increasingly significant role in electric machines, while the electric machines in their turn power the fans of PCs and cooling systems of datacenters.

Further information on the development of technical systems can be found in [2, 4].

---

[4] Ideality – the notion, regarded in the next section.

[5] More information can be found in Chapter 6.

Applications of Electric Machines as the Key Formative Factors or Driving Forces

Most existing systems created by humans emerge, evolve, and change in order to meet both expressed and unrecognized user needs. Generally, consumers desire more functionality and quality at lower costs and with fewer negative consequences. In TRIZ, there is a concept known as the "ideality of needs." In [5], the degree of ideality of needs is proposed to be represented by the following formula:

$$I = \alpha \frac{\sum_{i=1,j=1}^{\infty} Qn_i Ql_j}{\sum_{k=m}^{0} C_k + \sum_{l=n}^{0} \beta_l H_l} \Rightarrow \infty, \tag{2.1}$$

where I is the degree of ideality of needs, Qn is the quantity of needs, Ql is the quality of needs, C is the time and cost required to satisfy needs, H represents harmful effects ("penalty factors"), i, j, k, l are the ordinal numbers of variables, and α, β are adjustment coefficients.

Ideality is probably the most universal quality criterion for systems and the primary objective function. In practice, many specific (narrower) criteria are used. Let us consider several applications and their corresponding criteria.

1. **Tidal Turbine.** As shown in Chap. 1, the primary design criterion for turbines utilizing tidal and ocean currents is the levelized cost of energy (C_E), which is defined as:

$$C_E = C_\Sigma / E_\Sigma, \tag{2.2}$$

where C_Σ is all costs, E_Σ is the total energy generated during the turbine lifetime.

$$C_\Sigma = \sum_{t=1}^{n} \frac{(I_t + M_t)}{(1+r)^t}, \tag{2.3}$$

where I_t is the investment cost in year t, M_t is the operational and maintenance costs in year t, n is the system's life cycle.

$$E_\Sigma = \sum_{t=1}^{n} \frac{E_t}{(1+r)^t}, \tag{2.4}$$

where E_t is the electricity generated in year t, r is the discount rate.

The cost of the generator is included in the investment costs. The generator's mass and dimensions may affect the cost of other turbine components. The efficiency of the generator determines the amount of energy produced, and high reliability is crucial since repairing tidal turbines installed underwater is very expensive. In practice, the generator characteristics that influence C_E most are prioritized in the following order: high efficiency over a wide load range, high reliability, low cost, and optimal mechanical integration into the turbine's structure.

2. **Pump in a Water Supply System for Households and Utilities.** For such systems, the primary objective function is typically the lifecycle costs:

$$C_\Sigma = I + C_e + C_o, \tag{2.5}$$

where I represents the investment costs, C_e is the cost of consumed energy, C_o refers to the operational expenses over the pump system's lifecycle.

The efficiency of a motor driving a pump determines the amount of energy consumed, and the cost of the motor is part of the investment expenses. In practice, when trying to minimize C_Σ, high efficiency is much more important than the purchasing cost of the motor. Therefore, efficiency is the key objective function for electric machines in this application.

3. **All-Electric Aircraft (AEA).** The main requirements in designing any passenger aircraft—ensuring safety and operational cost-effectiveness—are addressed together. The objective function related to cost-effectiveness is the cost per passenger (a), which can be represented as follows:

$$a = \frac{C}{N_p \cdot L}, \tag{2.6}$$

where C represents the costs, N_p is the number of passengers transported, L is the distance passengers are flown.

In an AEA, part of the costs C will come from the energy consumed. The key characteristic of AEA, directly impacting the widespread adoption of this type of transport, is flight range. To carry the maximum number of passengers over the greatest distance, the AEA must have minimal weight. Additionally, safety considerations demand high reliability from the propulsion system. Thus, low weight of electric motor driving the propeller and motor's reliability are two the main criteria. High degree of mechanical integration of the electric motor into the AEA's body structure can further positively affect the aircraft's overall weight. High efficiency of the propulsion motor will reduce energy consumption costs, ease of heat dissipation, and increase the flight range by reducing battery drain.

Information from the three examples discussed is summarized in a more concise form in Table 2.1. We can see that the objective functions for electric machines vary significantly depending on the application. In the context of informational selection, these objective functions are of primary importance.

There exist applications where electrification is the result of optimization based on objective functions not directly related to economics. For example, the replacement of diesel engines with electric motors in propulsion systems of some ferries is currently driven more by environmental concerns, like pursuing the goal of reducing carbon dioxide emissions. In that application, the primary requirement for the electric motor is compactness, as the space in the engine room may be limited. If the electric motor meets this requirement, making electrification of this concrete ferry feasible, then the objective function for the motor will be more or less the same as for an all-electric aircraft (passenger transportation efficiency).

Table 2.1 The influence of applications and their objective functions on the objective functions of electric machines

Supersystem (higher-level system)	Main objective function of the supersystem	Expression	Main objective functions for electric machine in priority order
Tidal turbine	Levelized cost of energy	$C_E = C_\Sigma/E_\Sigma$	High efficiency, high reliability, low cost, mechanical integration
Water supply pump	Life cycle costs	$C_\Sigma = I + C_e + M$	High efficiency, low cost
All-electric aircraft	Passenger transportation efficiency	$a = \dfrac{C}{N_p \bullet L}$	Low weight, high reliability, high efficiency, high degree of mechanical integration, redundancy

To predict the evolution of electric machines, it is essential to understand the evolution of the supersystems and their applications. The short-term evolution of the system (electric machine) is determined by the dynamics of development, the history, and the resources of the system itself. In contrast, long-term evolution of the system is shaped by the supersystem (and the application of electric machine), and the longer the forecast period, the higher the level of the supersystem that influences and determines the evolution of the system (the electric machine).

For example, since the late twentieth century, wind turbines and tidal turbines have gone through a phase of determining the optimal structure of the power conversion system (drive train). In laboratories and on the markets, systems with gearboxes (with different numbers of stages) and direct drives have competed. This process is still ongoing, and all possible system variants are present in the market, although the share of direct-drive systems is steadily increasing. Clearly, the gear ratio of the gearbox determines the characteristics of the electric generator, and often its type as well. That is, the success of certain drive train structures decisively influences success of the corresponding electric generators and the level of investments in these particular types and topologies of electric generators.

Evolution Along the "Supersystem-System-Subsystem" Axes

Related systems, belonging to different hierarchical levels, evolve together and in a coordinated manner. This refers to the supersystem, system, and subsystem (Fig. 2.3).

If an electric machine is considered as "system", the "supersystems" in relation to electric machines include the electrical grid, machine control systems based on power electronics, the informational environment (today, this involves automated process control systems, various monitoring systems, and, possibly, the Internet of Things), the production environment, as well as applications—systems powered by the electric motors (e.g., a submersible pump, electric vehicle, elevator) or systems that drive the generators (e.g., turbines).

The subsystems of electric machines include the materials from which the machines is made, as well as various components and parts of the machines. In

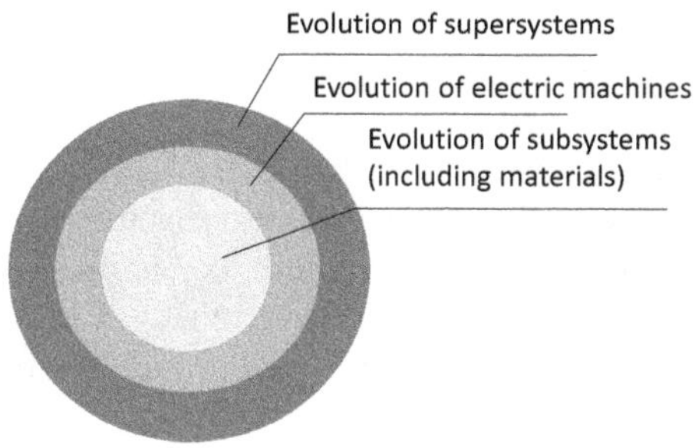

Fig. 2.3 Three hierarchical levels of evolution

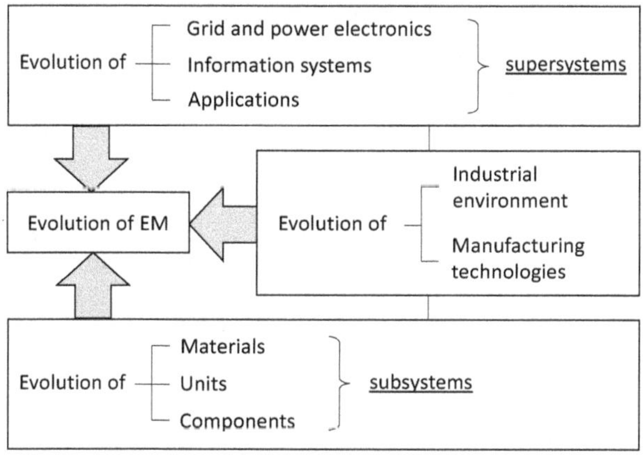

Fig. 2.4 Supersystems and subsystems influencing the evolution of electric machines (EM)

parallel with the development of new materials, new production technologies are emerging.

One could debate where to categorize new production technologies—whether they belong to the supersystem (alongside the production environment) or the subsystem (related to materials). For example, a coil printed on a 3D printer is a component of the electric machine, which means it belongs to the subsystem. At the same time, the 3D printer is clearly part of the production environment. A compromise interpretation is proposed in Fig. 2.4, where the production environment and production technologies are highlighted as a separate block that influences the evolution of electric machines and is connected to both the supersystem and the subsystem.

Supersystems: The Electric Grid, Power Electronics, and the Information Environment

The widespread integration of power electronics devices into networks began back in the 1970s, and today, the presence of electronic frequency and voltage converters, including high-power ones, has become commonplace in both the large-scale grids and micro-grids. The emergence and evolution of frequency converters have had, and continue to have, a significant impact on the development of electric machines.

First, a range of machine types has emerged, such as switched reluctance motors, synchronous reluctance motors, and permanent magnet synchronous machines, which cannot operate without power electronics converters. Second, electrical grids are becoming more "polluted" with high-frequency harmonics in current and voltage waveforms, and electric machines operating directly from the grid must have the necessary "immunity" to this distortion. Third, when using electric machines with frequency converters, changes in the machine design are required—particularly in the insulation system and bearings. It is well known that many technical systems evolve together with other systems, with the strength of the connection between such systems intensifying as they co-evolve. The co-evolution of electric machines and power electronics frequency converters is a good example of this, and one distinct aspect of this evolution is the gradual integration of the electronics and the electric machine into a single system, eventually merging into a unified module (in TRIZ terms this process can be referred to as "integration into the supersystem").

The evolution of a technical device or system can be described by changes in its key parameters over a long period of time. The development of power converters can be seen as the continuous improvement of quality indicators such as efficiency, specific power (by weight and volume), and cost. The increase in specific power of power electronics (PE) converters since 1970 is illustrated in Fig. 2.5, which also reflects the time interval (about 10 years) between the appearance of prototypes in laboratories and the release of products on the market. The volume and mass of PE converters are becoming increasingly insignificant compared to the volume and mass of the electric machines they control. The cost of PE converters is also decreasing relative to the cost of the machine. This is an important and noticeable trend, which will likely result in the full integration of PE converters and electric machines, at least in low-power systems, in the foreseeable future.

For integrating electric machines into the information environment to enable e.g. condition monitoring and predictive maintenance, sensors and data acquisition

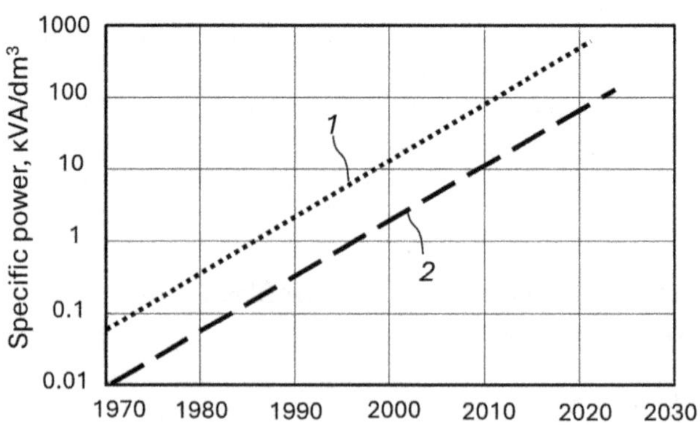

Fig. 2.5 Increase in the specific power of PE converters since 1970 [6]: Line 1—systems in the R&D stage, Line 2—systems already available on the market

devices are needed, including both contactless and integrated sensors within the machine. Research in this area is actively ongoing.

Manufacturing Environment

In line with the Evolution Patterns discussed above, the manufacturing environment is evolving toward reducing human involvement in the production process. Each year, the level of automation and robotization in electric machine manufacturing plants increases. Flexible automation systems are widely used, robots are utilized for assembly tasks. When designing a machine, it is important to consider the specific manufacturing processes and specifics of the plant where the machine is going to be produced.

Evolution of Electric Grids

The share of energy generation from renewable sources, such as wind and solar power, is steadily increasing. Energy generation is becoming more decentralized. Energy storage systems of various types are being integrated into the grid, leading to more elements with bidirectional/reversible energy flows. More autonomous networks are also emerging. The grid is becoming "smarter" to manage various instabilities. The growing presence of power electronics devices, including high-power ones, in the grid is increasing harmonic distortions, which can cause electromagnetic compatibility issues with some newly implemented systems. Electric machines are key components in many of the new power generation and energy storage systems, such as flywheel-based storage systems, so grid quality and stability must be considered when designing such electric machines.

Subsystem Evolution

The importance of subsystem evolution is illustrated in [2] where it is shown that the power output of electric turbogenerators increased from 0.8 MW in 1917 to 540 MW in 1972, while the overall design of the turbogenerators remained almost unchanged. This significant increase in power was made possible due to the development of materials and improved processing technologies, as well as advancements in turbogenerators' auxiliary systems, such as balancing, control devices, excitation, and protection systems. This illustrates the evolution of subsystems, primarily the materials used in the construction of electric machines.

Evolution of Materials

The relative importance of engineering materials changes throughout the evolution of humanity the graphical interpretation of this is shown in Fig. 2.6. These trends are also reflected in the domain of electric machines. We can see this, in particular, in the use of ceramic bearings, epoxy resins for consolidating windings, and composite materials in load-bearing structures of the machines.

Let us consider some examples of the evolution of magnetic and conductive materials. Among the main research directions for soft magnetic materials (electrical steels), we can highlight the following:

- Improvement of the steel properties
- Use of thinner laminated sheets
- The possibility of creating more complex core shapes (powder metallurgy technologies)

Improvements in steel properties include increasing saturation flux density, mechanical strength, maximum operating temperature, magnetic permeability, as well as reducing losses and reaching lower temperature coefficients. An example of the improvement of one of the properties is the trend of reducing losses in steel (values halved over 50 years) across various grades of JFE Steel (Fig. 2.7). However, it should be noted that improving some properties in practice is often possible only at the expense of deteriorating others.

Among the main research directions for hard magnetic materials (permanent magnets, PMs) are the increase of energy product (EP), improvement of mechanical properties, and performance at high temperatures. For PM, the critical system parameter that can describe its evolution is EP. Interestingly, the historical growth of EP (Fig. 2.8) resembles an S-shaped curve, which may indicate a development limit for rare-earth-based PMs is already reached. Some of the materials used to produce PMs are approaching their theoretical limits. Nevertheless, the level of EP achieved

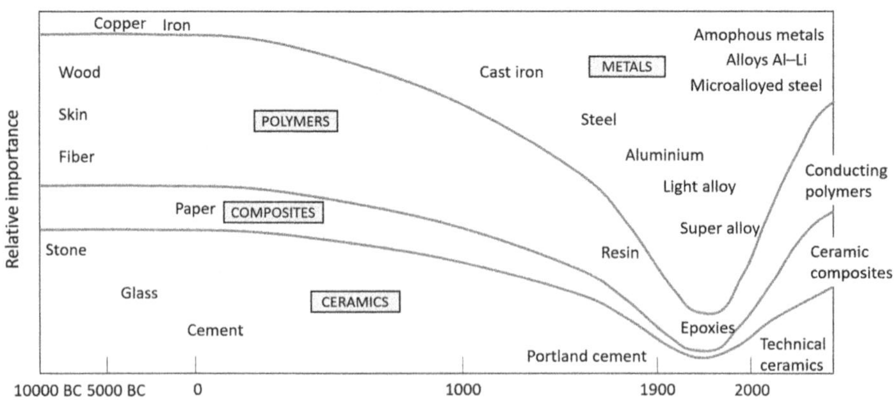

Fig. 2.6 Changes in the relative importance of engineering materials

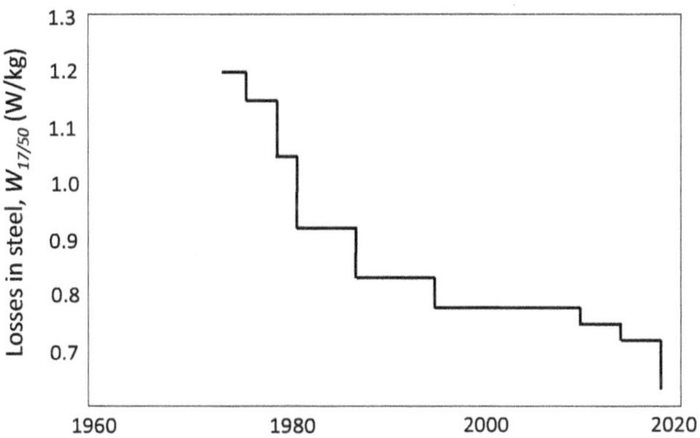

Fig. 2.7 Reduction of steel losses in the development of various grades of JFE Steel [7]

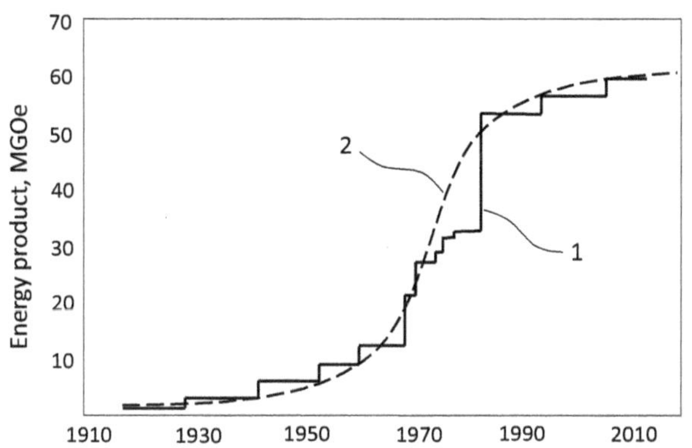

Fig. 2.8 Improvement in the energy product of PM materials: 1—actual values, 2—envelope curve

by the late twentieth century was already sufficient for the widespread adoption of PM-based machines. In the foreseeable future, new "non-rare-earth" types of PMs may emerge, possibly based on nanotechnologies.

In the field of conductive materials, carbon nanotubes (CNT) and superconductors deserve particular attention. For the practical application of CNTs, it is necessary to connect individual CNTs into fibers, which is a complex challenge. Various methods are currently being developed to create long fibers based on CNTs, which can then be woven into multistrand wires. However, the properties of the resulting multistrand CNT-based wires are still significantly worse than those of individual CNTs. As of the early 2020s, the electrical conductivity of CNT fibers has reached 15–20 MS/m (Fig. 2.9), meaning that the conductivity of copper has not yet been achieved.

The superconductors (SCs) BSCCO, YBCO, which appeared in the late 1980s, and MgB_2, which emerged in the twenty-fist century, are already used in electric machines, though primarily in prototypes for now. These SCs can be cooled with liquid nitrogen, unlike the first-generation Nb-based SCs, which required cooling

Fig. 2.9 Increase in electrical conductivity of CNT fibers over time [8]

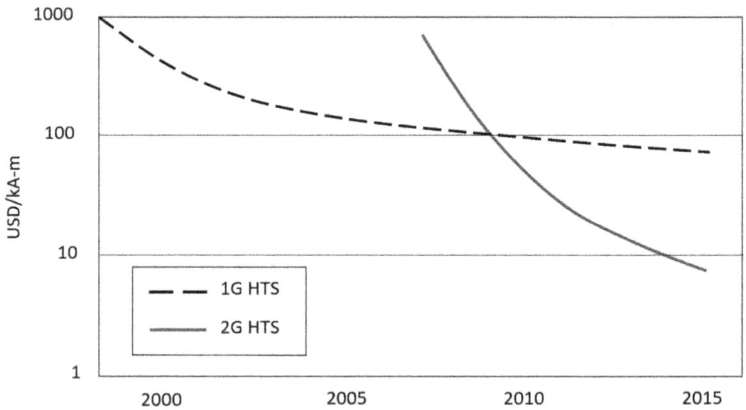

Fig. 2.10 Price trends for two generations of superconductors

with liquid hydrogen or helium. Currently, there are ongoing developments to create SCs that can operate at room temperature. Recently developed SCs such as H_2S, YH_6, YH_{10}, and LaH_{10} already work at temperatures above 200 K, but only under very high pressures—over 150 GPa [9]. These developments do not guarantee the appearance of SCs that operate at room temperature and normal pressure in the near future, but they make it quite likely in the foreseeable future.

An important factor that has long hindered the widespread adoption of SCs is their extremely high cost. In [10], historical data and projections for the price reduction of first- and second-generation SCs are provided. The rate and absolute magnitude of the price decrease are significant—practically by 1–2 orders of magnitude over 5–10 years (Fig. 2.10). The predicted price reduction is usually associated with the development of the SC market and the increase in production volumes. The forecast is outdated now, but it still supports the main idea well.

It would not be an exaggeration to call some of the improvements in active materials over the past decades radical. Alongside the development of active materials, there is also progress in insulation and structural materials. In particular, in the field of insulation materials, work is being done to improve electrical strength, thermal conductivity, lifespan, and resistance to high temperatures. Various additives, including nanocomposites, are being tested. For example, between 1995 and 2002, the electrical strength of polyethylene terephthalate (PET) increased from 300 MV/m to 520–600 MV/m [11].

We can roughly estimate the impact of the introduction of new materials that may appear in the future. Suppose a new conductive alloy is invented, with electrical conductivity twice that of copper. Using such a conductive material in electric machines would mean halving winding losses or maintaining the same losses while increasing the current by $\sqrt{2}$ times. If a new magnetic material with saturation flux density twice that of modern laminated magnetic steel is also invented, the flux density in the air gap could be increased by about two times. As a first approximation, we can assume that the volume of the electric machine is inversely proportional to the current density and the flux density in the air gap. The use of both of the aforementioned materials would lead to a reduction in the machine's volume by $2\sqrt{2} \sim 2.82$ times. The weight could also be reduced by approximately the same proportion. This would be a significant improvement, leading to radical changes in applications where the mass and size of the machines are critical, such as in aircraft and exoskeletons.

Evolution of Design and Manufacturing Methods and Tools

Fig. 2.11 shows the major milestones in the development of design methods and computational capabilities. In recent years, tools have emerged for solving multiphysics and coupled problems. Neural networks and machine learning have started to be used, and the datacenter infrastructure for distributed computations has appeared. In the near future, it will likely become possible to conduct virtual testing of electric machines.

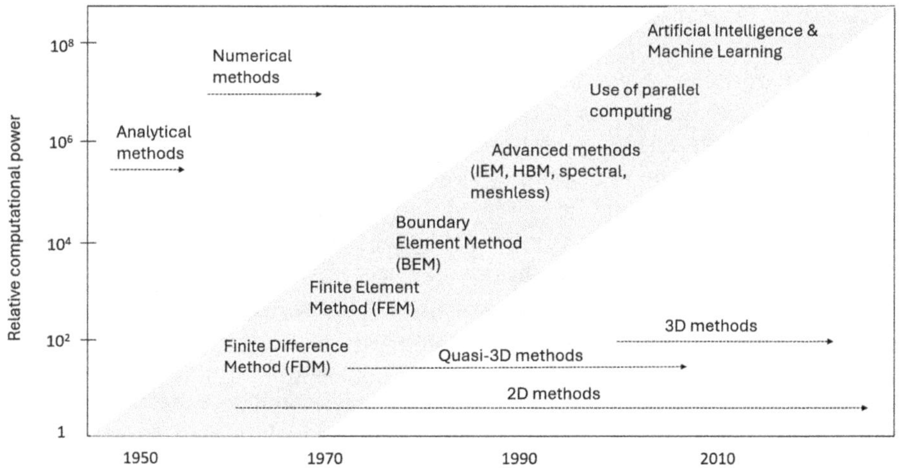

Fig. 2.11 Evolution of design methods and computational capabilities

In the field of electric machine manufacturing, increasingly advanced CNC machines and automated production complexes are offering new possibilities. Overall, there is a widespread increase in the degree of automation and robotization. Advanced methods and tools for measuring and assessing the quality of components and materials (such as permanent magnets) are also emerging. An example of new manufacturing technologies includes additive manufacturing and lithography [12], which, for instance, allows the "printing" of windings with a high copper fill factor for certain types of micromachines.

Directions of Evolution of Electric Machines

The main directions of evolution can be outlined as follows:

- Increasing maximum achievable power and torque
- Improving controllability
- Increasing maximum achievable rotational speed
- Enhancing energy conversion efficiency
- Increasing specific power and torque (improving compactness, reducing mass)
- Increasing the degree of mechanical integration with relevant surrounding structures (integration into applications—supersystems)
- Miniaturization (appearance of progressively smaller machines designed for practical purposes)
- Emergence of new machine types displacing the old types

Let us illustrate progress in some of these directions, taking first the increase of maximum achievable power.

Increase of Power Capability

The most powerful machines are turbogenerators. Over the past 60 years, the power of the largest turbogenerators has increased almost tenfold (see Fig. 2.12) [13]. Today, humanity has reached power levels of above 2 GW (for 4-pole turbogenerators). Some scientists think that this is close to the limit of modern technology. To quote [14]: "Power can only be increased up to certain limits. Calculations show that surpassing the upper limit, constrained by the 2500 MW turbogenerator, whose rotor rotates at 3000 rpm, will not be possible because this limit is primarily determined by the strength characteristics: the stresses in the mechanical structure of a higher-powered machine increase to the point where centrifugal forces will inevitably cause rotor failure." Additionally, transporting such large machines (over 500 tons) presents significant challenges. However, there is confidence that in the future, with the emergence of new active and structural materials, it will be possible to significantly improve the compactness of turbogenerators and surpass the 2.5 GW limit (for 2-pole machines with the speed of 3000 rpm). 4-pole machines with the speed of 1500 rpm will likely cross this threshold earlier than the 2-pole ones.

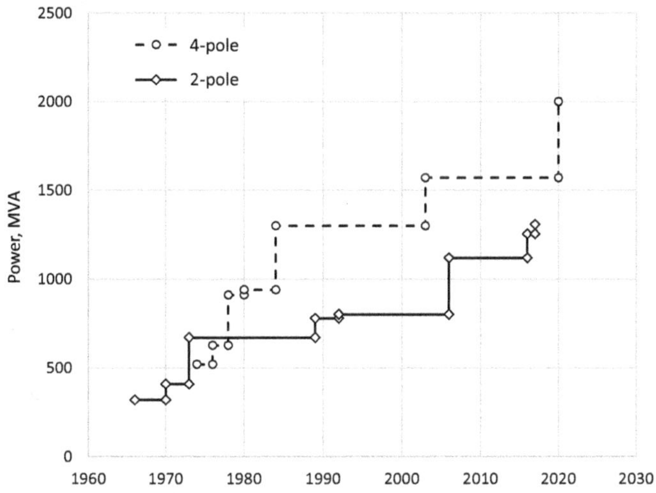

Fig. 2.12 Historical increase in the power of turbo generators

Increase of Torque Capability

As an example of growth of maximum achievable torque in electric machines, let us consider machines with high-temperature superconductor (HTS) excitation windings. Fig. 2.13 shows that almost immediately after the emergence of HTS, there was a steady increase in the maximum achievable nominal torque of such machines. Today, the torque level has exceeded 10 million Nm.

Miniaturization

There is also ongoing work in different companies in the direction of miniaturization. For example, at Orbray, a company that manufactures micromachines, the minimum machine size they could produce decreased from 10 mm to 0.6 mm between 1973 and 2014 (Fig. 2.14).

Increase of Maximum Speed Capability

Thanks to the capabilities of power electronics in generating high-frequency currents, the maximum achievable rotational speed of machines is increasing. In low-power machines, rotational speeds of one million rpm have already been reached.

It's important to note that record high power levels in machines are achieved at relatively low rotational speeds (1500 rpm), while record rotational speeds are reached in machines with low power (less than 1 kW). Fig. 2.15a shows applications that outline the boundaries of current capabilities in terms of power and rotational

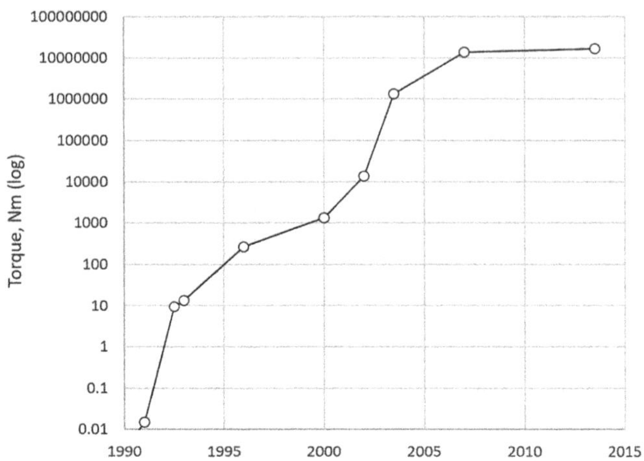

Fig. 2.13 Historical increase in torque for HTS machines, data from [15]

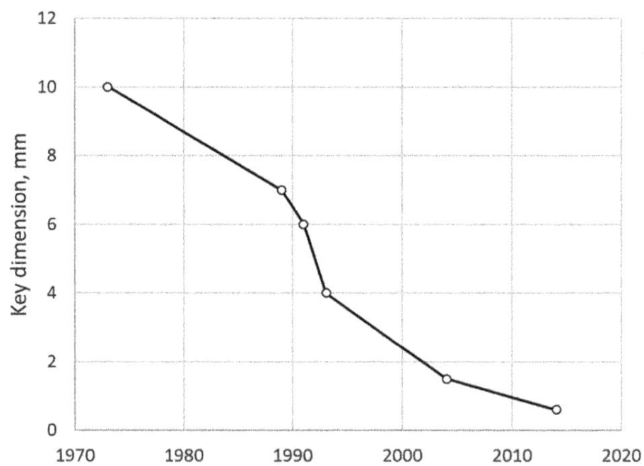

Fig. 2.14 Reduction of the minimum possible machine size [16]

speed. We can envision evolution as an attempt to push the boundary in Fig. 2.15b upwards and to the right and down and to the left.

Increase of Controllability and Efficiency

Let us note that power electronics significantly increase the controllability of electric machines overall.

As an example of the efficiency increase in machines during historical development due to incremental improvements in materials and optimization, we can cite the growth in efficiency of traditional Siemens generators with a capacity of 200 to 500 MW (Fig. 2.16).

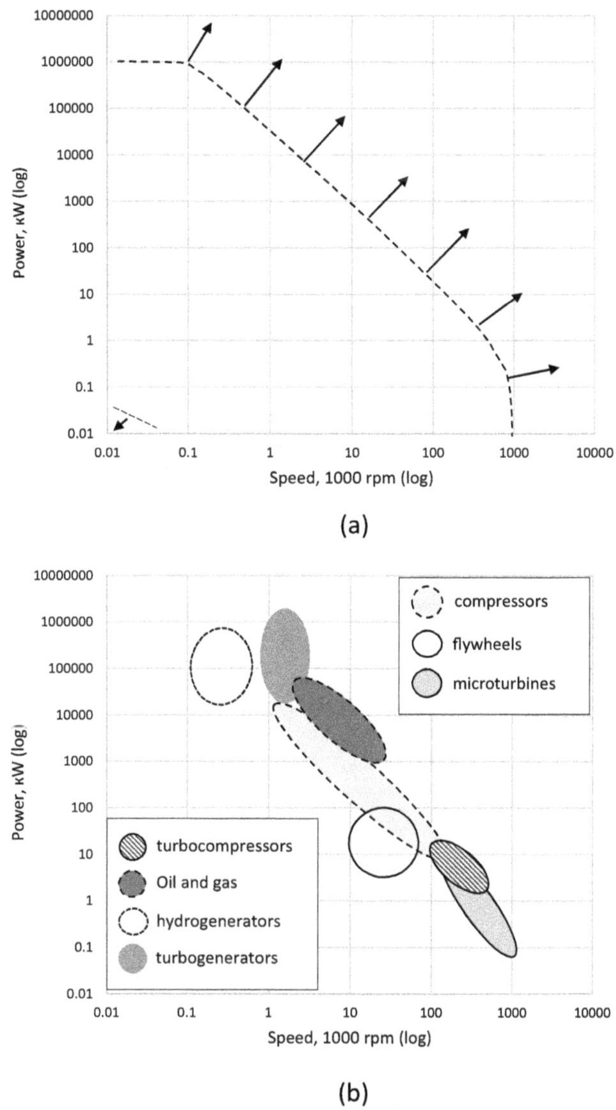

Fig. 2.15 Applications of machines with high power and rotational speed (**a**) and schematic representation of the "feasibility limits" being stretched (**b**)

Improving Compactness and Reducing Mass

In practice, compactness (a common term associated with volume) and the machine's mass are usually directly related. "High compactness" of a machine is often reasonably understood as "low mass." This is the sense in which the term "compactness" will be used below. In Chap. 1, the constant C_M was introduced. The lower the value

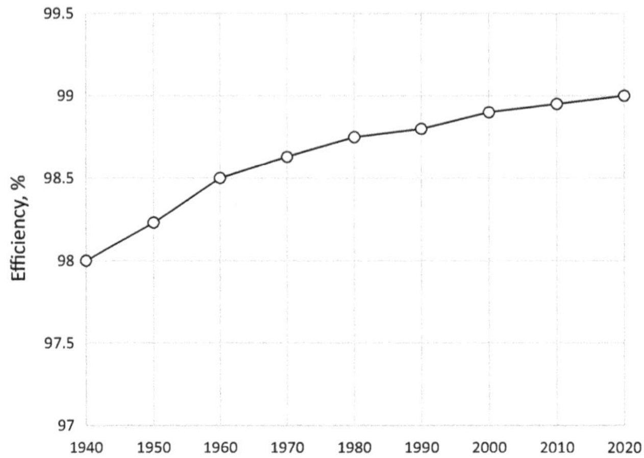

Fig. 2.16 Trend of efficiency improvement in Siemens generators [17]

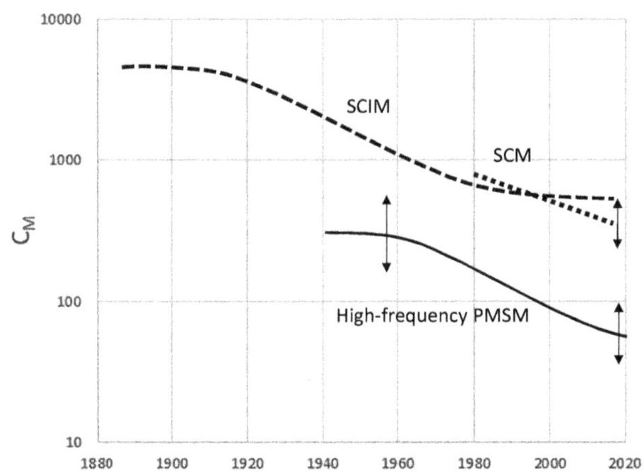

Fig. 2.17 Trends in the reduction of the machine constant

of C_M, the more compact the machine. Fig. 2.17 shows the evolution of the compactness of high-frequency permanent magnet synchronous machines (HF-PMSM), superconducting machines (SCM) and squirrel cage induction machines (SCIM) [18]. Vertical arrows indicate the ranges in which the values of C_M for this type of machine typically fall. We can see that HF-PMSM is significantly more compact than SCIM and SCM. This is primarily the result of using high-frequency power supply and the effect of utilizing a large number of poles.

Table 2.2 presents a comparison of a high-frequency (HF) system with a PM motor design for a drone propulsion [19] against a traditional system with an induction motor and a frequency converter (FC-IM) designed for general industrial

Table 2.2 Comparison of the conventional and the high-frequency systems

	Conventional industrial system (ABB)	High-frequency UAV system (T-motor)
Electric machine (EM)	**M3BP 80 MC**	**P80 kv100**
Rated voltage, V	400	40
Rated power, кW	1.1	1.1
Rated speed, rpm	2900	2934
Number of poles	2	42
Rated frequency, Hz	50	1050
Efficiency, %	81.8	~ 85
Protection	IP55	IP55
Diameter, mm	160	91.6
Length, mm	225	40
Volume, cm³	5760	336
Mass, kg	18	0.650
C_M	900	33
Frequency converter (FC)	**ACS580–01-04A0–4**	**Flame 80 A HV**
Rated voltage, V	400	48 (DC side of the inverter)
Rated power, кW	1.5	>2
Rated current	3.8 A	80 A
Efficiency, %	~ 98%	~ 97%
Dimensions (W × H × D), mm	222 × 303 × 125	84 × 35.5 × 19.5
Volume, cm³	8408	58
Mass, kg	5.1	0.109
System (EM + FC)		
Volume, cm³	14168	394
Mass, kg	23.1	0.759

(It should be noted that ABB is the world leader in induction motor technology for industrial applications. This comparison is not a comparison of companies who are among the best ones in their respective product areas, but comparison of the system types)

purpose [20]. The difference in weight and compactness is very significant[6]. In comparison with the HF system, the FC-IM system appears like being from a previous generation of technical systems. It can be concluded that systems from different generations coexist in the market today. Fig. 2.18 shows the traditional system (a) and the HF system (b), with the scale of the drawings chosen to illustrate the size differences between the corresponding electrical machines and frequency converters. This crucial advantage of HF systems with PMSM could potentially lead to new

[6] It should be noted that intensive air cooling created by the UAV propeller and the lower voltage plays an important role in compactness of the HF-PMSM system.

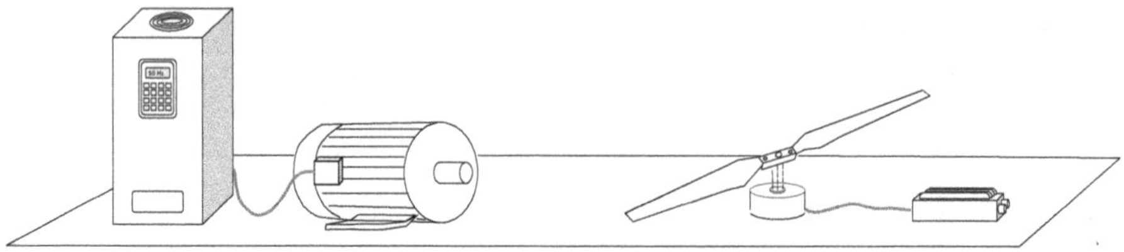

Fig. 2.18 Examples of conventional system (**a**) and high-frequency system (**b**)

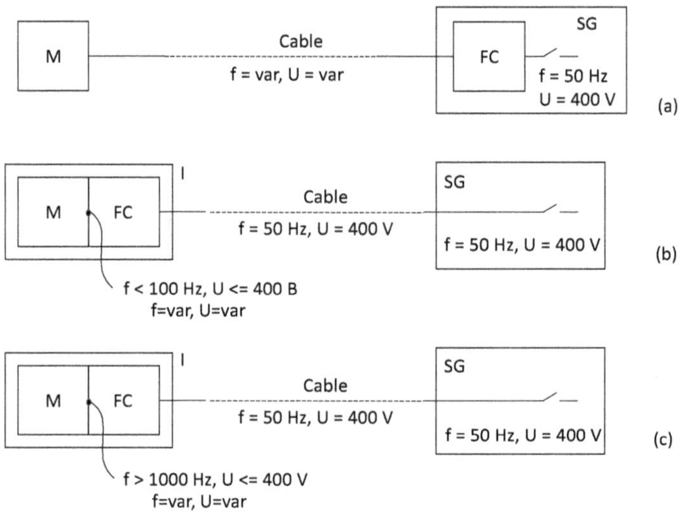

Fig. 2.19 Evolution of system topologies: *M* machines, *VFD* variable frequency drives, *SG* switchgear, *I* integrated module

standards in the industry. Machines currently considered "special-purpose" could move into the category of "general industrial" machines in the future.

Fig. 2.19 presents a forecast[7] for the potential development of variable speed drive systems. In Fig. 2.19a, the most common topology is shown, with the frequency converter (FC) located in a switchgear cabinet and the motor connected to the FC via a long cable. Over the past 20 years, integrated drives have been introduced, where the FC and the motor form a single module (Fig. 2.19b). In the future, it is likely that the advantages of high-frequency systems will be utilized, and the frequency between the FC and the motor (such as, e.g., a PMSM-type motor) will be significantly increased (Fig. 2.19c). It is quite possible that the motor in such a system will have multiple galvanically isolated multiphase windings.

The evolution of compactness naturally affects not only permanent magnet machines but other types of machines as well. For example, in [14] data is presented indicating that, through improvements in design, use of forced cooling, and the use of insulation materials that tolerate higher temperatures, it was possible to increase the specific power of DC machines by three times over a period of 50 years

[7] One of many other possible forecasts.

(approximately from 1935 to 1985). It is also reported in [14] that the specific material consumption in turbogenerators decreased by more than three times between the 1950s to the 1990s.

Increasing the Degree of Mechanical Integration

One direction of evolution is the increased degree of mechanical integration with surrounding structures (integration into the application—higher-level systems, in TRIZ terms). In typical permanent magnet generators for direct-drive wind turbines, the mass of the generator's supporting structures can account for up to 50–60% of its total mass. Through mechanical integration, structural mass can be significantly reduced. This principle is well illustrated by the Sway wind turbine generator (Fig. 2.20). By significantly increasing the diameter, the circumferential speed increases, which positively affects the generator's mass—essentially creating an "inbuilt gearbox." The Sway generator was an axial-flux PMSM without iron cores in the stator (ironless PMSM), allowing the structural elements supporting the active parts to be lightened since in axial-flux ironless machines there are no axial attraction forces between the stator and the rotor. The wind turbine blades themselves largely serve as the supporting structure for the rotor.

Let us note that the high compactness of HF-PMSM also opens up new possibilities for mechanical integration with mechanisms which are the parts of the application (transition into a higher-level system in TRIZ terms).

Transition to New Types of Electric Machines

One of the characteristics of the evolution process is the displacement of certain types of machines by others. The number of patent applications for a specific technology is an indirect indicator of the level of investment and the number of developers involved in the advancement of a certain technology.

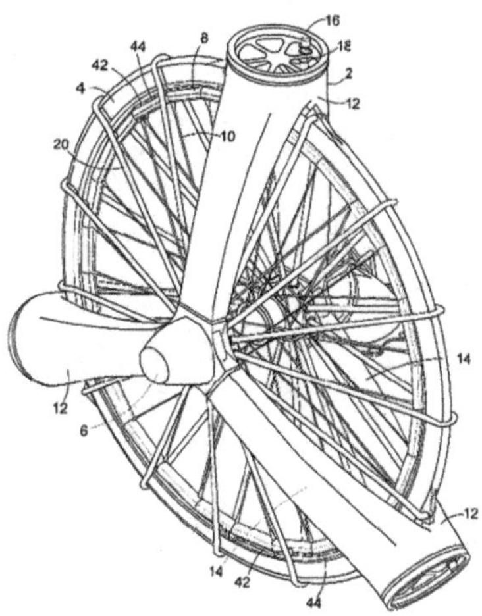

Fig. 2.20 Sway wind turbine generator: illustration from patent CN103987962A2

Fig. 2.21 shows the historical dynamics of patent applications for four types of electric machines in absolute (a) and relative (b) numbers. The number of inventions related to permanent magnet synchronous machines (PMSM) has been growing over the last 20–25 years, while for induction machines (IM) and direct current machines (DCM), it has remained at the same level. Interest in switched reluctance machines (SRM) is significantly lower than in PMSM. In Fig. 2.21b, it is clearly seen that the relative shares of IM and DCM are decreasing. It is interesting to compare the curve for the change in C_M for IM (Fig. 2.17) with the declining share of IM patents—it becomes evident that investments in IM are decreasing compared to investments in other types of machine, and consequently, this type of machine is not developing as intesnively as in the past. Against this backdrop, the rapid development of PMSM is obvious.

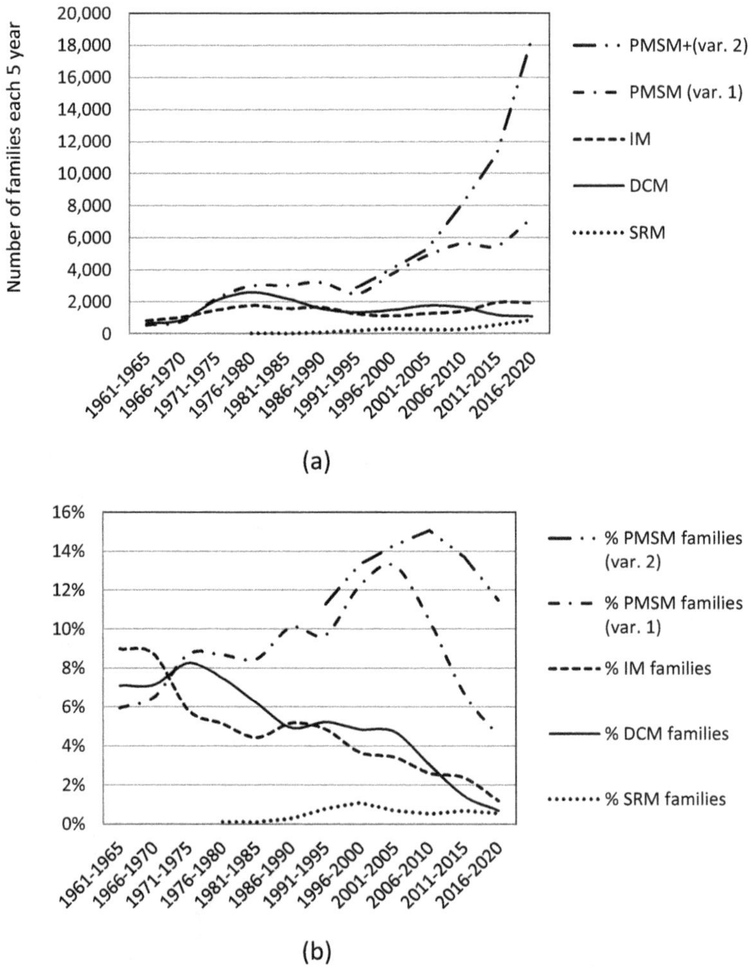

Fig. 2.21 Historical dynamics of patent applications

Appendix 3 provides more detailed statistics on patents for various types of machines, including leading countries and companies. This data can offer some insights into global competition in the field of electric machines in a historical context.

In Chap. 1, Fig. 1.19 presented a logarithmic map that links specific power and torque with nominal power and rotational speed. The map reflects the fact that electric machines are a class of devices that largely follow logarithmic laws when it comes to the relationship between mass, power, and rotational speed [21, 22]. We can attempt to illustrate the evolution of the compactness of some of the machine types discussed earlier using this map, as shown in Fig. 2.22. A given electric machine technology typically exists and develops within a certain range of powers and speeds, occupying a corresponding area on the logarithmic map. Figure 2.22 outlines the approximate areas of the current development of superconducting machines (SCM) and high-frequency PM machines (HF-PMSM). It is likely that in the near future, we will witness competition between SCM and HF-PMSM in the power and speed range of 0.1–1 MW and 1000–3000 rpm. As prices for HTS materials decrease, it may become more justified to produce SCM for power levels below 1 MW. Meanwhile, HF-PMSM are already beginning to fill niches in applications with power above 100 kW. Overall, we can see that the development of electric machines as a class of systems is uneven, driven by the introduction of different types of electric machines in various applications occupying some niches, which can potentially grow.

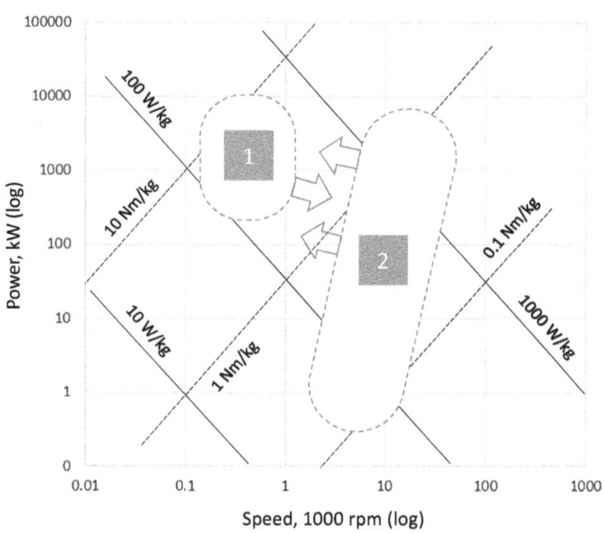

Fig. 2.22 Examples of compactness evolution: 1—SCM, 2—HF-PMSM

Opposing Forces and Constraints

Evolution is characterized not only by driving forces but also by *opposing forces* and *constraints*, which together determine the direction of evolution (Fig. 2.23). Any complex technical system "resists" changes in one way or another. Opposing forces are often created by people who either experience psychological inertia or fear worsening their situation due to the introduction of something new. Methods to overcome this include demonstrating the advantages of the new breakthrough solution in real-world applications and then working on reduction of its cost.

Any evolving class of systems periodically encounters limitations and boundaries. Some types of constraints with examples from the field of electric machines are provided in Table 2.3. In simpler technical systems, the limit (an insurmountable constraint) can be the exhaustion of alternative development options. Systems

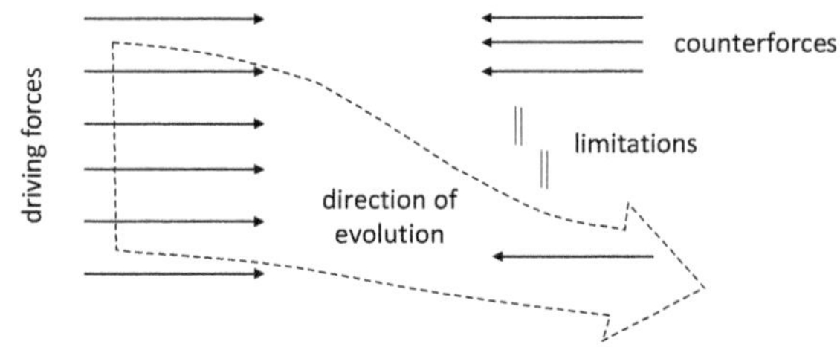

Fig. 2.23 Illustration of driving forces, opposing forces, constraints, and the direction of evolution

Table 2.3 Examples of constraints

Constraints related to…	Examples from the field of electric machines
… laws of physics, chemistry, and biology	Mechanical limitations due to high circumferential velocities in machines with ultra-high rotational speeds, where centrifugal forces become so great that using rotors made up of multiple interconnected elements becomes problematic
… the imperfections of other components in the supersystem	Insufficient capacity and high weight of batteries currently slows down the adoption of all-electric aircrafts and their electric motors. Interestingly, the same issue prevented the adoption of Jacobi's engine in electric-powered passenger boats back in the 1830s
… the imperfections of subsystems and components	Brushes in commutator assemblies cannot reliably operate at high rotational speeds, limiting the use of DC machines to relatively low speeds
… excessive increase in design complexity	Excessive mechanical complexity can be a barrier for "multilayered" (multiple air gap) electric machines
… the finiteness of nonrenewable natural resources	Rare-earth materials used to make permanent magnets are a limited resource
… environment and pollution	Current production methods for rare-earth permanent magnets are not environmentally friendly
… negative effects on humans	Noise, vibrations
… competition with other similar systems	Some market niches (e.g., municipal water supply pumps) are already occupied by certain types of electric machines (squirrel-cage induction motors), which may be slightly inferior to new motors (e.g., synchronous reluctance motors) in performance but are already optimized in terms of cost

typically reach such a limit within several decades after their emergence. Electric machines are a highly diverse and branched class of technical systems, so they are unlikely to reach this limit anytime soon. However, certain types of electric machines may well "hit the ceiling" of their development and be replaced by alternative options.

A constraint can also be interpreted as the absence of something, for instance, a missing component. In such cases, the emergence of the missing component is seen as the removal of a limitation. For example, the advent of power electronics removed limitations on creating machines operating at ultra-low and ultra-high rotational speeds. The evolution of a technical system can be viewed as a process of gradually removing various constraints.

In [2], it is shown how the power of a turbogenerator can be limited in practice: the ratio of the generator's power to the power generation site's power is limited by a simple empirical rule—one generator's power should not exceed 10% of the total power generation site's power, so that in the event of a failure, it can be replaced by the system's reserve power, which typically amounts to 15–20%.

Let us note a few additional characteristics of the evolution of technical systems mentioned in [4], which are fully applicable to electric machines, and provide the corresponding examples:

- Changes in different areas of technology can significantly influence one another. One invention can either promote other inventions or hinder and even prevent them. Similarly, one technical system can either support the emergence of new systems in other areas or impede their development and even block their appearance. For instance, the increased reliability and cost reduction of internal combustion engines (due to wide adaptation and mass production) act as obstacles to the adoption of electric vehicles and corresponding electric machine technologies.
- Evolution is a multidimensional process. Evolution along certain axes may follow different patterns than along others. Furthermore, different subsystems can evolve at different speeds, influencing each other's evolution by either accelerating or slowing it down. For example, improvements in the brushes of DC machines, which extend their lifespan, slow the replacement of DC drives with AC drives.
- There are two interesting effects in the development of technology: divergence and convergence. Divergence refers to the differentiation of systems into specialized ecological niches, where each system evolves within its own niche. After divergence, convergence may occur, where systems come closer together, with local advancements being transferred between relatively similar applications. For example, permanent magnet synchronous machines (PMSMs) can be used in vastly different applications, such as marine vessels and aircraft. The requirements in these applications are distinct, leading to different PMSM designs (divergence). At the same time, the experience of integrating PMSMs into the outboard propulsion units of marine vessels can be used to create generators for tidal turbines (convergence).

- As a system evolves, changes accumulate within it, and these changes do not simply accumulate but transform the system, giving it new properties, ultimately creating a new system (a "rebirth" takes place). An example of this could be various "hybrid" topologies of electric machines, combining the principles of reluctance machines (uneven air gaps), synchronous machines (permanent magnets), and induction machines (damper windings).
- The criterion of a system's quality and competitiveness is its success in the market. In a healthy economy, the principle of positive selection operates, meaning that systems with the highest level of ideality are chosen. In times of crisis, a mechanism of negative selection may come into play: systems with the lowest cost may be chosen without proper assessment of their characteristics and quality. For instance, when the price of permanent magnets surged fivefold in 2011 due to political circumstances (with prices quickly returning to normal levels by 2013), a number of promising projects with direct-drive PMSM were halted, and preference was given to geared drive-trains, and even to machines without permanent magnets and corresponding system topologies, despite the superior characteristics of the machines with permanent magnets.

Standards as Limiters and Guides

The design and operation of most electric machines are governed by various standards and directives. Here are three examples:

1. IEC 60034 regulates:
   (a) Dimensions and mounting interfaces of electric machines
   (b) Certain characteristics, such as acoustic noise levels and efficiency classes
   (c) Specific procedures, like the procedure for testing insulation systems
   (d) Guidelines, such as those for designing and operating AC motors specifically designed to be powered by frequency converters
2. The Low Voltage Directive regulates:
   (a) Protection from hazards arising from the use of electrical equipment
   (b) Electromagnetic compatibility (EMC)
   (c) Restrictions on the use of certain hazardous substances
3. The Waste Framework Directive outlines measures to protect the environment and human health by preventing or reducing the negative impact of production waste, as well as measures to increase resource efficiency. This directive is related to recycling of electric machines.

Standardization is important for both the design and operation of systems or sites using electric machines, as well as for electric machine developers and manufacturers. Standards define a certain discrete space within which new equipment is created. So, what is the purpose of standardization? First, standards, directives, and legal requirements ensure that electric machines meet safety requirements for both people and equipment. Second, they ensure the smooth integration of new electric machines into existing industrial units and lines, regardless of the country of origin. Third, they provide predictability for the future replacement of the machine, even many years after the first installation.

Initially, manufacturers are guided by international, national, and industry standards, but the specification or standard of a particular operator or user may be more demanding than, for example, an industry standard. This can give the operator an advantage but may complicate the electric machine manufacturer's work.

A separate topic is military specifications and standards. Although they are not covered in this book, it is worth mentioning that electric machines in military applications often operate under similar conditions to civilian ones. However, there can be differences, such as requirements for resistance to impact forces. Another significant difference is that military technology may use prohibited, rare, or expensive materials that are generally not available in civilian applications.

Final Comments

It can probably be said that while the main driving force of evolution in the nineteenth century was inventive activity, which led to the creation of the fundamental types of electric machines and the elimination of dead-end development paths, the twentieth century's evolution was primarily driven by the improvement of material properties, various subsystems of machines, and the emergence of power and control electronics. In the twenty-first century, the evolution of materials will likely continue to play its important role, alongside the introduction of fundamentally new manufacturing technologies. Throughout all periods, the emergence and development of new applications have been the most critical factor in the advancement of electric machines. Even in the early days of Electromechanics, electric machines were developed for specific applications, such as Jacobi's motor (1838) for powering a boat, and Davidson's motor (1842) for powering a locomotive.

Thus, the main driving forces of the evolution are: (1) the emergence of new applications and the evolution of existing ones, (2) new materials, and (3) new manufacturing technologies.

Electric machines follow the same laws of evolution as other technical systems. Understanding these laws is essential, and the works of TRIZ specialists will be beneficial to electrical engineers in this regard.

References

1. B. J. G. van der Kooij, The invention of the electromotive engine, 2015., ISBN: 1503095878
2. D. Sahal, *Patterns of Technological Innovation* (Addison-Wesley, 1981), 381 p.
3. M. Orlov, *Foundations of classical TRIZ*, 2nd edn. (M.: Solon-Press, 2006), 432 p.
4. B. Zlotin, A. Zusman, Patterns of evolution: recent findings on structure and origin, in *TRIZCON'2006, Milwaukee, WI, USA*, (2006)
5. V. Petrov, *Technology of innovations.*, ISBN 965–7127-01-7 (Tel Aviv, 2007), 90 p.

6. J.W. Kolar, U. Drofenik, J. Biela, M.L. Heldwein, H. Ertl, T. Friedli, S.D. Round, PWM converter power density barriers. IEEJ Trans. Indust. Appl. **128**(4), 9–29 (2007)
7. T. Takamiya, K. Hanazawa, T. Suzuki, Recent development of grain-oriented electrical steel in JFE steel. JFE Tech. Rep. **21**, 1 (2016)
8. Carbon conductors for lightweight motors and generators, DE-EE0007865, Rice University, University of Maryland, DexMat, Irvin Global Industries, April 2017—Oct 2019
9. A. Extance, *The race is on to make the first room temperature superconductor* (Chemistryworld, 2019) www.chemistryworld.com
10. W. Tong, Direct drive superconducting wind generators, in *Wind Power Generation and Wind Turbine Design*, (WIT Press, 2010), 741 p.
11. S. Li, Y. Shihu, Y. Feng, Progress in and prospects for electrical insulating materials. IET J. High Volt. **1**(3), 122–129 (2016)
12. B. Dehez, F. Baudart, Y. Perriard, *Analysis of a new topology of flexible PCB winding for slotless BLDC machines*, vol 2017 (IEEE International Electric Machines and Drives Conference (IEMDC), Miami, FL, 2017), pp. 1–8
13. Y. Satake, K. Takahashi, T. Waki, M. Onoda, T. Tanaka, Development of large capacity turbine generators for thermal power plants. Mitsubishi Heavy Indust. Tech. Rev. **52**(2), 1 (2015)
14. I. Kopylov, *Electric Machines*, 2nd edn. (Textbook for BSc students, Urait, 2012)
15. AMSC presentation, Concepts for high power wind turbines introducing HTS technology. World Green Energy Forum 2010, Gyeongjucity, Korea, November 17–19, 2010
16. Orbray's website. https://orbray.com/en/product/dccorelessmotor/motorhistory.html
17. F. Werfel, Supraleitung und Innovationene in der Energietechnik, Internationale Fachtagung Innovative Glanzlichter fuer das Solarzeitalter aus Sachsen, June 14, 2012
18. A.V. Matveev, Analyzing the types and series of induction machines on the basis of an alternative machine constant. Electrichestvo **7**, 41–56 (2020)
19. T-motor's website. https://uav-en.tmotor.com
20. ABB's website. www.abb.com
21. A.V. Matveev, Driving Forces and Lines of the Evolution of Rotating Electrical Machines. Part 1. Elektrichestvo **1**, 44–54 (2021)
22. A.V. Matveev, Driving Forces and Lines of the Evolution of Rotating Electrical Machines. Part 2. Elektrichestvo **2**, 4–15 (2021)

The Evolution Actors, Tools, and Processes

3

The process of evolution can be seen as an ongoing stream of innovations. The specific actions that drive evolution forward are, naturally, carried out by people—individuals (scientists, engineers, developers, designers, inventors), groups, and organizations (universities, research institutes, companies, corporations, the military-industrial complex, governments, and international associations). Today, the role of individuals is limited. True "lone wolves" are rare, perhaps found only among consultants. Even solo startup founders quickly build a team around them. The role of "solo inventors" is largely confined to idea generation (although this is a crucial role!). When a new, promising idea emerges and becomes known, public research centers and private corporations across various countries almost immediately start allocating funds for further development and competition between large teams begins. In any case, nowadays, the improvement of products or solutions begins almost as soon as their first versions appear.

A creator of something new, aiming to disrupt the status quo and initiate a step in what he or she sees as the "evolutionarily correct" direction, must be aware of the full complexity of the innovation ecosystems and understand the goals and operational mechanisms of all players involved. The main groups of participants in the evolution process operate within four environments: (1) academia, including research institutes; (2) industry, including sector-specific and cross-sectoral committees; (3) governments and supranational bodies; and (4) the private capital financial sector. One individual may play roles in two or more of these environments. For instance, a university professor might serve as an expert for a government body, a specialist from an industrial enterprise might also teach at a university, and students could work on their own startup in their free time, and so on.

As mentioned in Chap. 2, there are typically multiple possible evolutionary paths for a system from its current state to the next. Often, the path that materializes is the one initiated earlier and into which the most financial and human resources have been invested. Thus, the issue of investment is crucial. Investments can be public or private, or a combination of both. Private investments may come from companies,

A. Matveev, *Shaping the Future of Electric Machines*,
https://doi.org/10.1007/978-3-031-95248-7_3

funds, or individuals. In essence, groups with political or financial influence shape the directions of research and development, the level of funding, and consequently, the number of specialists working in the chosen direction. These groups create the conditions for innovation (e.g., through grant mechanisms) and control the process (e.g., through certification mechanisms).

Goal-Setting as a Management Tool

The United Nations has outlined the Sustainable Development Goals (SDGs) for humanity. Government agencies (to a large extent) and private corporations and funds (to a somewhat lesser extent) base at least part of their planning on these goals. Through negotiations, policymakers from different countries commit to achieving specific targets within the broader framework of the UN's goals. Grant-making organizations use these goals as a basis for directing investments. The European Commission is one example of such an organization. Within the UN's list, many SDGs stand out as particularly relevant to Electromechanics and electric machines; examples of such goals are presented in Table 3.1.

Figure 3.1 illustrates how the European Union's political priorities impact the development of hybrid and fully electric aircrafts[1] by setting concrete objectives to reduce carbon emissions and fuel consumption, as outlined by the supranational body, the Advisory Council for Aviation Research and Innovation in Europe (ACARE) [1]. As a result of such goal-setting and targeted funding, numerous concepts for hybrid and fully electric aircraft have been developed in recent years. The

Table 3.1 GDG and Electromechanics

Sustainable Development Goals (SDG)	Electromechanics providing the solutions
Ensuring access to affordable, reliable, sustainable, and modern energy for all	Small generators for distributed generation, flywheel energy storage
Taking urgent action to combat climate change and its impacts	Generators for renewable energy installations, such as, e.g., wind turbines
Making cities and human settlements inclusive, safe, resilient, and sustainable	Electric transport
Ensuring sustainable consumption and production patterns	High-efficiency motors in industrial applications that minimize energy consumption
Conserving and sustainably using the oceans, seas, and marine resources for sustainable development	Electric propulsion
Good health and well-being	Electromechanical prosthetics and mobility aids

[1]The terms "fully electric" and "all-electric" aircraft are closely related and often used interchangeably.

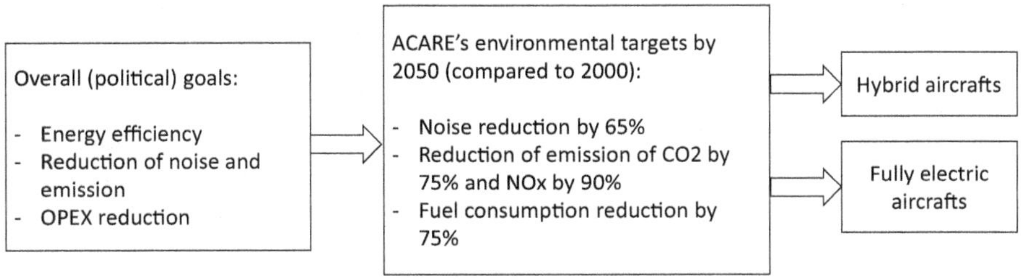

Fig. 3.1 Influence of selected political priorities on the electrification of aviation

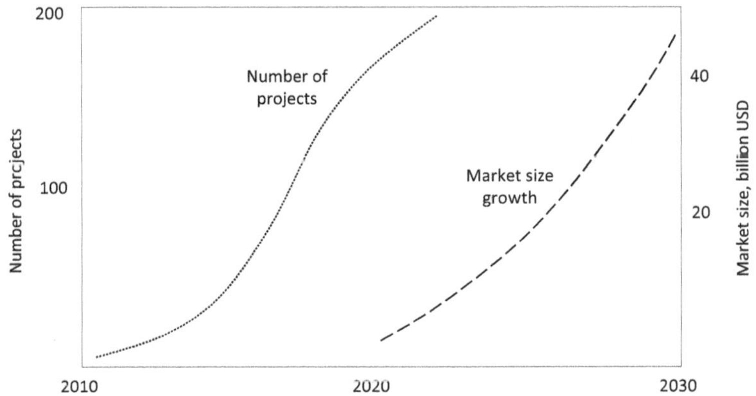

Fig. 3.2 Growth in the number of projects followed by market growth

explosive growth in aviation electrification projects is shown in Fig. 3.2. It is noteworthy that the related market growth follows with a certain time lag.

Awareness of the goals listed in the strategies of various organizations funding scientific research and engineering development will be useful for those seeking support for new ideas and innovations aligned with them.

Key players in various commissions and committees—policymakers and officials—rely on expertise from the scientific and engineering community. These expert groups holds significant influence over the evolution of technical systems, including electric machines.

Financial Support as a Management Tool and the Expert Community

One of the most important mechanisms for guiding innovation directions is the investment and grants provided by commercial and governmental institutions (Fig. 3.3). Grants can be public, governmental, private, or corporate.

The evolution management process through funding has three main stages:

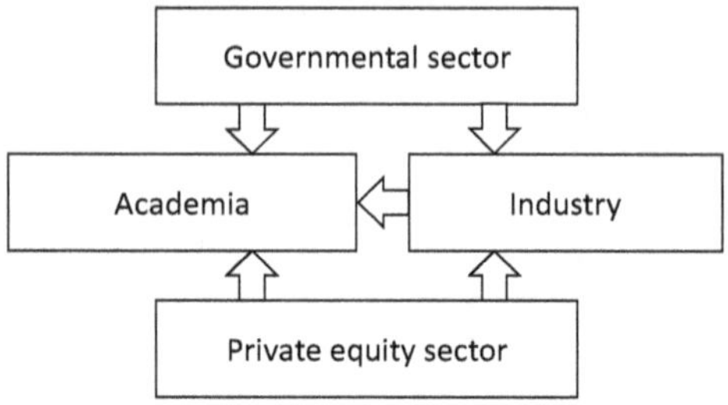

Fig. 3.3 Direction of investments (arrows indicate the financial flows)

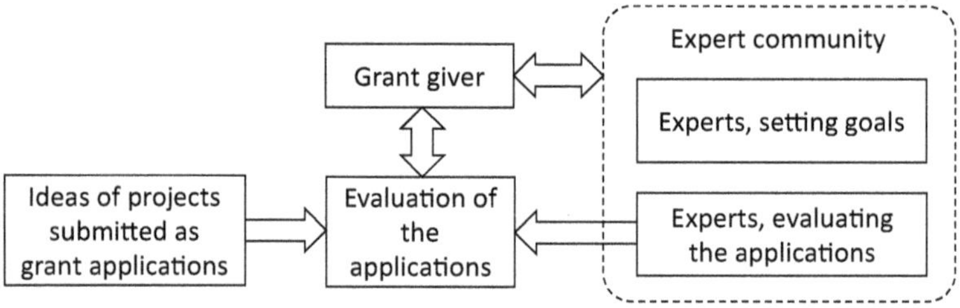

Fig. 3.4 Role of the expert community

1. Defining and establishing the broad areas of research and development (e.g., energy security or environmental protection).
2. Identifying more specific topics within these areas (e.g., offshore wind farms or hydrogen-powered aircraft) and the expected outcomes of projects.
3. Implementing a selection process to determine which projects receive financial support (often through a rigorous filtering system).

Funding of new developments can also take the form of loans and investments, which is more common for private commercial organizations. Regardless of the type and source of funding, similar procedures are used, including assessing the alignment of a project with the goals of the grantor or investor, and expert evaluation of the entire project portfolio of the investor to determine how naturally new projects might fit within it. Criteria for assessment are selected, and projects are evaluated individually and in groups by experts. This process includes estimating the likelihood that the team standing behind the proposed innovation, with the proposed plan and resources, will deliver valuable results. Given that experts conduct the assessments, the role of the expert community is essential both in setting development goals and in evaluating projects, as illustrated in Fig. 3.4.

The Role of Universities and Research Institutes in Technological Evolution

Universities primarily bear the responsibility of educating the next generation of scientists, engineers, and technologists, as well as preparing future leaders in technology and innovation who will drive progress across various industries and shape the course of technological evolution. In addition to this, universities play a role in technology development in several more direct ways:

- Universities create an environment that encourages creativity and innovation. They inspire students and researchers to think critically, take risks, and push the boundaries of what is possible.
- Unlike many industrial enterprises, universities are not solely focused on short-term profits. They can invest in long-term research projects that may take years or even decades to yield fundamental discoveries and technological breakthroughs. Universities encourage interdisciplinary research, bringing together experts from diverse fields to work on complex problems. This approach often leads to breakthroughs that would be difficult to achieve within a single discipline.
- Universities contribute to the spread of knowledge by publishing research findings and making them accessible to the public. This openness facilitates information exchange and accelerates technological advancement.
- The academic environment can also serve as an incubator and accelerator for ideas originating outside the university. Universities provide a space for external researchers to develop their concepts and create new knowledge relevant to their fields.
- Regardless of where an idea originates—within academic circles or outside—technology transfer to industry eventually takes place, often through partnerships with industry on research projects. Many universities have technology transfer offices that support the commercialization of research outcomes. Technology transfer may involve licensing technologies to companies or launching new enterprises based on university-developed innovations.
- Universities also play a role in examining and discussing the ethical, social, and political implications of emerging technologies. They participate in discussions on the responsible development and use of technology, helping to shape ethical standards and regulations. This ethical foundation is a critical element in managing technological evolution.

In Fig. 3.5, the main outputs of universities are shown; these outputs, if they meet quality standards and address current needs, can be very beneficial for implementing innovations. Key actors in the academic environment include researchers, professors, graduate students, and staff responsible for innovation and technology transfer.

Alongside universities, research institutes play a significant role in innovation. Research institutes may belong to either the academic community or the industry, depending on their specific orientation and objectives. While universities focus on student education, research institutes are oriented toward scientific research and innovation. They help translate scientific research into tangible products, services, or technologies.

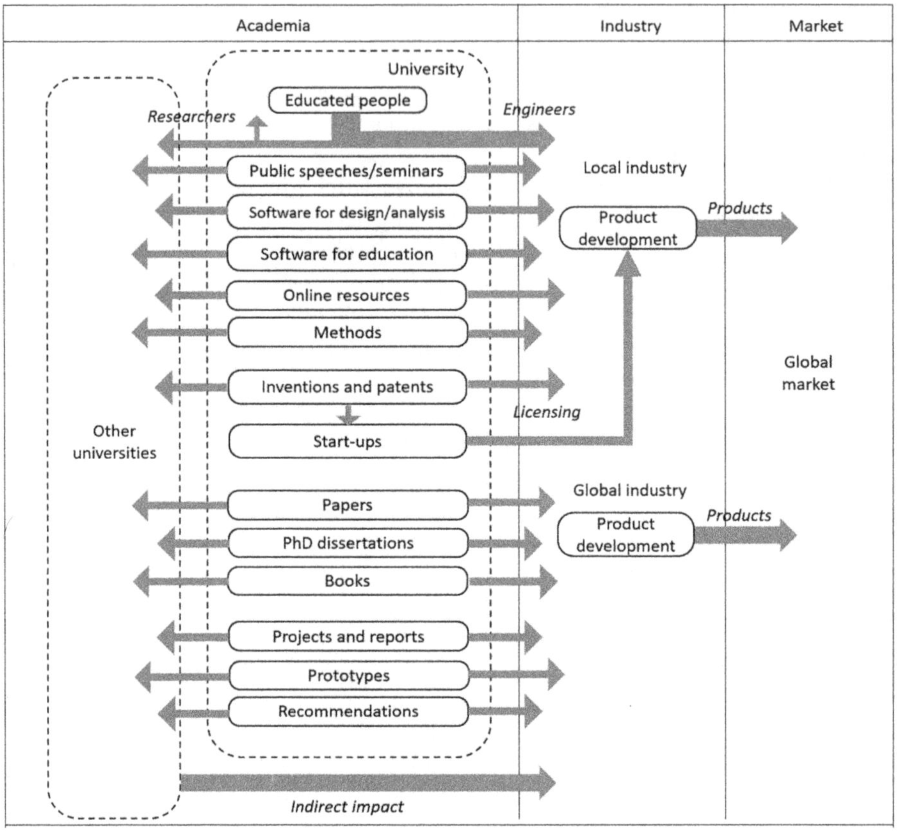

Fig. 3.5 University activity outputs

Actors and Processes in Industry

Success in industry largely depends on innovation. In the context of this book, innovation can be defined as the practical realization and implementation of a new or improved electric machine or related system, or a process for manufacturing of a new electric machine. Such an innovation may be based on one or more new solutions—let's refer to the conceptualized ideas behind these solutions as "innovation proposals." While most innovation proposals are filtered out by experts, some progress to prototyping, and a few, after successful testing, become part of new products or processes, evolving into true innovations. Most innovations are

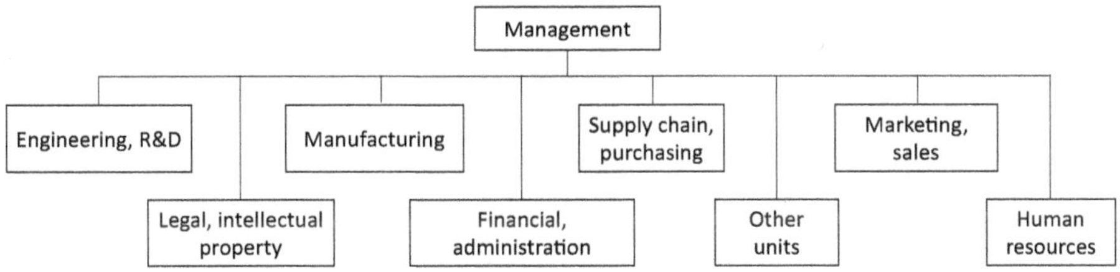

Fig. 3.6 Typical industrial company structure

implemented in industry, so additional attention should be given to the related processes. When bringing an idea to realization within an industrial enterprise (a company), it is important to understand how the company's structures operate (Fig. 3.6). Notably, innovation work can involve all departments of the company.

Let's assume an innovation proposal has emerged. To decide whether to pursue the proposed direction, several questions should be addressed, including:

- To what extent are the technologies, potentiall required in the realization of the new idea, are mature, tested, and available, including any key components used in the new solution?
- Is the company capable of taking an innovative leap, considering factors such as the company's size, experience, and competencies?
- What are the potential risks associated with the innovation pathway?
- What is the value of the new development for the end-user? It's crucial to have a clear understanding of the impact of potential implementation in advance.

Answering these questions may involve nearly every department within the company. However, the core strategic planning work will primarily be conducted by management in coordination with engineering, production, R&D, marketing, and sales departments. Below are several examples illustrating practical aspects of exploring these questions and a range of possible tools.

Example 1 Assessing the Maturity Level of a New Development in the Context of Modern Technology. The Gartner Hype Cycle is a graphical representation and methodology designed to track the adoption, maturity, and expectations surrounding new technologies, trends, and concepts over time. It illustrates the trajectory of technologies as they progress from early stages to widespread adoption. This is a valuable visualization tool for decision-makers and innovators, helping them anticipate and navigate the challenges and opportunities related to new technologies. By understanding which stage of the Gartner Cycle a technology is in, stakeholders can make more informed decisions about when and how to invest, adopt, or deploy new technologies, managing risks and maximizing benefits. An example of the Gartner Cycle, with various technologies relevant to electrical engineering systems plotted on the curve, is shown in Fig. 3.7.

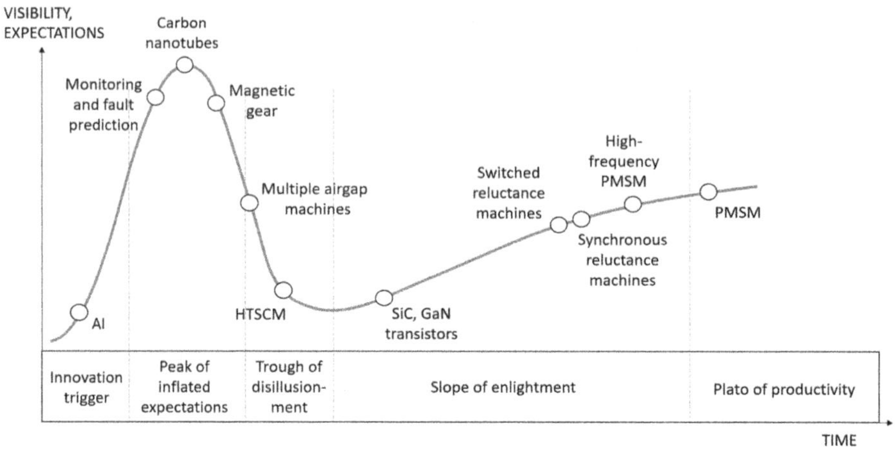

Fig. 3.7 Gartner hype cycle curve with some selected technologies[2]

In addition to monitoring cycles and trends (as discussed in Chap. 2), it is also important to track potential low-probability but high-impact technological events, such as the emergence of a material with radically improved properties or radically improved production technology. Such events can lead to transformative changes, making it essential to identify, evaluate, and leverage them promptly.

Example 2 Strategic Step Planning. For a company, especially a small one, decision-making will always prioritize business sustainability, with a tendency to avoid abrupt shifts. Figure 3.8 illustrates a helpful tool offering recommendations based on the experiences of multiple generations of industrial entrepreneurs. In the solution space shown in Fig. 3.8, moving one cell from the current state is considered feasible, while moving two cells is usually not recommended (marked with an "X").

The simplest option, labeled "1", involves expanding the use of proven technology into a similar application in a related market. For instance, moving from the pump market where the company has established a position with certain electric drive system technology to the industrial fan market using the same drive systems will be this kind of a move. However, attempting to enter the market of propulsion systems for electric aircraft with the same electric machine technology and electronics as used for pumps, would not be recommended.

A safe option labeled "2" might involve adding to its portfolio a product or a system sharing common system elements with the existing systems. For example, for a company with induction motors (IM) and synchronous reluctance motors (SynRM) in the portfolio, adding permanent magnet synchronous motors (PMSM) could be a viable option, since the frequency converter topology is the same for all the three machine types and mechanical elements of the IM or SynRM (housings, shafts, etc.) can be reused in PMSM series of products.

[2] Author's subjective opinion which may be wrong and should not be taken as any kind of advice.

Move “3,” which combines both “1” and “2,” is also acceptable but may present more complexity.

> One of the main questions is whether the proposed innovation is likely to provide sustainable value

Example 3 Assessing the Risks of Resource Investment in Specific Technologies. Any new idea or technology will undergo a risk assessment before manufacturers invest resources in its development. This is a natural process, and innovation proponents should be prepared for it. Here we demonstrate risk assessments in the context of competition between companies focusing on different machine types—PMSM and SCIM. Example of possible risk assessments are presented in matrices in Fig. 3.9, where the top-right corner represents unacceptable risk, and the bottom-left corner represents minimal, negligible risk. These risk assessments are subjective and intended only as illustrative examples for a 10–20 year horizon.

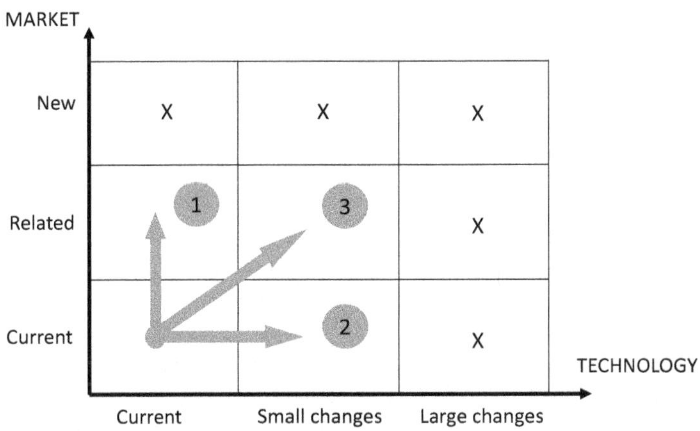

Fig. 3.8 Recommendations for startups and small companies

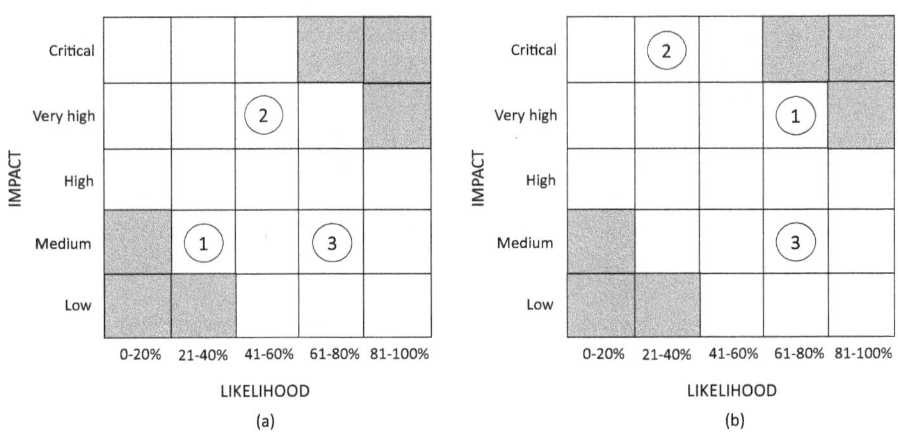

Fig. 3.9 Example of risk assessment matrices for SCIM (**a**) and PMSM (**b**)

Let's start with an example for an SCIM manufacturer (see Fig. 3.9a). Possible risks include:

- "1"—Emergence of stronger, non-rare-earth magnets with lower costs, which could let PMSM suppliers enter applications traditionally occupied by SCIM (estimated probability: 21–40%, impact severity: medium).
- "2"—Increased energy efficiency standards for machines (e.g., adoption of IE5–6 efficiency classes as new standards and gradual banning IE2 class motors), which would require major SCIM series redesigns (probability: 41–60%, impact severity: very high).
- "3"—Expansion of synchronous reluctance machines (SynRM) in applications with variable-speed drives (probability: 61–80%, impact severity: medium).

To mitigate these risks, SCIM manufacturers may need to invest in additional marketing to highlight the specific advantages and applications of SCIM in contexts of the market niches they occupy.

For a PMSM manufacturer (see Fig. 3.9b), potential risks might include:

- "1"—A surge in permanent magnet prices by several times, the scenario already experienced in 2011 (probability: 61–80%, impact severity: very high).
- "2"—Restricted access to rare-earth metals due to political factors (probability: 0–20%, impact severity: critical).
- "3"—Growing criticism of permanent magnets on environmental grounds, as their production is associated with pollution of the environment (probability: 61–80%, impact severity: medium).

One way to manage these risks could be diversifying the product portfolio to include motors that do not rely on permanent magnets. Another approach could be investing in the development of non-rare-earth magnets.

Example 4 Developing a Value Proposition. Marketing and sales departments play a direct role in gathering and analyzing the market and applications related information, one of their key tasks is the formulation of "value propositions"—the specific benefits or advantages that a new or improved electric machine can offer for a particular application. The potential success of a new development often depends on the sales representative's expertise, ideally based on his or her engineering background.

To illustrate, let's develop a value proposition for using a new electric motor in drones, where the motor has higher efficiency and reduced weight compared to competitors. The competing criteria and trade-offs involved in selecting an electric motor for drones were discussed in Chap. 1. Building on that, we'll briefly show how this value proposition can be visually represented for clarity.

In Fig. 3.10, the solid lines represent curves that show the impact of battery size on a drone's flight time and payload capacity. Three battery capacity options are illustrated: X, 1.5X, and 2X, where X represents the basic battery's capacity or charge. The graph shows that with a smaller battery (and thus lower battery weight), the drone can carry more payload, but this comes at the cost of reduced flight time.

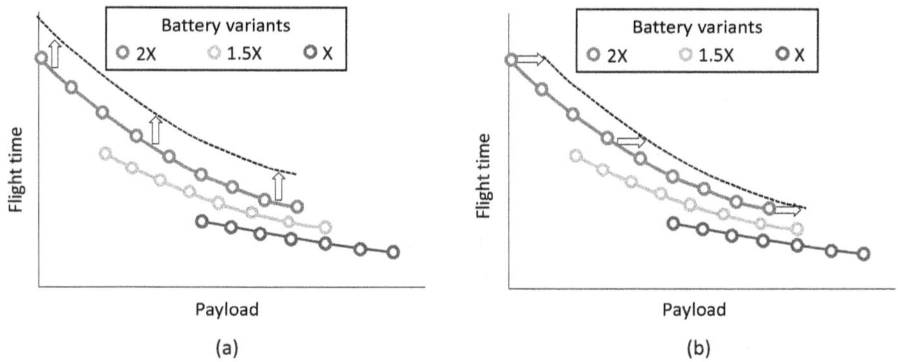

Fig. 3.10 Qualitative graphical interpretation of the value proposition

When using a higher-efficiency motor, the battery discharges more slowly, extending the drone's flight time. The curve's shift with the new motor shown for the 2X battery case reflects this benefit. A possible value proposition could be: "Our technology lets drones fly farther" (Fig. 3.10a). If each motor weighs Y grams less and there are eight motors on the drone, that frees up 8Y grams, which could be allocated to increased payload capacity. An alternative value proposition could be: "*Our technology enables drones to carry heavier loads*" (Fig. 3.10b). Naturally, a combined proposition is also possible, such as: "*Our technology allows drones to fly farther and carry more*". For delivery drones, implementing new motor technology means covering a broader geographical area and delivering heavier packages.

After the Decision has been Made. If a decision is made to pursue an innovation proposal, certain factors become crucial in the next stage:

- When implementing the innovation, it's ideal to achieve broad adoption of the new technology across multiple applications.
- For radically new solutions, it's essential to look beyond just the product, technology, or individual solution.
- Ultimately, price remains a critical factor determining a product's market success.

Experience shows that innovations thrive best in the "right environment." Management and HR teams need to foster a company culture that supports innovation at all levels and in all stages. This culture should encourage openness to new ideas and tolerance for potential setbacks. Communication channels between team members, across all levels of the hierarchy, should be well-established. Overall, there should be a strong commitment to continuous improvement across all areas of the company's operations.

In the following practical examples, we'll explore which company departments are involved and highlight potential internal and external collaboration strategies.

Example 5 Developing Flexible Technology Platforms. A new concept is more likely to be adopted if it has the potential to become part of an existing technology platform or to serve as the foundation for a new one. A technology platform is a foundational framework that provides a set of components, tools (for development and production), and expertise that enables the design, deployment, and operation of specific solutions. This foundation simplifies and accelerates the development of various products and systems in a given field. Platforms can be scaled, customized, and configured to meet specific requirements. They often foster an ecosystem of developers and partners, whose integrated solutions can expand the platform's capabilities and reach.

An example of a technology platform based on surface-mounted PMSM is illustrated in Fig. 3.11. The main elements of this platform include:

- Rotors with surface-mounted permanent magnets (including solutions for retention and protection of the magnets).
- Concentrated winding of different variants.
- PMSM control systems (including sensorless vector control).

Figure 3.11 also shows example products that can be developed based on the technology platform: a turbine-generator, marine thruster, winch, generator with a gearbox, and pump, created for the respective applications or markets: hybrid aircrafts, marine vessels, renewable energy extraction, and the oil and gas installations, respectively.

Within the platform, the R&D team who have developed the platform, should be capable of managing the key characteristics of the machines. As an illustration, let's consider modifying the parameters of a permanent magnet generator for wind turbine by varying the axial length and radial thickness of active parts. This generator has the following specified characteristics: rated power of 4.3 MW, voltage of 690 V, rotational speed of 15 rpm, rated torque of 2738 kNm, and outer diameter of the

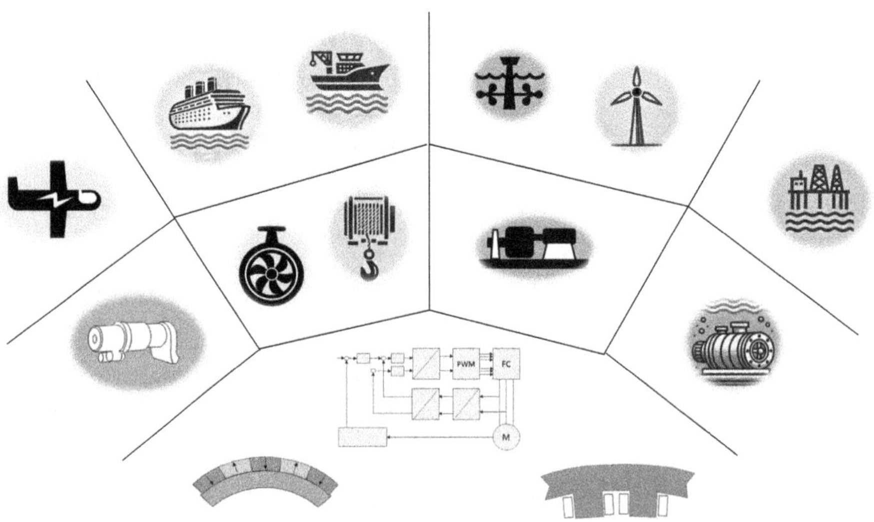

Fig. 3.11 Technology platform and applications

Table 3.2 Design variants for PMSG

Parameter	Var. 1	Var. 2	Var. 3	Var. 4	Var. 5
Inner diameter of active parts, m	5.57	5.57	5.39	5.39	5.39
Active stack length, m	1.64	1.35	1.16	1.37	0.97
Mass of stator iron, ton	8.7	7.1	13.3	15.7	11.2
Mass of rotor iron, ton	5.3	4.4	6.5	7.7	5.5
Mass of PM, ton	7.5	6.2	4.2	5.0	3.5
Mass of winding, ton	5.3	4.5	5.7	6.5	4.9
Mass of active parts, ton	27.4	22.7	30.3	35.6	25.6
Cost of stator iron, thousands of Ecu (imaginary currency)	104.0	85.7	160.1	188.9	134.2
Cost of rotor iron, thousands of Ecu	63.9	52.7	78.1	92.1	65.4
Cost of PM, thousands of Ecu	301.3	248.4	168.6	198.2	140.7
Cost of winding, thousands of Ecu	48.1	40.5	51.0	58.6	44.1
Cost of active parts, thousands of Ecu	517.3	427.2	457.8	537.7	384.3
Efficiency, %	95.3	94.7	95.9	96.4	95.2
Losses, кW	202	228	176	155	206
Cosφ	0.94	0.91	0.81	0.85	0.75
Reactive power, кVA	284	430	1014	764	1439
Machine constant C_M (Calculated using only mass of active parts)	200	165	221	260	186

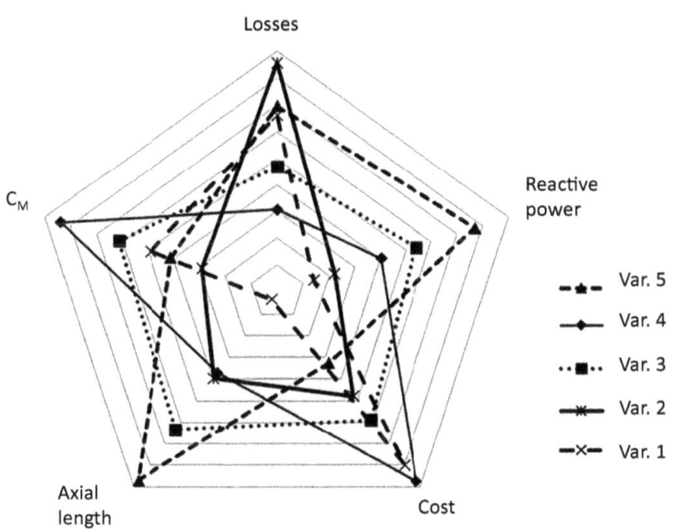

Fig. 3.12 Normalized characteristics

active parts of 5.85 m. Five design variants are listed in Table 3.2. The key characteristics of the five variants are compared in Fig. 3.12, where the parameters are normalized so that more favorable values appear closer to the center of the diagram.

None of the variants in this example is the best for all the five selected characteristics. Depending on the application and system requirements, these options will be evaluated as "more" or "less" suitable. To ensure flexibility within its technological platform, it is essential to have a well-developed CAD system.

> An innovation has higher chances of success if it is coming as a part of a technological platform

Example 6 Collaborating with Customers to Achieve Breakthrough Results. Close collaboration with the customer, or even co-development, can significantly increase the likelihood of breakthrough outcomes. This is especially common in projects where electric machines are integrated into a specific mechanism or assembly. As an example, let's consider how switching from an induction motor (SCIM) to a permanent magnet synchronous motor (PMSM) in a subsea pump unit installed at great depth on the seabed could substantially improve the pump's performance. To withstand pressures of several hundred atmospheres, the motor cavity in the pump is filled with a specialized "barrier" fluid. The maximum achievable rotational speed and power of the motor are limited by friction (viscous) losses, P_l, in the barrier fluid. These losses depend on the rotational speed n, rotor radius R, and axial rotor length L, according to the following formula: $P_l \equiv n^3R^4L$. When designed for the same specification, PMSM will have a more compact rotor than SCIM. A very rough theoretical estimate can be 10–20% reduction of both the length and the radius as illustrated in Fig. 3.13. Transition from a SCIM to a PMSM design in the pump unit will then considerably reduce the magnitude of friction losses thanks to the smaller rotor. This opens for an increase of the possible rotational speeds and power of the pump. However, this innovation is impossible without adapting the

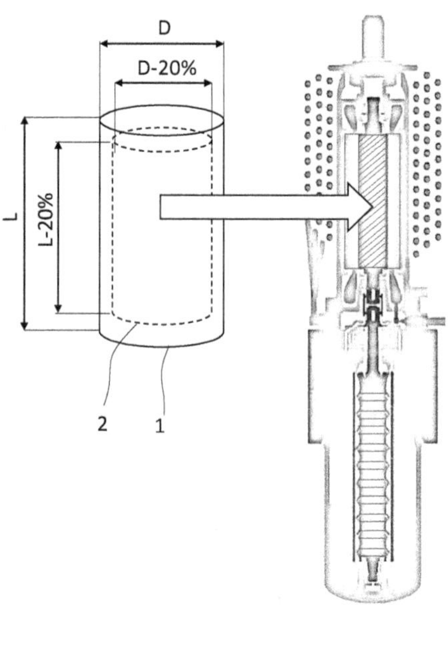

Fig. 3.13 Illustration of the effect of electric machine type on the rotor size in a subsea pump unit (1—SCIM rotor, 2—PMSM rotor)

entire pump assembly and even the entire power and control system for the pump motor. In this specific case, the possibility of using a PMSM will also depend on the ability to start and control the PMSM via a long cable.

> Co-development can significantly increase the likelihood of breakthrough innovation

Example 7 Achieving Target Production Volume and Competitive Product Price. The process of scaling up the production requires close collaboration between the procurement, production, and engineering departments. Chap. 1 provided pricing models for electric machine products available on the market. The models are derived for the established product. Achieving the position of an established player is a long journey. To illustrate the way from prototypes to mass production let's take as example a multiple-pole, 1 kW PMSM that can be used in, e.g., household appliances or small vehicles. This machine operates at high frequencies, so thinly laminated sheets are used for the stator. The machine housing is aluminum with cooling fins. The rotor design includes NdFeB magnets. Figure 3.14 illustrates the costs and investments involved in the development and production of this machine. The goal is to reach mass production within 2 years. The graph shows how the final product price varies with the planned production volume, where each step involves a tenfold increase in volume. In this example, investments are made in multiple stages, total-

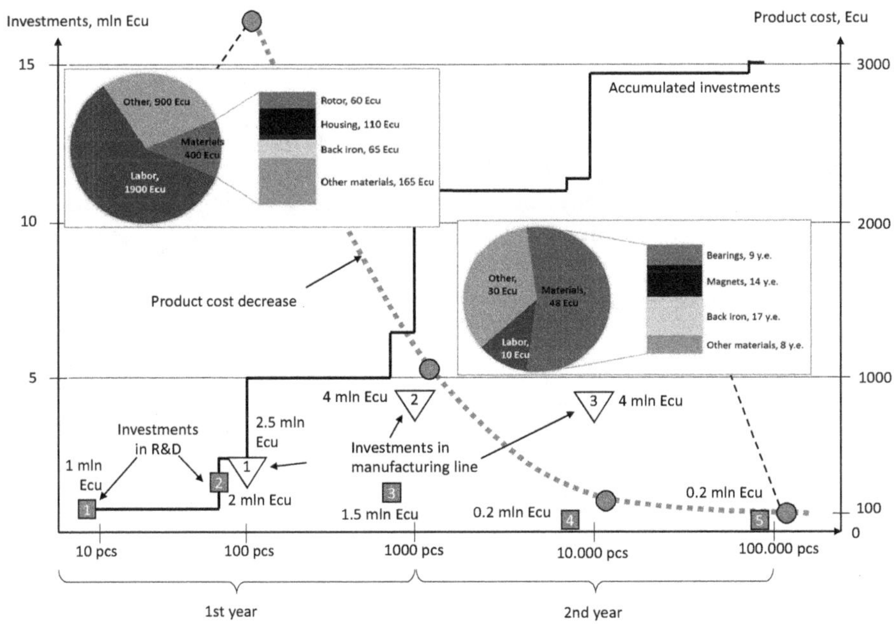

Fig. 3.14 Investments and product cost reduction

ing 15 million Ecu. Investments are split between R&D and production infrastructure (a robotic line).

In this example, R&D investments are planned to be in five stages:

1. For a prototype series of 10 pcs—one million Ecu.
2. For a trial series of 100 pcs—two million Ecu.
3. For a series of 1000 pcs—1.5 million Ecu.
4. Design adjustments for a series of 10,000 pcs—0.2 million Ecu.
5. Design adjustments for a series of 100,000 pcs—0.2 million Ecu.

Investments for establishing the robotic production line are done in three stages:

1. For the trial series of 100 pcs—2.5 million Ecu.
2. For the series of 1000 pcs—four million Ecu.
3. For the series of 10,000 units—four million Ecu.

As we can see in Fig. 3.15, increasing production volume brings about significant changes, particularly:

- Radical reductions in costs and product price—up to a 30-fold decrease!
- A substantial shift in the proportion of manual labor (see the pie charts).
- Changes in the three most expensive materials and the proportion of their contributions to product cost (see the pie charts).

The price of a prototype and a final product can vary by orders of magnitude

When comparing the cost and price of new design to a competing machine, it's crucial to understand the product life cycle stage of the competitor's machine. Electric machines should be designed not only for specific design performance but also to suit specific production technologies (Fig. 3.15). For example, in the case of micromachines, automated production is essential for achieving cost targets, thus designing for manual operations makes no sense.

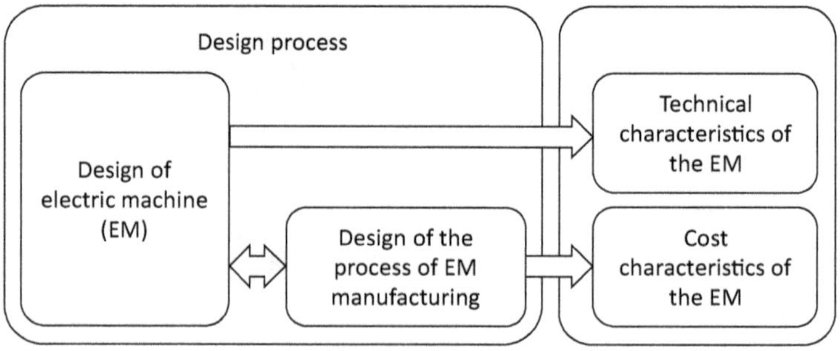

Fig. 3.15 Design considerations and goals

The Influence of Inventors and Their Inventions

Inventors and their work are often the starting points for all innovations, sparking evolutionary changes. Today, electric machines continue to evolve mainly through a continuous flow of incremental improvements, though major breakthroughs still remain possible. Looking back, the main breakthrough in electric machines history, in terms of invention and the adoption of key solutions still in use, took place in the late nineteenth century, with some fewer important inventions emerging throughout the twentieth century. Here, we'll review some historical examples, examining the patterns of emergence of inventions and their life cycles.

Induction Machines with Wound Rotor and Squirrel-Cage Rotor

The first multiphase motor that could be classified as induction motor was proposed by Walter Baily in 1879. Baily used a stationary electromagnet with four poles, wherein switching the current of the poles created a moving magnetic field in a copper disk placed beneath the poles. The field induced eddy currents in the disk, which interacted with the magnetic field of the poles, causing the disk to rotate in the direction of the moving field. Essentially, this was an induction motor with a solid rotor (in modern terminology).

In 1885, Galileo Ferraris demonstrated an induction motor using two pairs of electromagnets to create a rotating magnetic field. Ferraris' motor more closely resembled modern machines, as the electromagnets surrounded a cylindrical rather than a disk-shaped rotor. More importantly, Ferraris proposed generating a rotating magnetic field by supplying two sinusoidal currents with a 90° phase shift.

In that same year, Nikola Tesla demonstrated a very similar motor. Two years later, on October 12, 1887, Tesla filed a patent application for an "Electro-magnetic motor" and was granted patent US381968 on May 1, 1888. The solution presented in this patent (Fig. 3.16) was a multiphase induction motor with slip rings, representing one of the variants for an induction motor with a wound rotor.

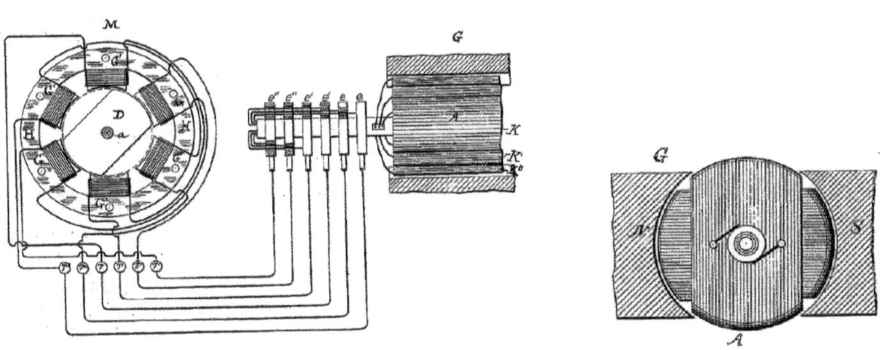

Fig. 3.16 Illustrations from US381968

A year and a half after Tesla's patent, Mikhail Dolivo-Dobrovolsky patented the concept of the squirrel-cage rotor, a design with bars and end rings (patent application CH1532A, filed October 26, 1889). The squirrel cage from this patent (Fig. 3.17a) is nearly identical to the modern version. Dolivo-Dobrovolsky's rotor design proved to be the optimal solution and has been used in squirrel-cage induction motors since then. Another year later, in 1890, Dolivo-Dobrovolsky also patented the concept of an induction motor with a wound rotor and rotor circuit resistance regulation (patent application CH3062A (Fig. 3.17b), which remains close to the form in which this type of electric machine is still used.

Considering the above, we can conclude that credit for the invention of modern induction motor designs can be shared among Baily, Ferraris, Tesla, and Dolivo-Dobrovolsky. The creation of induction motors in forms close to its contemporary designs was a monumental breakthrough, but as we see, authorship is collective and the invention process was stretched over time. Starting with Arago's experiments in 1824 as a reference point, we can outline the key milestones as shown in Fig. 3.18.

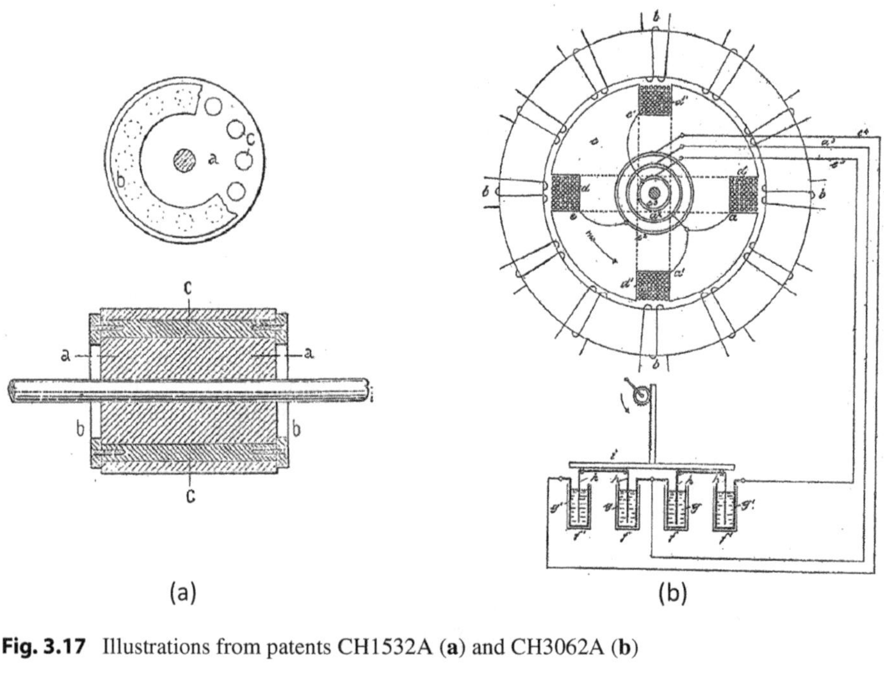

Fig. 3.17 Illustrations from patents CH1532A (**a**) and CH3062A (**b**)

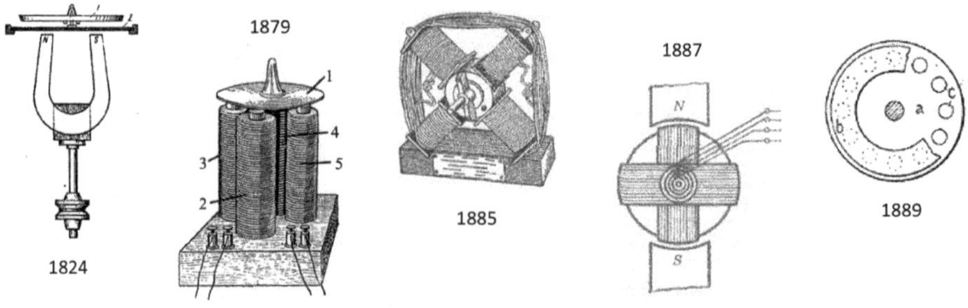

Fig. 3.18 Evolution of the induction motor concept

The Three-Phase Transformer

Beyond inventing the squirrel cage and the induction motor with a rotor circuit resistance regulation, Dolivo-Dobrovolsky is renowned for his work in advancing the concept of the three-phase network and demonstrating its practical advantages, as well as inventing the three-phase transformer (1892, patent CA39585A, Fig. 3.19). Few in the history of electrical engineering can claim so significant inventions as Dolivo-Dobrovolsky. His inventions mentioned above, along with the three-phase generator, became foundational components of the three-phase power generation and distribution system that has been the primary solutions already for over 130 years.

Rolling Rotor Motors

The twentieth century also brought some intriguing inventions. One of them was the *motor with a rolling (and wave) rotor* (Fig. 3.20), invented by A.I. Moskvitin (patent "Slow-speed electric motor," SU68211A1, 12.15.1944). In the latter half of the twentieth century, motors with rolling rotors (MRR) found niche applications in certain production mechanisms such as dispensers, clock mechanisms, manipulators, telemetry control systems, rheochord and wire potentiometer winding machines, and arc furnaces. These motors were mass-produced in a variety of models in France, England, Japan, the USSR, and the USA. Today, these motors are rarely used. The MRR example illustrates that the lifespan of some implemented inventions can be finite, with MRRs remaining relevant for only about 30–50 years.

Stepper Motors (StM)

The first motor that could be described as a "stepper motor" appeared in the 1830s, consisting of an electromagnet that drove a ratchet wheel. It's difficult to definitively identify who developed a version closest to the modern StM, but the name Frank

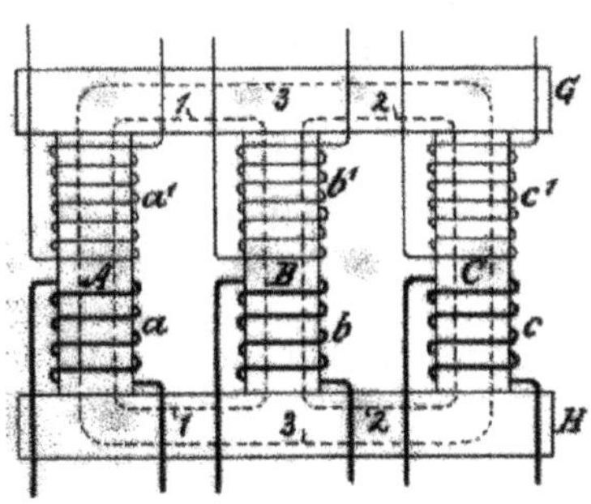

Fig. 3.19 Illustration from CA39585A

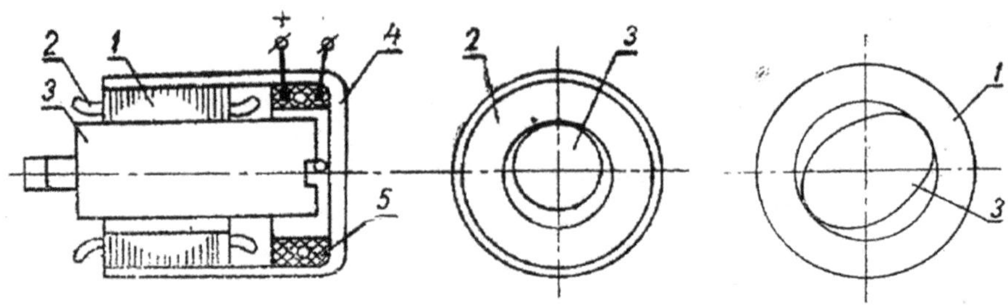

Fig. 3.20 Illustrations from SU68211A1: machines with rolling (**a**) and wave (**b**) rotor

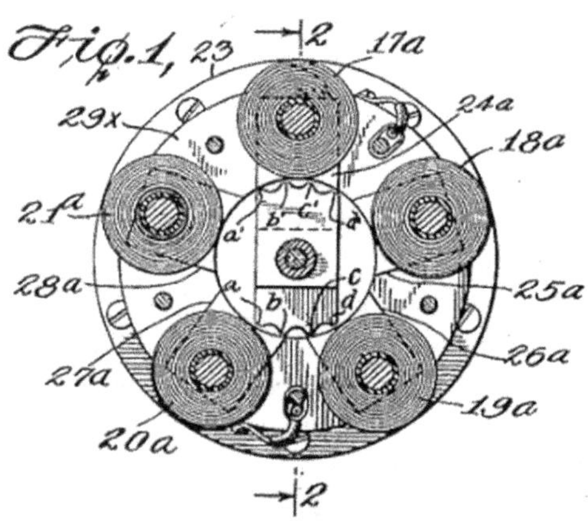

Fig. 3.21 Illustration from US1408555A

W. Wood is frequently mentioned. In 1918, Wood filed patent US1408555A for an invention titled "Art of producing step-by-step movements." This patent describes a stepper motor with five coils on the stator (Fig. 3.21). By controlling the polarity of the stator coils, a stepping motion could be achieved, and it was also possible to change the direction of rotation. Notably, in this patent application, Wood references his earlier invention from 1914, US1100860A, "Electric indicator system," which describes a similar system. Practical applications of stepper motors began on warships in the 1920s. Modern StM designs were developed in the 1950s, gained popularity in the following decade, and are still widely used today. Although it may be an overstatement to say that F. Wood invented the stepper motor, it is fair to conclude that Wood made a significant contribution to the development of this type of motor.

Electromechanical Amplifier (EMA)

EMA was invented by Ernst Alexanderson in 1940 and remained in use until nearly the end of the twentieth century, before being replaced by electronic systems. This means the EMA was used for nearly half a century.

Switched Reluctance Motors (SRM)

A machine resembling SRM was invented already in the first half of nineteenth century—that machine used mechanical commutator. The emergence of power and control electronics in 1970s recalled the old concept to life and the modern SRMs have got acceptance in many applications.

Halbach Magnetic Assembly

The Halbach magnetic assembly (Fig. 3.22) is a special configuration of permanent magnets characterized by the almost complete absence of a magnetic field on one side due to the unique arrangement of the individual magnet pieces. The principle was first proposed by James M. Winey in 1970 for an ideal case of continuously rotating magnetization, induced by a single-sided strip coil (patent application US3674946A, titled "Electromagnetic transducer"). John C. Mallinson further applied this effect in 1973 while working to improve the properties and quality of magnetic tape recording. In the 1980s, Klaus Halbach developed a magnetic assembly based on this effect, designed for intense electron beam radiation in accelerators. This configuration was subsequently named after Halbach. The Halbach assembly has been used in electrical machines since the early 1990s and remains popular today, particularly in high-specific-power and high-torque applications, such as coreless permanent magnet synchronous machines.

Magnetic Gears and "Pseudo-Direct Drives"

The first electromagnetic gearbox was likely proposed by A. Neuland over 100 years ago in 1916, in patent US1171351, titled "Apparatus for transmitting power." This solution (Fig. 3.23) did not use permanent magnets.

In 1941, H. Faus proposed a gearbox design in patent US2243555 ("Magnet gearing") that resembles a conventional mechanical gearbox in structure, but with the teeth replaced by permanent magnets (Fig. 3.24). In 1972, N. Laing revisited the coaxial design (Fig. 3.23), similar to Neuland's, but incorporated permanent magnets, as disclosed in patent US3645650 (Fig. 3.25).

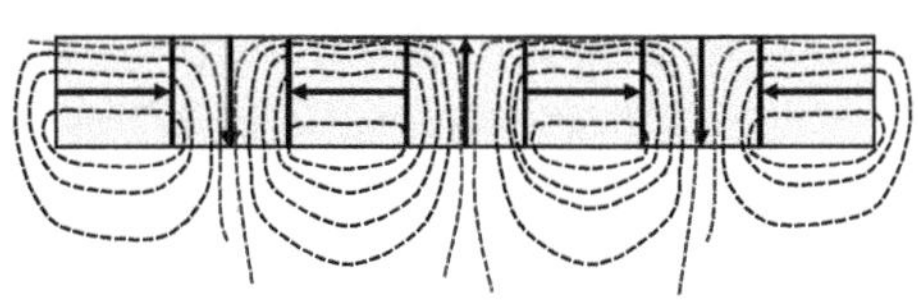

Fig. 3.22 Magnetic field of the Halbach array

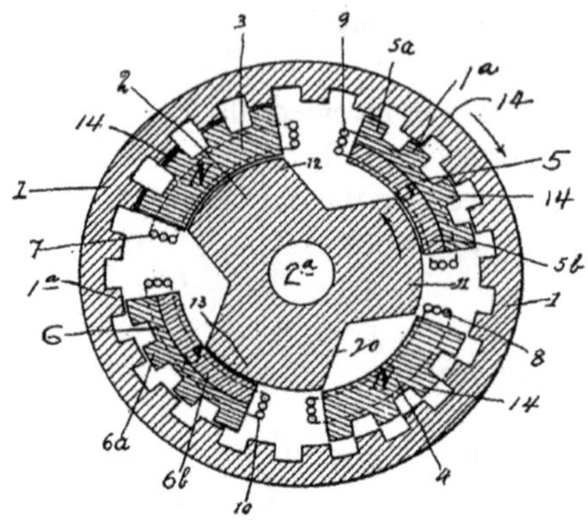

Fig. 3.23 Illustration from US1171351

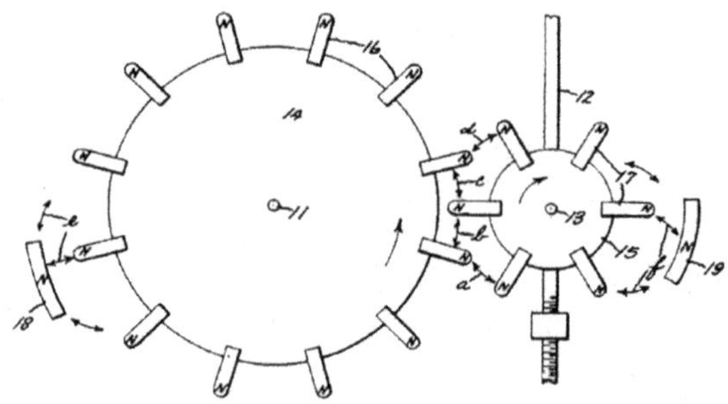

Fig. 3.24 Illustration from US2243555

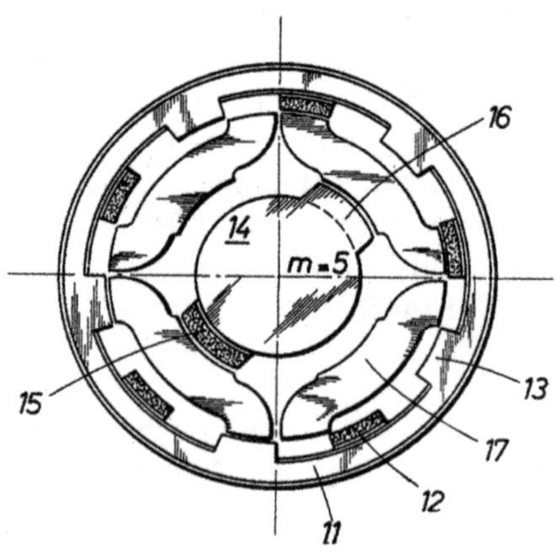

Fig. 3.25 Illustration from US3645650

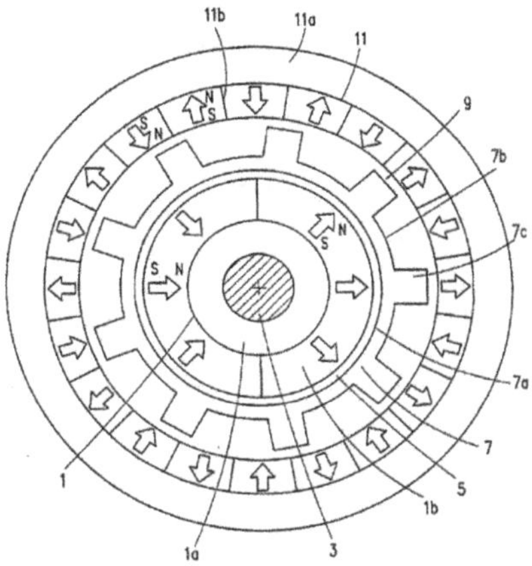

Fig. 3.26 Illustration from US5633555A

Fig. 3.27 Illustration from US2010207472A1

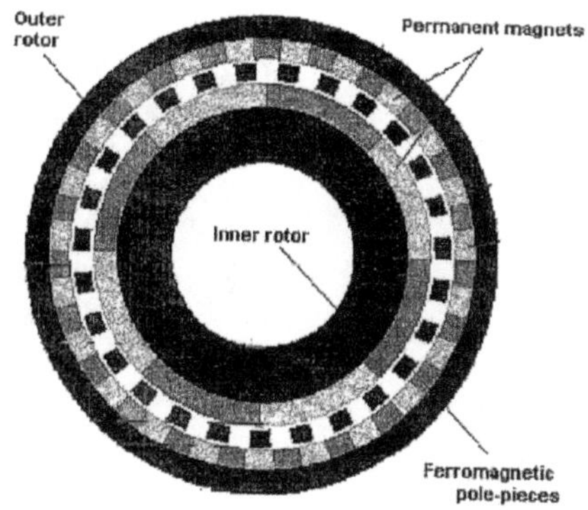

In 1997, B. Ackermann took the next step in patent application US5633555A ("Magnetic drive arrangement comprising a plurality of magnetically cooperating parts which are movable relative to one another"), enhancing the toothed and magnetic structures of magnetic gearboxes (Fig. 3.26).

Finally, in 2001, K. Atallah and D. Howe described a version of a magnetic gear in their article [2] (Fig. 3.27 from patent application US2010207472A1) that proved competitive in practical applications. Patents related to magnetic gears have been appearing for several decades now, and Fig. 3.28 outlines key milestones in the development of these systems. No single inventor can truly be credited as the sole creator of modern magnetic gears. However, we can conclude that the transition to practically viable versions occurred at the turn of the twenty-first century, thanks to the work of Ackermann, Atallah, and Howe. In 2006, Atallah, together with J. Rens, proposed a

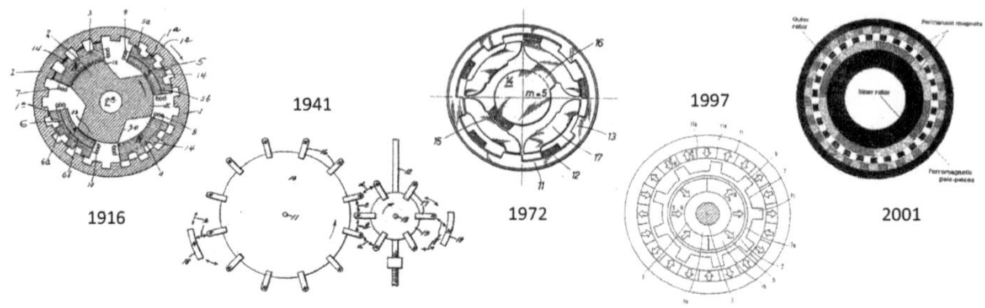

Fig. 3.28 Evolution of the concept of magnetic gear

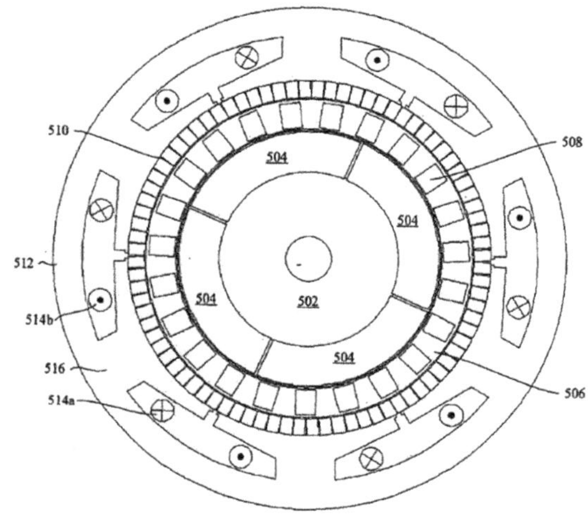

Fig. 3.29 Illustration from GB2437568A

machine with an integrated magnetic gear in patent application GB2437568A, titled "Electric machine having elements interacting in a magnetically geared manner." This design shown in Fig. 3.29 has become known as a "Pseudo-direct Drive" (PDD).

Interestingly, as early as 1968, F.M. Yuferov, in patent GB1196213 for a "Low-speed synchronous electric motor," proposed a machine with electromagnetic reduction that shares several characteristics with the PDD (Fig. 3.30). This again highlights the complexity of assigning singular credit for technological breakthroughs.

Some Identified Innovation Patterns

The examples of induction motors and magnetic gears illustrate a recurring pattern: at a certain point, a series of significant inventions arises from an already known material. As we saw with the induction motor, Arago's effect was known as early as

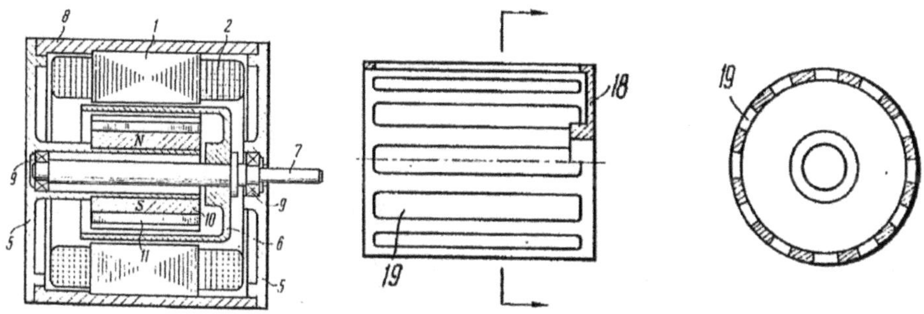

Fig. 3.30 Illustration from GB1196213

1824, but a breakthrough did not occur until 1879–1889. Similarly, with magnetic gears, attempts were made as early as in 1916, but practical solutions only emerged between 1997 and 2001. The skill of outstanding innovators in the electric machines field can lie in their ability to find some forgotten concept, find new ways to address the problems which prevented the concept from realization, and, finally, turn the intriguing ideas into practical, "working" solutions.

> One of the patterns is finding some old concept and bringing it back to life using the newest technological achievements

Life Spans of Selected Inventions Within the Context of Supersystems and Subsystems

Figure 3.31 presents the life spans of selected inventions in the field of electric machines, as well as some evolutionary changes at the levels of supersystems and subsystems. Influences from supersystems include various devices related to electronic converters for controlling electric machine (such as power switches, integrated circuits, and microprocessors), while influences from subsystems include permanent magnets and superconducting materials.

Not all promising concepts stand the test of industry and market acceptance, however. Here is a couple of historical examples of promising ideas that are still struggling to become true successful innovations:

First Example—High-Voltage Generators. In the 1970s, in USSR a new type of hydrogenerator with a voltage of 121 kV was developed and built, designed to connect directly to a 110 kV power transmission line without a step-up transformer. The stator windings was of cable-type and used paper-oil insulation. This hydrogenerator, with a power output of 14.5 MW, was installed at the Skhodnya Hydroelectric Power Plant in Moscow. It successfully underwent testing when connected to the regional grid, demonstrating the feasibility of creating hydrogenerators with unprecedented voltage levels. In the 1990s, in Sweden there were developed hydrogenerators and turbogenerators with cable windings of the "Powerformer" type, resulting in dozens

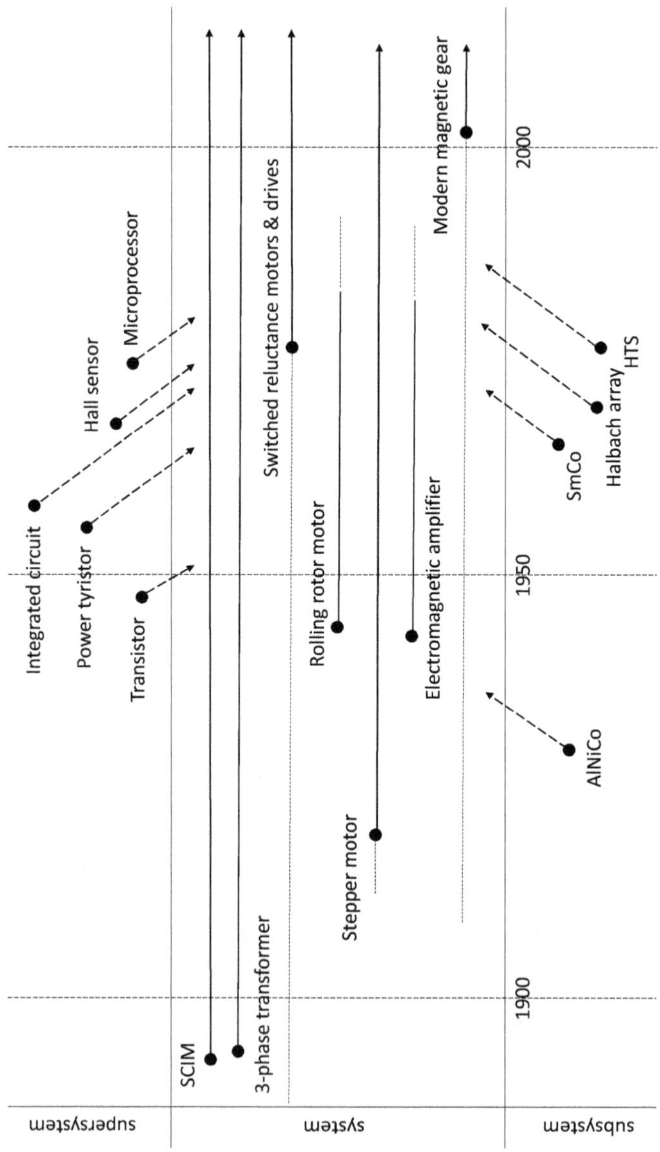

Fig. 3.31 Life spans of selected inventions within the context of supersystems and subsystems

of patents for various aspects of their design. One such hydrogenerator operated at 45 kV, with a power of 11 MVA and a rotation speed of 600 rpm, and, like its Soviet predecessor, it was connected to the grid without a step-up transformer. Its stator windings used cross-linked polyethylene insulation. A turbogenerator with a voltage of 136 kV, power of 42 MVA, and rotation speed of 3000 rpm was connected to the grid in 2000. Another hydrogenerator, with a voltage of 155 kV, power of 75 MVA, and rotation speed of 125 rpm, was commissioned in 2001. Despite the successful demonstration, these impressive machines did not achieve broader adoption.

Second Example—Large Coreless Generators. Coreless generators, featuring two rotor assemblies with magnets and a single coreless stator were considered to be the perspective concept for gearless wind power converters. Such generators with large diameters (up to 20 m) were under development for wind power converters in several projects in the 2010s (one such concept is shown in Fig. 2.19). However, several companies working on similar designs ceased operations, leading to project closures.

"Serial" Inventors

Earlier we referred to several significant inventions by Nikola Tesla and Mikhail Dolivo-Dobrovolsky. Both of these outstanding electrical engineers held dozens of inventions, though history tends to remember only their most impactful contributions. In the early 2020s, the Espacenet database [3] recorded approximately 40,000 patent applications annually related to electrical machines. This indicates that the potential for new inventions in electromechanics remains vast, despite the field being over two centuries old.

The history of invention includes many "serial" inventors who have filed hundreds or even thousands of patents, often in fields outside electromechanics. Some of the prolific inventors are still active today. For example, contemporaries Shunpei Yamazaki and Kia Silverbrook each hold several thousand patents[3]. The personal qualities and work methods of serial inventors have been studied through interviews. Typically, such inventors exhibit the following traits: resilience to criticism, high energy, dissatisfaction with the status quo, curiosity, and a willingness to adopt and adapt unconventional ideas. The following notable points were identified about serial inventors:

- Skills and abilities
  - The ability to think beyond the "barriers" of different technical fields and scientific disciplines.
  - The skill to mentally combine various objects and visualize new solutions in their imagination.
  - Studying patents in their own field and related areas.
  - Developing an understanding of what qualifies as sufficiently novel to merit patenting.
  - Acquiring skills from adjacent fields of knowledge.

[3] Here, perhaps, one might assume that they are not working alone, but have some sort of a support team.

- Their principles
  - Pursuing goals without abandoning them, even after multiple failures.
  - Tackling challenges independently rather than passing them to others.
  - Refusing to accept the existence of insurmountable obstacles.
  - Recognizing the critical importance of working with colleagues to achieve results.
- Sources of ideas
  - Setting significant objectives and striving to achieve them.
  - Actively seeking problems that require solutions.
  - Engaging with others to identify problems and generate innovative ideas.

Outstanding inventors provide valuable lessons for personal development necessary for being active players in the evolution of electrical engineering and electrical machines. In this field, as well as in other areas of technology, fostering serial inventiveness is crucial for sustained innovation.

Economic Impact of Inventions and Innovations

It would be interesting to quantify the flow of innovations and its dynamics in the field of electric machines in recent years. Unfortunately, the author was unable to find such data, and this work provides only an approximate assessment based on indirect indicators. Let us turn to a related industry—automotive. Modern cars contain dozens of electric machines, making automotive a related industry to electric machine building.

In [4], it is shown that companies like Opel and Volkswagen had around 100,000 innovative proposals annually in the early 2000s. These proposals ranged widely—from minor improvements to inventions with significant economic impact. The average economic benefit per proposal was about €1000, while the effectiveness of so-called "outstanding" innovations was assessed at no less than €400,000. Interestingly, "outstanding" innovations accounted for 15% of the total economic impact, while 85% came from mass innovations. This indicates that the importance of minor improvements should not be underestimated.

During the same years, Opel and VW filed several hundred patent applications annually, amounting to less than one patent per 100 innovative proposals. A similar ratio (1:100) between submitted ideas and patent applications likely applies to electric machines industry as well.

Given that approximately 40,000 patent applications are filed annually in class H02K "Dynamo-electric machines", and assuming the same 1:100 ratio of innovative proposals to patent applications, this suggests an impressive figure of four million innovative proposals annually! Naturally, the vast majority of these ideas involve only minor improvements, and these ideas are in most cases not processed to any realization.

It is also worth noting that the value of the same idea can vary depending on the city, country, or region where it is implemented. Feasibility depends on the availability of materials, components, manufacturing capabilities, traditions of engineering schools, established standards, system solutions, and many other factors.

The Role of Patents

Some proposed new solutions may reach the level of inventions, meaning they possess novelty and an inventive step[4]. In such cases, the implementation of these solutions may involve application for protection of intellectual property rights (filing patent applications), which can, in certain cases, provide exclusivity on the relevant market or opportunities for licensing.

Patents are often used as tools in competitive markets. For example, the patent US5083039B1 for a "Variable speed wind turbine," filed in 1991 by Robert Richardson and William Erdman, can be considered one of the most commercially valuable patents related to electric machines. Initially filed on behalf of US Windpower, the patent was later acquired by General Electric (GE) through a series of mergers and acquisitions. Recognizing its significance, GE effectively leveraged the patent in competitive battles within the US wind energy market.

The first claim of the patent US5083039B1 is as follows:

> *A variable speed wind turbine comprising: a turbine rotor including at least one blade mounted to a rotatable shaft; a multiphase generator having a stator and a rotor coupled to the turbine shaft for rotation therewith; a power converter including active switches operable for controlling stator electrical quantities for each phase of the generator; torque command means associated with turbine parameter sensors for defining a torque reference signal; and generator controller means coupled to the power converter for controlling the active switches including field orientation means responsive to the torque reference signal for defining a desired quadrature axis current in field coordinates, and switch control means for controlling the active switches to produce stator electrical quantities that correspond to the desired quadrature axis current.*

The described approach represents what some specialists today could identify as a fairly standard scheme for stator field control in a generator using feedback loop, as illustrated in Fig. 3.32. This solution gained significant popularity in wind power industry starting in the 2000s. Since vector control of induction machines had been known since the 1970s, some specialists did not recognize this adaptation as fundamentally new, even though it was adapted for a specific domain—wind power harvesting. Nevertheless, the patent was granted and remained in force until 2011.

During the 2000s, many competitors challenged the originality and novelty of the invention, particularly as GE began leveraging the patent to block access to the

[4] Terms of the patent law.

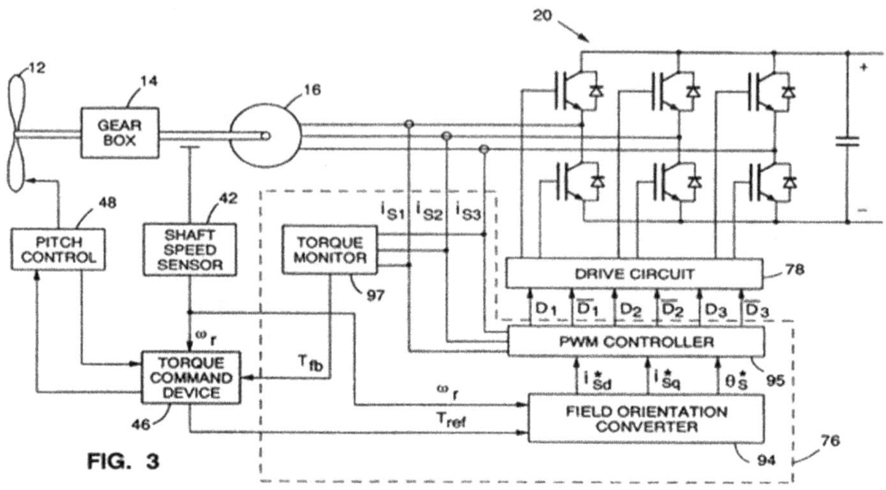

Fig. 3.32 Illustration from US5083039B1

US market for suppliers of wind turbines, such as Mitsubishi and Enercon. GE succeeded in defending its position in court, essentially reaching an exclusive status on the market for high-power, variable-speed wind turbines in the USA. Some competitors were forced to adopt less efficient control schemes to avoid infringing on GE's patent or to come to some deal with GE.

Opinions may differ regarding this case, but it undeniably highlights the skill of the specialist who crafted the text and claims of the patent. None of GE's competitors managed to circumvent the patent. Typically, patent applications are drafted by patent attorneys, whose expertise in patent law and intellectual property sometimes plays a pivotal role in the evolution of technical systems. Strategies for intellectual property protection, including patenting, form a distinct field of knowledge with extensive literature, such as [5].

Patent Searches

One practice accessible even to nonspecialists is patent search. This practice involves examining existing patent applications and granted patents, the texts of which are available on the websites of national, regional, and international patent offices. Patent searches are conducted for various purposes:

- Ensuring a technical solution is novel before submitting a patent application. Neglecting this step may result in the patent office finding during the examination process that the idea lacks novelty and thus cannot receive legal protection in the form of a granted patent.

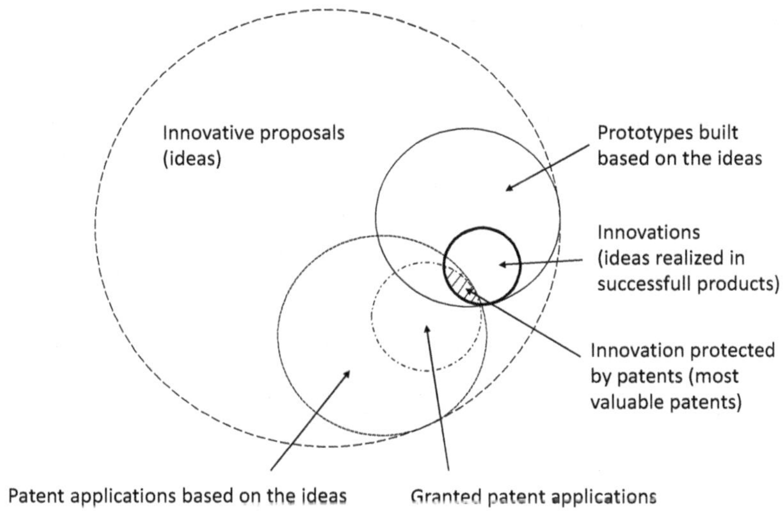

Fig. 3.33 Overlapping pluralities to the process of innovation

- Developing a knowledge base in a specific field, identifying promising research directions, and forecasting the development of science and technology.
- Identifying solutions that can be applied in one's products or systems (e.g., from expired or abandoned patents).
- Ensuring that the introduction of a new product to the market does not violate existing intellectual property rights covering similar technical solutions.

Inventions, Patents, Prototypes, and Products

It's essential to recognize that not all innovations are accompanied by patent applications and granted patents. This concept is illustrated in Fig. 3.33, where innovative proposals (ideas) are represented by a large circle, which contains other subsets. The set of patent applications based on these ideas includes a smaller set of granted patents. Notably, the subset of prototypes built based on the ideas only partially overlaps with the subset of ideas for which patent applications are filed, indicating that the processes of prototyping and patenting often proceed independently. Similarly, innovations and granted patents are only partially intersecting subsets, meaning that in most cases, that either the specialists responsible for intellectual property fail to predict which ideas will prove most impactful or the ideas are not patentable because of the existing prior art. According to data from [6], only about 5% of patents hold strategic value for their owners. Figure 3.33 also highlights that ideas realized in successful products (innovations) make up a very small portion of all submitted ideas.

Using the Patent Statistics for Certain Estimates

The number of patent applications is an indirect indicator of the level of investment, the number of specialists involved, and the success of technological development in any particular industry, region, or time period. Tables 3.3 and 3.4 presents the results of a series of searches in the open patent database www.espacenet.com for the H02K class (dynamo-electric machines). The number of patent families listed in the table is shown for five-year periods. Leading countries and companies (by the number of patent applications) are also indicated for the same periods for reference. In addition, largest world economies (by GDP) are indicated for each period. It is important to note that a single patent application may be filed in several countries (forming a patent family), so the total number of patent families is less than the sum of patent applications in individual countries.

Let's interpret the data in the table. Starting from the 1970s, we observe the economic and technological rise of Japan, which bypassed Germany and became the global leader for more than three decades. This period coincides with Japan's entry into the top three largest economies, eventually reaching second place (after the USA) in terms of GDP. In the early 2000s, China surpassed Japan and the USA to firmly take the lead in the patent production within the class H02k. Interestingly, despite being the largest global economy, the USA was not the leader in the number of patent applications in the field of electrical machines during the considered period (the last 65 years). Over seven decades, the number of patent applications in the H02K class has increased about eight times (Fig. 3.34). The explanations for such radical growth may include the increase in the Earth's population from 3 billion to 8 billion during this period, the greater involvement of almost all countries in the patenting process, as well as China's anomalous activity in the last 15 years.

By comparing available data on the sales of electric machines, which show constant growth, with the rate of increase in patents for various electric machines inventions, one can conclude that Schmookler's hypothesis of "demand-pull," according to which the flow of inventions in a particular industry correlates with demand for the products of this industry holds true for the electric machines industry.

With patent statistics, one can roughly estimate the number of specialists involved in innovation activities and investments. According to [7], in the USA during the 2000s, there were filed 5 patents for every 100 scientists and engineers and the same number of patents for every ten million dollars invested in R&D. We will assume these ratios (1 patent per year for every 20 scientists, 1 patent per year for every two million dollars in investment) as the baseline. Currently, about 40,000 new patent applications for electric machines are filed globally every year. Using these ratios, we estimate that there are 800,000 scientists and engineers working on electric machine-related topics worldwide, with approximately 80 billion dollars in R&D investments in this technological domain. Considering that only about 5% of all patents have a real impact [6], i.e., representing actual innovations, we can assume that the annual number of useful innovations in the electric machines field worldwide exceeds a thousand per year.

Table 3.3 Results of the search on class H02K

	1961–1965	1966–1970	1971–1975	1976–1980	1981–1985	1986–1990
Number of families	9070	12,025	25,230	34,430	35,610	31,880
Leading countries	Germany—3130, USA—2580, France—2400	Germany—4235, Japan—4070, USA—3316	Japan—17,900, Germany—4120, USA—3310	Japan—26,200, Germany—3730, USA—3230,	Japan—27,900, Germany—3440, USA—3270	Japan—22,900, USA—3830, Germany—3570
Leading companies	Siemens—700, Licentia—640, GE—270	Siemens—550, Licentia—470, GE—280	Hitachi—930, Siemens—390, GE—270	Hitachi—2280, Tokyo Shibaura—1120, Mitsubishi—840	Hitachi—1710, Matsushita—1260, Mitsubishi—1150	Toshiba—981, Mitsubishi—965, Hitachi—870
Largest economies	USA, USSR, Germany	USA, USSR, Germany	USA, USSR, Japan	USA, USSR, Japan	USA, USSR, Japan	USA, Japan, USSR

Table 3.4 Results of the search on class H02K (contd)

	1991–1995	1996–2000	2001–2005	2006–2010	2011–2015	2016–2020
Number of families	25,710	30,490	36,880	53,910	82,100	161,000
Leading countries	Japan—17,050, USA—4600, Germany—4290	Japan—18,100, USA—7500, Germany—6300	Japan—19,260, USA—11,470, China—9160	China—26,640, Japan—18,640, USA—13,280	China—54,200, Japan—19,100, USA—17,200	China—132,400, Japan—18,370, USA—17,680
Leading companies	Toshiba—965, Matsushita—900, Mitsubishi—582	Matsushita—1010, Denso—640, Hitachi—630	LG Electronics—860, Denso—810, Matsushita—770	Matsushita—1000, Denso—860, Toyota—830	Mitsubishi—1350, Toyota—890, Siemens—850	Zhuhai Gree—2057, Mitsubishi—1325, Denso—959
Largest economies	USA, Japan, Germany	USA, Japan, Germany	USA, Japan, Germany	USA, China, Japan	USA, China, Japan	USA, China, Japan

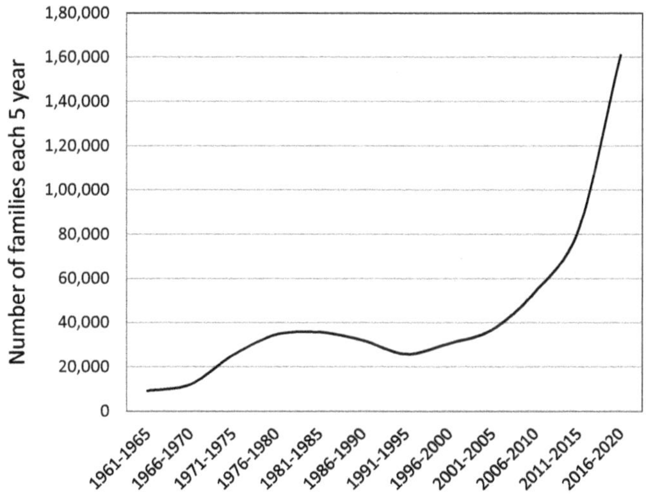

Fig. 3.34 Dynamics of patent applications in the H02K class

Thus, we can draw the following conclusion: the work of hundreds of thousands of specialists worldwide results in hundreds of improvements in electric machines and directly related systems every year.

Every year the work of hundreds of thousands of specialists worldwide results in hundreds of significant improvements in electric machines and directly related systems

Concluding Comments

To summarize the chapter, we can once again list the key players: policymakers and officials, investors, technology and market experts, researchers, technology scouts, visionary entrepreneurs, customers for new equipment, managers at various levels, salespeople, inventors, R&D engineers, and patent attorneys. The success of any innovation requires nearly all of these players, and it is essential that each fulfills their role effectively.

Here we conclude the attempt made in the first three chapter to at least outline the main mechanisms, processes, and players involved in the evolution of electric machines. This description is, of course, subjective and incomplete, and the author hopes that in the future, it will be possible to study and examine this topic more deeply and thoroughly, as active participation in the evolutionary process, and ultimately managing that evolution, is impossible without a comprehensive understanding of all the key processes.

References

1. www.acare4europe.org
2. K. Atallah, D. Howe, A novel high-performance magnetic gear. IEEE Trans. Magn. **37**(4), 2844–2846 (2001)
3. www.espacenet.com
4. M.A. Orlov, T.R.I.Z. Easy, *Universal practical course* (M.: Solon-Press, 2011), 384 p.
5. V. Irish, *Intellectual property rights for engineers* (The Institution of Electrical Engineers, 1994)
6. S.S. Harrison, P.H. Sullivan, *Edison in the boardroom revisited: how leading companies realize value from their Intellectual property*, 2nd edn. (John Wiley & Sons, 2011)
7. B. McKern, *Managing the global network corporation* (Routledge, 2003), 304 p.

"Anatomy" of Electric Machines as the Object of Evolution

4

To create something radically new, it may be necessary to begin by shifting paradigms and discourses. This chapter introduces a new classification of electric machines as an alternative to the established discourse, along with a method for organizing knowledge about electrical machines.

In the near future, the combination of unconventional thinking and the use of Artificial Intelligence (AI) could become a key tool in the creation of new technical solutions. AI refers to the capability of software systems to perform creative functions traditionally regarded as possible only for humans. It is well-known that there already exists software that simulates intellectual activity, surpassing human intelligence in specific narrow fields. AI is gradually finding its place in engineering disciplines, with numerous studies demonstrating its application in system stability control, prediction, and optimization. Examples in the field of Electromechanics can be found in [1–3]. This progress raises the intriguing and timely question of when AI will begin to be applied to the design and development of electrical machines. While there are no reliable forecasts about when AI will become prevalent in Electromechanics, it is reasonable to assume that development will follow the stages illustrated in Fig. 4.1.

At present, it is difficult to predict the exact type of AI that might be used in the design of electric machines. Possible types include expert systems, knowledge bases, and reasoning systems that mimic processes such as thinking, reasoning, and speech. Also, artificial neural networks and evolutionary computations, which model intelligent behavior based on biological elements, are being actively developed and tested. One noteworthy approach is based on the premise that elements of human thought are not numbers but components of fuzzy sets. AI can process fuzzy sets significantly better than humans. For instance, while an experienced designer might be able to keep track of dozens of critical machine parameters, AI could handle an order of magnitude more parameters and combine them effectively, potentially enhancing the speed and quality of decisions in the design process. It is plausible that *combinations* of different types of AI will be used in the future.

A. Matveev, *Shaping the Future of Electric Machines*, https://doi.org/10.1007/978-3-031-95248-7_4

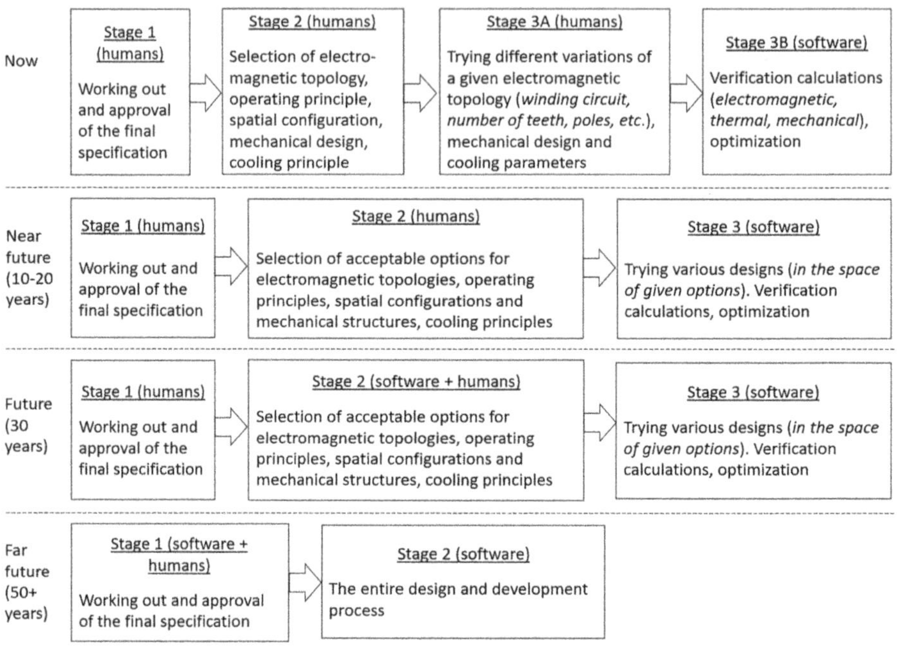

Fig. 4.1 Predictions for human involvement in key stages of electric machine development

It is worthwhile to explore the types of logical "reasoning" AI could be trained to perform and to propose a classification of electric machines as a "space" for this reasoning to operate within. This new classification could serve as a framework for AI-driven innovation in electric machine design and development.

To illustrate the reasoning process that could potentially be "taught" to AI, let's consider the following task: achieving rotational motion with a certain torque using a set of electromagnets and ferromagnetic bodies. The reasoning could unfold as follows, step by step:

1. It is known that a ferromagnetic body is attracted to an energized electromagnet. This principle should be "communicated" to the AI by defining a specific interaction pair: "*electromagnet–ferromagnetic body.*"
2. If a ferromagnetic body is attached to a cantilever fixed at a certain pivot point (as shown in Fig. 4.2a), the process of the body being attracted to the electromagnet will result in the cantilever rotating around the pivot point.
3. Once the attraction is complete (as shown in Fig. 4.2b), the rotation will stop.
4. By adding another interaction pair positioned relative to the first pair as shown in Fig. 4.2c, the motion can continue. This is achieved by deenergizing the first electromagnet and energizing the second one.

Once AI "masters" this reasoning process, a follow-up challenge could be presented: how to optimally distribute electromagnets and ferromagnetic bodies on cantilevers around a central axis of rotation, and how to activate and deactivate electromagnets to ensure continuous rotation. AI could leverage its ability to

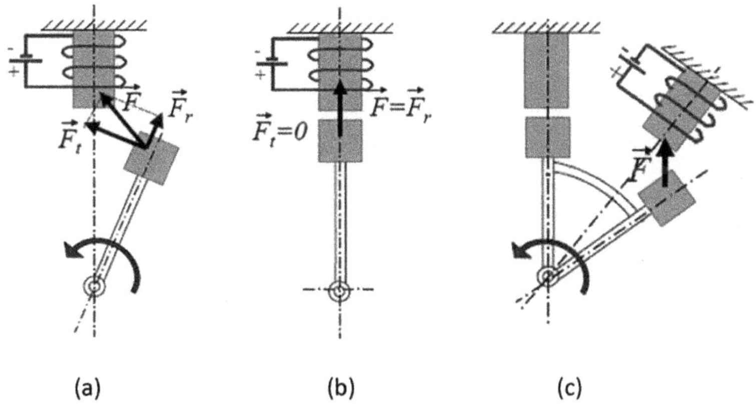

Fig. 4.2 Illustrating the logic of creating rotational motion: (**a**) energizing the first electromagnet, (**b**) aligned position, (**c**) energizing the second electromagnet

recognize spatial relationships between elements and assess their interactions. If successful, AI would effectively learn the operating principle of a switched reluctance motor (SRM). SRM is likely to be the simplest example for AI to explore in the context of electric machine design.

In this example, a "building block" is used—the interaction pair "*electromagnet–ferromagnetic body*"—which could be considered a "basic building element" for implementing the logic behind this type of electric machines. The proper selection of such logical building blocks will likely be particularly important for advancing AI applications in the design of electric machines.

Let us leave AI for a while and investigate electric machines classification, the primary subject of our discussion in this chapter. Classification is a system for grouping research objects based on shared characteristics. In science, classification involves dividing the scope of a concept according to a specific criterion, where the broader concept (class) is divided into types (subclasses), and those types are further divided into subtypes, and so on.

Existing Classifications of Electric Machines At the highest level, the classification of electric machines can arguably be based on the type of field in which energy conversion occurs. According to this criterion, electrical machines can be divided into *inductive* (energy conversion occurs in a magnetic field), *capacitive* (energy conversion occurs in an electric field), and *inductive-capacitive* [4]. Virtually all electric machines used in industry and households today are inductive, and our discussion will focus exclusively on this class of machines.[1]

The result of classifying electric machines using almost any set of criteria is not a tree with a "trunk" and "branches," but rather a multidimensional matrix. Let us illustrate this with an example. Figure 4.3 shows a classification based on the type of power supply (current type), while Fig. 4.4 shows a classification based on the

[1]Capacitive and inductive-capacitive machines should be considered in a broader discussion of electric machines evolution, e.g., when going into micro and nano scale.

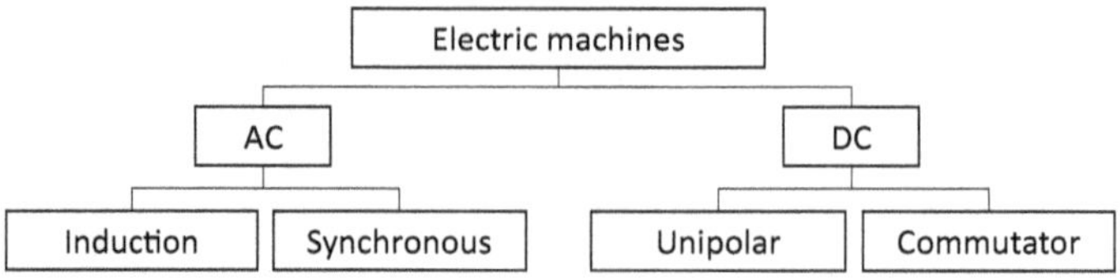

Fig. 4.3 A section of the classification tree based on current type

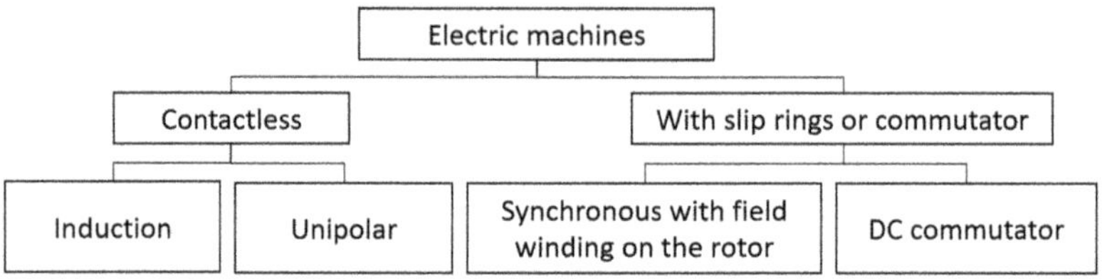

Fig. 4.4 A section of the classification tree based on energy transfer to the rotor

type of energy transfer to the rotor. When examining parts of the classification trees in Figs. 4.3 and 4.4, it becomes clear that, under the first criterion, induction and unipolar machines fall into different groups, while under the second criterion, they fall into the same group. Similarly, under the first criterion, synchronous and commutator machines are in different groups, but under the second criterion, they are grouped together. This highlights that when different classification criteria are applied, an individual electric machine does not necessarily have consistent "neighbors by subtype" in the branches of the classification trees.

To see how some leading minds of the academia saw classifications in the late twentieth century, let us consider classifications presented in two classical textbooks on electric machines—[4, 5]. Specifically, the following classification is provided in [4]:

- By current type: DC and AC machines, with AC machines further divided into asynchronous (non-synchronous), synchronous, commutator machines, and transformers.
- By direction of energy conversion: motors, generators, and reactive power compensators.
- By design of active parts: salient-pole and non-salient-pole machines.
- By rotor arrangement: machines with inner and outer rotors.
- By the number of moving parts: single-dimensional and multi-dimensional machines.

An original approach to classification is presented in [5], which starts from the statement that "*a necessary condition for an electric machine to perform electromechanical energy conversion is the change of self- or mutual inductances of the coils as the rotor turns*". Correspondingly, the question is posed on how an electric machine must be designed so that the self- or mutual inductances of its windings depend on the angular position of the rotor and change periodically as the rotor rotates. As a result, three fundamental designs (three branches of classification) are identified: machines with one winding on the stator and one winding on the rotor (D-1), machines with one winding on the stator and a toothed magnetic core on the rotor (D-2), and machines with two windings on the stator and a toothed magnetic core on the rotor (D-3). An additional classification criterion in [5] is the type of magnetic core—*toothed* or *smooth*—which corresponds to the division into salient-pole and non-salient-pole machines in [4].

Certain conclusions can be drawn about the practically applied classifications of electric machines during different time periods based on the content of textbooks from those time periods. Textbooks play a critical role as they shape the understanding of entire generations, establish fundamental paradigms, and define the language of the field. In the twentieth century, it was common to structure general electric machine textbooks into sections on transformers, induction machines, synchronous machines (with electromagnetic excitation), DC machines, and a broad category of other machines collectively referred to as "special machines." Thus, in practice, another implicit criterion was often used: the popularity of certain types of machines (market share, extent of usage, number of applications), which relates more to economics or structure of the industry than to technical attributes. Classifications frequently reflect the popularity of certain electrical machine types in the academic world at the time of writing the textbook or article. For example, in [6], a twenty-first-century book, electrical machines are divided into induction (including single-phase), synchronous (including synchronous reluctance and permanent magnet synchronous machines), and switched reluctance machines.

At times, it seems that the selection of machine classes covered in textbook chapters partly reflects the personal experience and scientific interests of the authors.

Critique of Existing Classifications Let us point out some shortcomings of existing classifications.[2] For example, the classification approach in [5], which includes three configurations (D-1, D-2, D-3), can be criticized for excluding certain well-known machines. Examples include synchronous reluctance machines, various hybrid machines, and induction machines with solid or massive rotors. Additionally, it is difficult to draw a clear boundary between "toothed" and "smooth" magnetic cores, given the existence of semiclosed slots and open slots with magnetic wedges. Furthermore, rotors with perfectly smooth surfaces but varying magnetic reluctance along different axes (as in synchronous reluctance motors) blur the distinction further. Similarly, the strict division between asynchronous and synchronous machines is not always justified. For instance, "synchronized asynchronous" and "asynchronized synchronous" machines are known examples. Certain machines with $q < 1$ (where q is the number of slots per pole per phase) do not fit the classical definition of synchronous machines as they operate on higher harmonics. Switched reluctance machines, which function with a "jumping" field, are neither synchronous nor asynchronous since they lack slip. In short, existing classifications do not always capture the full diversity of electrical machines.

[2] From the subjective perspective of the author.

Another commonly used classification criterion—energy conversion direction and machine purpose, dividing machines into motors, generators, and reactive power compensators—also has its limitations. For example, in electric vehicles, the main propulsion motor frequently transitions into regenerative braking mode during standard urban driving. As a result, the main propulsion electrical machine in an electric vehicle cannot be strictly classified as either a motor or a generator. Similar examples abound in other applications.

When it comes to classification by the type of current (AC or DC), there are examples that complicate the traditional division into AC and DC machines. For instance, in classical synchronous machines, the stator is powered by alternating current while the rotor uses direct current. Similarly, in homopolar inductor machines, the stator can have two windings supplied with currents of different types. Assigning such machines to either AC or DC categories becomes debatable. Furthermore, the advent and widespread adoption of power electronics fundamentally change the perception of which type of current supply (or network) is available for use. This is especially evident in cases where electronics is integrated with the active parts of the machine. The shapes of the currents and voltages generated by power electronics can vary widely—rectangular, trapezoidal, and so on—making the concept of "alternating current" somewhat ambiguous. Additionally, discussions are already underway regarding autonomous networks with variable frequency and voltage, further complicating classification by the type of current.

One more criteria—the flux direction in the machines—can also be criticized: it will be shown below, the division of machines into those with radial and axial flux is not always meaningful and often represents an unacceptable oversimplification.

Returning to contemporary literature, it can be said that the diverse classifications proposed not only fall short of perfection but often contain errors. Some publications even allow for so-called "conceptual overlaps" (e.g., "induction, DC, ... and disk machines"). There is also poorly conceived classification examples reflected in machine names, such as the term "brushless DC machine"—this machine, in fact, is powered by trapezoidal AC currents and thus qualifies as an AC machine, yet it traditionally retains its misleading name. Summarizing the above, one can say that the situation with electric machines classifications undoubtedly deserves critical scrutiny.

Questioning the Paradigm and Language Several questions can be posed:

- *What is the purpose of electric machine classification?*
- *What have conventional classifications led us to?*
- *What realities do they reinforce?*
- *How modern and relevant are they?*
- *Do they contribute to the advancement of Electromechanics?*

Clearly, existing classifications are invaluable and even indispensable for communication between specialists, particularly in the implementation and deployment of electric machines and accompanying systems. However, for inventors and designers, the established language and associated thinking patterns often present more of a limitation than an advantage. The terms used in classifications highlight only a selection of some "islands" in a "sea" of possibilities, obscuring the broader space of alternatives.

For an inventor, the goal is not to name something or fit it into a predefined "slot" but to think freely. An inventor's task, to some extent, is to scatter a mosaic assembled by others and reassemble it into something new. In the case of AI, the need for classifications may diverge even further from what current systems can provide. AI is likely uninterested in human terminology altogether. Human-defined terms and classifications can even act as constraints—"anchors" or "boundaries." In theory, AI could "discover" intriguing "hybrids" that entirely lack adequate descriptions within the established professional language.

The discussion above can be concluded by saying that there is a pressing need to revisit the established paradigms and redefine the language used in the classification of electric machines. This raises the following question: "*Can the diversity of electric machines be described through a practically useful and applicable classification—suitable for inventors and AI—that is free from the shortcomings of existing classifications?*" This chapter aims at proposing such a classification—one that is exhaustive, comprehensive, and logically consistent for both inventors and AI. The new classification will define the boundaries of the multidimensional space of possibilities in the domain of electric machines, encompassing all potential configurations of the machines.

> There is the need for exhaustive, comprehensive, and logically consistent classification to be used by inventors and for teaching and using AI.

The Radical Proposal To achieve the goal, the new classification proposes to discard several familiar classification criteria, such as:

- The type of current powering the machine.
- The synchrony of rotor and field rotation (thus eliminating the division into synchronous and asynchronous machines).
- The direction of energy conversion (e.g., motor vs. generator).
- The direction of flux relative to the axis of rotation (e.g., radial or axial flux machines).

Additionally, the concepts of "toothed" versus "smooth" cores will take on a different meaning in the new classification.

In the new classification, the following six key classification criteria are proposed:

1. By the motion of active parts (APs) and the type/method of energy transfer to these APs.
2. By the composition of active elements and the type of power supply (magnetic field sources).
3. By the mechanism/method of generating interaction forces between the APs.
4. By the spatial configuration/orientation of the interacting APs.
5. By the number of AP pairs (number of air gaps).

6. By the arrangement of the main magnetic flux contour relative to the air gap plane and direction of motion.

None of these criteria is inherently superior to or more fundamental than any other. This results in a nonhierarchical, "flat" matrix-style classification where there are no overarching genera, classes, or similar divisions. Any of the six criteria can be combined with any variants or combinations of the other criteria.

It is important to note that certain aspects of design unrelated directly to electromagnetic processes are not addressed by this new classification. Traditional classification attributes, such as those based on construction elements and mounting methods (e.g., B3, B5, B35, etc.), protection levels (e.g., IP21, IP54, etc.), and cooling methods (e.g., IC01, IC411, etc.), remain unchanged. Additionally, in specific cases, characteristics such as the aggregate state of active parts, the type of motion (e.g., linear, rotational, or combined), and others can still be applied.

The Basic Element and New Classification Features The new approach to classification requires defining what is known as the *basic element*. Since energy conversion occurs in the air gap, the air gap itself can serve as the "starting point" for classification. The basic element can be defined as a pair of interacting active parts (APs) and the plane of the air gap between them, "extracted" from the machine, as shown in Fig. 4.5. This gives rise to the *Pictogram* (Fig. 4.5, right), which serves as the fundamental visualization element and logical building block.

Pictograms can be used, for instance, to illustrate the presence of two or more active parts in the machine, separated by air gaps, the supply of current to the windings (single-phase or multi-phase), and the types of currents involved. It is important to note that the pictograms do *not* represent the spatial configurations of machines (rotating, linear, etc.). Figure 4.6 illustrates several examples of pictograms:

- Figure 4.6a: A general case showing three moving, interacting active parts, each powered by a different (unspecified) type of current.
- Figure 4.6b: Two active parts, one stationary and powered by an undefined set of currents, while the second receives no external power supply

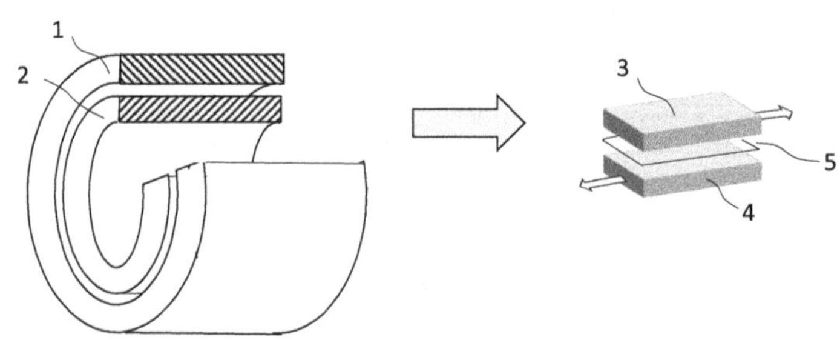

Fig. 4.5 Active parts with an extracted segment and their Pictogram (1—stator, 2—rotor, 3—stator element, 4—rotor element, 5—air gap plane between the stator and the rotor elements)

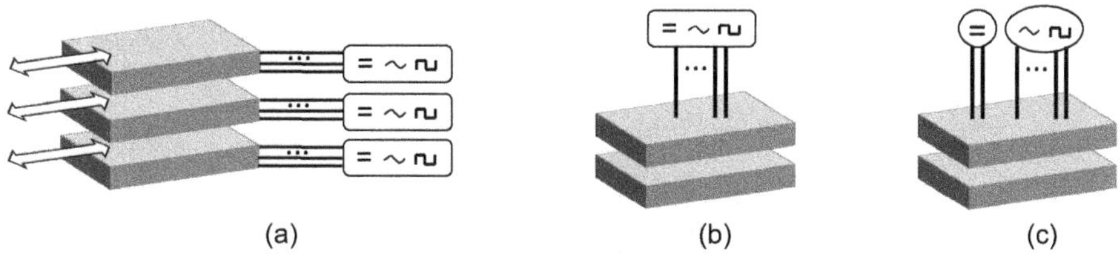

Fig. 4.6 Examples of using pictograms

- Figure 4.6c: A more specific configuration in which the stationary active part has two windings: one powered by direct current and the other by alternating current. Now let's consider the application of the proposed criteria in detail.

Criterion 1: "By the motion of active parts (APs) and the type of energy transfer to these APs". The vast majority of electric machines consist of one stationary AP and one moving AP. However, machines can have more than two APs. For instance, one AP may move while two others remain stationary, or conversely, two APs may move while a third is stationary (the moving APs can also be mechanically connected). Electric machines with two APs that both move relative to stationary surrounding objects are relatively rare.

Energy transfer to the stationary AP is typically direct and galvanic. For the moving AP, energy can be transferred through contact rings or commutators or magnetic field in the primary air gap, such as, e.g., rotary transformers (electromechanical cascades).

We can attempt to systematize these variations using pictograms. As an example, Table 4.1 lists all possible combinations of motion and energy transfer for two active parts. The row labeled "1M" (single moving AP) includes almost all the most common machines. Machines in the "2M" row (two moving APs) are rarer. Slip rings or a commutator are represented with a universal symbol—a circle divided in half, featuring concentric arcs on the right and diverging rays on the left. External energy sources are depicted as circles with a sinusoid inside. The topology in the table's cells can be referenced using alphanumeric indexes, such as "2M-1C" (two moving APs with energy transfer via contact rings/commutators to one of the APs) or "1M-2C" (one moving AP with energy transfer to both APs). For the "2M-EMC" configuration (two moving APs with energy transfer through a rotating transformer), it is also possible to transfer energy to both APs (this variant is not shown in Table 4.1). In the "2M-NT" variant, one of the moving APs could include an integrated energy storage system, which is another exotic option. Evidently, the number of APs can exceed two. Table 4.1 can easily be expanded as needed by following the same logic.

Criterion 2: "By field sources, field configurators, and kinds of power supply". Field sources can include windings with currents (located on iron cores or without cores) and permanent magnets. Windings can be powered by various types of currents. *Field configurators* are cores that serve as conductors of magnetic flux and shape the flux. These cores can have complex shapes, including holes or channels.

Examples of machines with key role of field configurators include switched reluctance machines, salient-pole synchronous machine, and synchronous

Table 4.1 Pictograms representing variants of motion and energy transfer

			Energy transfer to the moving APs			
			Collector or rings with brushes			Electromechanical cascades
			To one AP	To two APs	No energy transfer	
			«1C»	«2C»	«NT»	«EMC»
Motion of AP	Both APs move 	«2M»	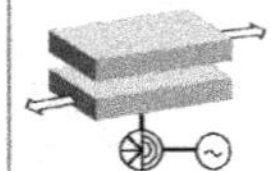	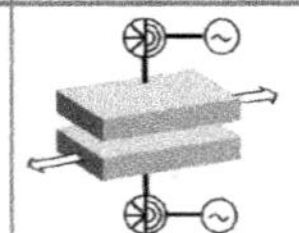	 Consumer or source of energy on one AP or both APs	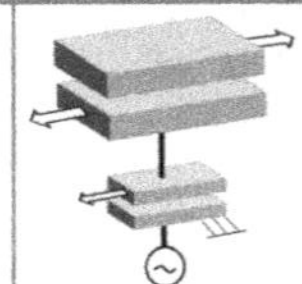
	One AP moves 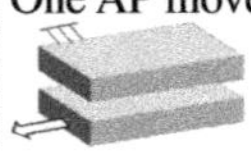	«1M»	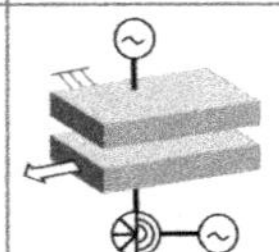			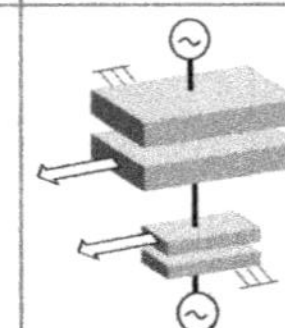

reluctance machine. In these machines flux is shaped due to the pronounced "magnetic irregularities" of the cores. Any toothed core can be considered as a field configurator, however, when the machine has many slots, especially closed or semiclosed, it can be regarded as not having any magnetic irregularities. Slotless machines have cores which are not toothed; thus they do not shape flux and do not have magnetic irregularities. Ironless machines do not have cores, so they have neither magnetic irregularity, thus, nor field configurator.

Examples of pictograms for some well-known machine types are shown in Figs. 4.7 and 4.8. Field configurators bringing in magnetic irregularities are symbolically represented by white broken lines. A short-circuited winding (such as a squirrel cage or damper winding) is depicted as small white circles. Permanent magnets are shown with arrows. Pictograms for some machine types without energy transfer to the rotor are shown in Fig. 4.7. Pictograms for machines with direct energy transfer to the rotor are presented in Fig. 4.8.

Pictograms can be used to visualize all the dimensions of the space of variants corresponding to one of the classification criteria. For example, Table 4.2 provides all possible variants of machines with a single multiphase winding powered by

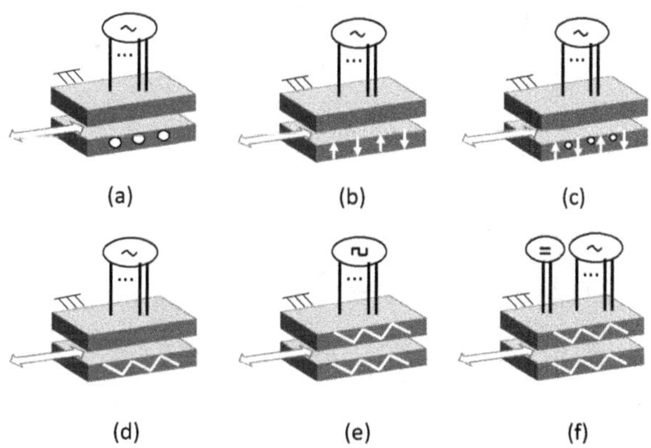

Fig. 4.7 Pictograms of: (**a**) squirrel cage induction machine, (**b**) permanent magnet synchronous machine (PMSM), (**c**) PMSM with damper winding, (**d**) synchronous reluctance machine, (**e**) switched reluctance machine, and (**f**) homopolar inductor machine

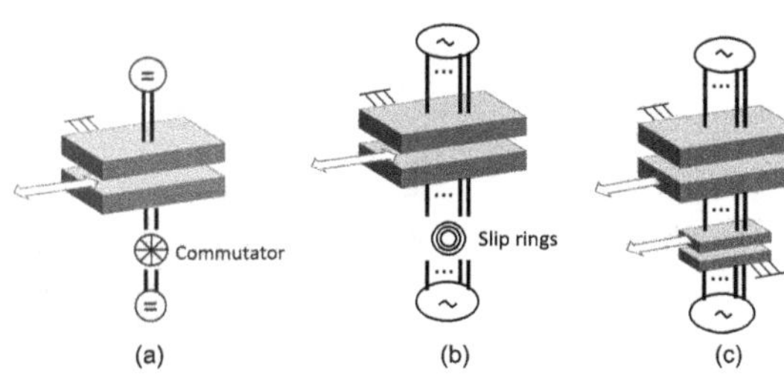

Fig. 4.8 Pictograms of: (**a**) direct current machine, (**b**) double-fed induction machine (with a wound rotor), and (**c**) synchronous machine with a shaft-mounted exciter

Table 4.2 Machines with one moving AC with a multiphase winding with sinusoidal current

			Stationary part			
	General case 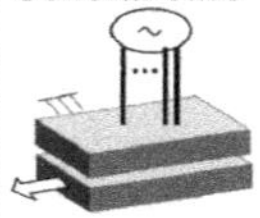		Smooth surface 	MI 	Magnets 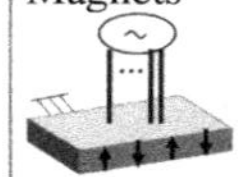	MI + magnets 
			A	*B*	*C*	*D*
Moving part	Solid 	*1*		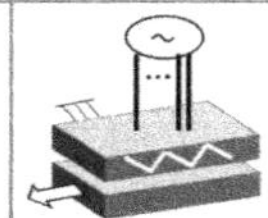	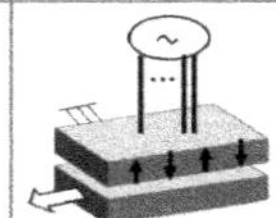	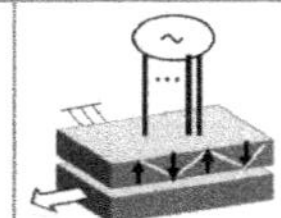
	SCW 	2				
	MI 	*3*	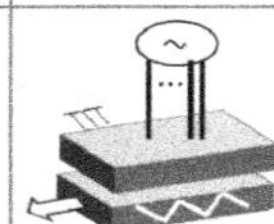		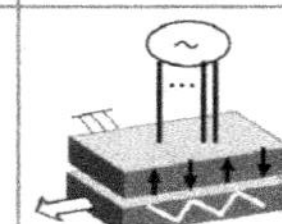	
	Magnets 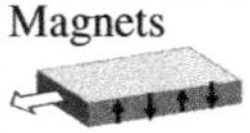	*4*	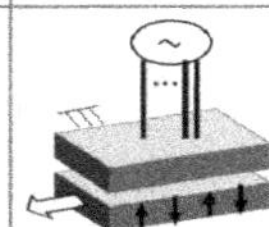		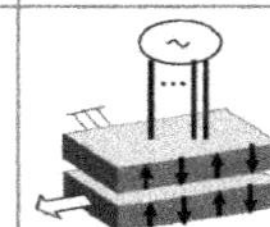	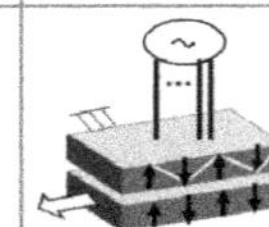
	SCW + MI 	*5*		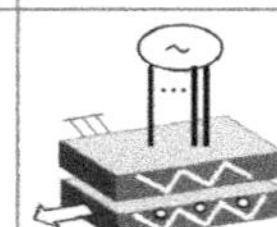		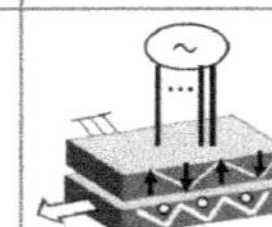

	SCW + magnets 	*6*	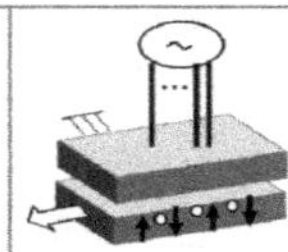	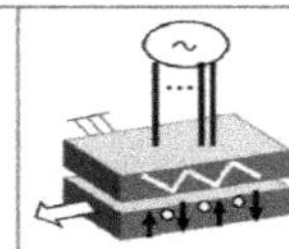		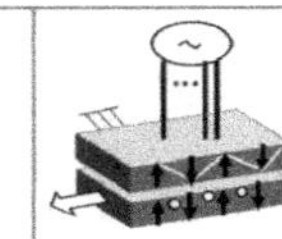
	Magnets + MI 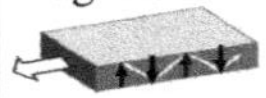	*7*	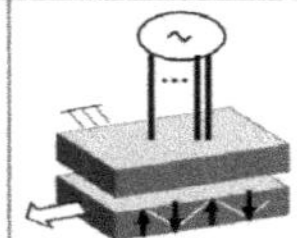	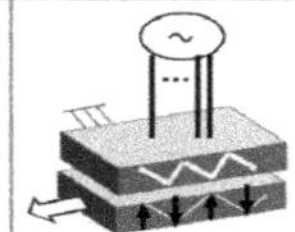		
	SCW + magnets + MI 	*8*		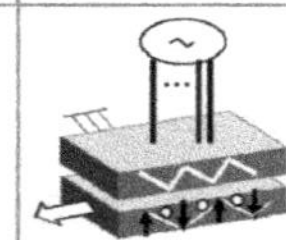	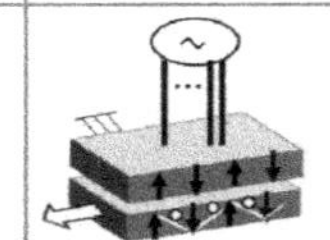	

Fig. 4.9 Eight configuration options for the moving active part

	1	2	3	4	5	6	7	8
SCW		x			x	x		x
MI			x		x		x	x
magnets				x		x	x	x

Fig. 4.10 Four configurations for the stationary active part

	A	B	C	D
MI		x		x
magnets			x	x

Table 4.3 Principles of force generation and specific cases

No	Principles of force generation[a]	Specific cases
1	A ferromagnetic body aligns itself in an external magnetic field	An electromagnet attracts a ferromagnetic body
2	An electromagnet aligns itself in an external magnetic field	An electromagnet attracts/repels another electromagnet, a permanent magnet attracts/repels an electromagnet
3	A permanent magnet aligns itself in an external magnetic field	An electromagnet attracts/repels a permanent magnet
4	Ampere's force acts on an element carrying an electric current	Force acts on a conductor with current in a field created by electromagnets or permanent magnets

[a] It is important to note that the forces acting on the stator and the rotor of the same machine can be of different type

sinusoidal current, for the case of one stationary and one moving active part, without energy transfer to the moving part ("contactless" machines). The table is structured as follows: first, configurations for the moving active part are determined. The active part consists of field sources and field configurators, if any. A short-circuited winding (squirrel cage or damper winding) is denoted as "SCW." Magnetic irregularity (field configurator) is denoted as "MI." Figure 4.9 shows all combinations of field sources and configurators for the moving active part.

Secondly, configurations for the stationary active part (AP) are determined. It is assumed that placing a short-circuited winding (SCW) on the stationary AP is pointless,[3] so we have only four variants in Fig. 4.10. The main winding, which is an integral part of any configuration, is always present in all configurations of the stationary active part in Table 4.3, so the main winding is not shown for simplicity of the pictogram depiction.

Some configurations from Table 4.2 are recognizable and well-known:

- Induction machine with solid rotor—in cell "A-1".
- Squirrel-cage induction machine—"A-2".
- Synchronous reluctance machine—"A-3".
- Switched reluctance machine with sinusoidal supply—"B-3".
- Flux-switching PM machine—"D-3".

[3] Though this variant can be investigated in some exotic cases.

- Permanent magnet synchronous machine (PMSM)—"A-4".
- PMSM with a damper winding—"A-6".

In Table 4.2 the supply currents are sinusoidal. Similar tables can be constructed using a supply current that is rectangular, trapezoidal, or of a more complex waveform instead of the sinusoidal only. Accordingly, in all the pictograms, the symbol in the ovals will change. Another possible group of configurations includes separating (as in Fig. 4.7f) windings into those powered by DC and those powered by AC. Such separation helps specify the topologies into narrowed subclasses.

In general, the type of supply (single-phase or multiphase) and the type of current significantly determine the machine's winding design, thereby constraining the winding and the core design space once the configuration is chosen. However, it is important to note that even with the same field sources and configurators, the winding options can be still highly diverse.

Most configurations visualized in this manner are likely impractical or nonviable. This applies both to Table 4.2 and to any tables created in a similar way. Nevertheless, creating such tables can be useful for inventors and designers. In particular, pictograms or their components can be generated or drawn by AI or inventors and then combined with the design options defined by the criteria described below. Then, such combinations can be analyzed in detail.

The new classification is created with AI training and application in mind.

Criterion 3: "By the mechanism of force generation between the active parts". It is proposed to categorize the ways of generating electromagnetic forces into four principles, described in Table 4.3. Whether all possible cases can be reduced to these four principles and their combinations is a question worthy of scientific debate.

Let's go through the principles (presented by their specific cases) and then consider their combinations.

Principle 1 is "An electromagnet attracts a ferromagnetic body". It is used, for example, in inductor machines, motors with rolling rotors, switched reluctance motors, and synchronous reluctance machines.

Principle 2 is "An electromagnet attracts/repels another electromagnet". It applies, for example, to synchronous machines with electromagnetic excitation and partially (as will be shown later) induction machines. In Fig. 4.11, the stator of a synchronous machine can be viewed as an "inverted" electromagnet (stator's field lines are closing inside and not outside the source like in usual electromagnets). When the field rotates, it "pulls" the rotor along, and the rotor can also be viewed as "electromagnet."

Principle 3 is "An electromagnet attracts/repels a permanent magnet". Permanent magnet synchronous machines (PMSM) operate according to this principle. In Fig. 4.12, a PMSM is shown where, as the field moves from position "A" to position "B," the mutual positioning of the excited teeth with windings (acting as electromagnets) and the three permanent magnets (forming a single pole) changes. In

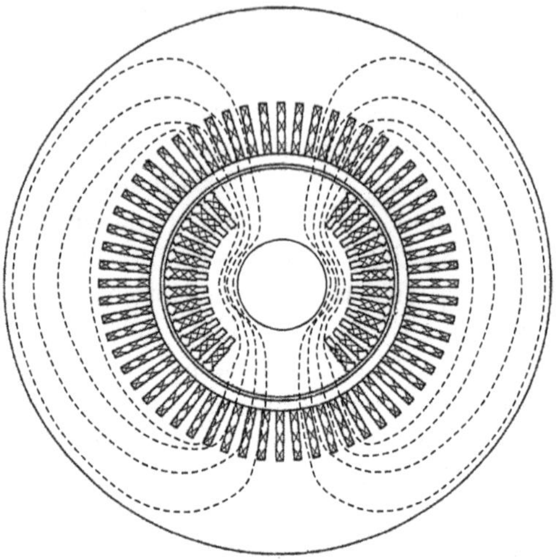

Fig. 4.11 Synchronous machine with electromagnetic excitation

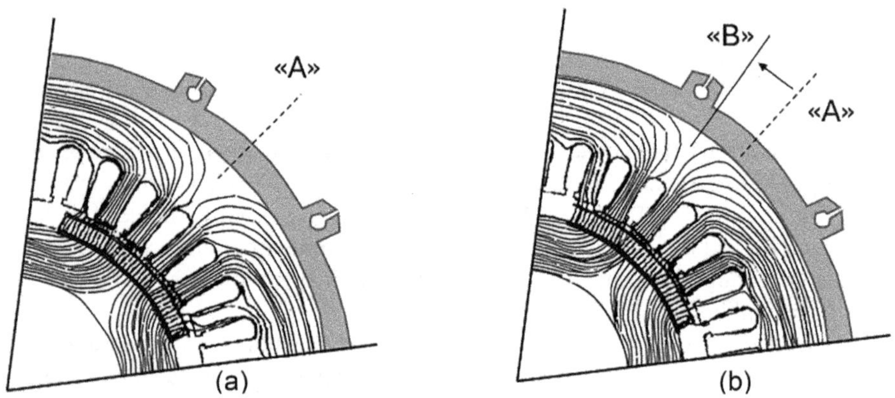

Fig. 4.12 Field rotation in a PMSM: (**a**) initial field position "A", (**b**) position "B" after the field displacement

position "A", the force of attraction is primarily radial. In position "B", a tangential component appears, thereby implementing the rotation. Interestingly, in PMSMs with fractional slot windings, both attraction and repulsion of permanent magnets are present. The task of designing the winding structure in such machines is to ensure that the currents in the slots generate a field direction in the teeth corresponding to the direction of field of the passing magnets.

Principle 4 is "Ampere's force acts on a conductor with current". Machines such as PMSMs with slotless windings (Fig. 4.13) and induction machines with solid (or massive) rotors operate according to this principle.

Squirrel case induction machine (SCIM) is a special case. Its operating principle is often described as follows: a three-phase system of stator currents generates a rotating magnetic field, which, intersecting the closed-loop conductors of the rotor winding, induces currents in them. These currents, in turn, interact with the stator field to produce torque. This description assumes that the operating principle of SCIM is based on Ampere's forces acting on the rotor winding conductors. However, there is a discrepancy in this description of the operation principle: most of the magnetic field travels through the ferromagnetic teeth of the rotor (as shown in Fig. 4.14a), not through the slots containing the aluminum conductors of the cage. Therefore, the field does not directly intersect the conductors in the slots, and the currents in the conductors cannot interact directly with the stator field. More field would go into the slot if the teeth were saturated. An interesting study was presented in [7], where an idealized model of a machine with rectangular teeth was used to analyze the areas where tangential forces producing torque are applied. The following results were obtained:

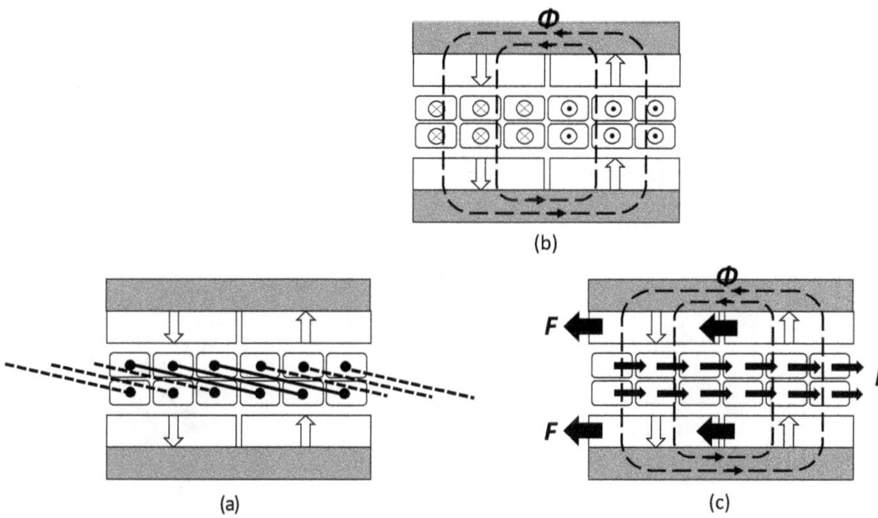

Fig. 4.13 PMSM with slotless winding: (**a**) winding structure, (**b**) current distribution, and (**c**) field lines and forces

- When neglecting magnetic saturation of the iron teeth, the distribution of tangential forces acting on different slot parts and elements was as follows: 3% of force was acting on the current-carrying conductor and 97% of force was acting on the tooth boundaries. No force was acting on the tooth body.
- When taking magnetic saturation of the iron teeth into account, the distribution changes to the following: 34.4% of the force is acting on the current-carrying conductor, 27.6% on the tooth boundaries, and 37.9% on the tooth body.

Based on this research, another explanation of SCIM operating principle avoiding the discrepancy highlighted above can be proposed:

- Logical step 1: The multiphase system of stator currents creates a rotating magnetic field, which intersects the short-circuited rotor windings and induces currents in them.
- Logical step 2: The induced currents generate their own magnetic field (Fig. 4.14b), which interacts with the stator field. This interaction causes the stator field to "twist" due to the rotor currents, and the resulting field may appear as shown in Fig. 4.14c. This field generates torque.
  So, we have a "two-stage" process here.

So it can be assumed that approximately one-third of the forces act on the slot walls (F_{ts}), one-third on the current-carrying conductor (F_c), and one-third on the tooth body (F_{tb}) as illustrated in Fig. 4.15. It is worth noting that in "bottle-shaped" slots, the field distribution appears slightly different from the simplified model in Fig. 4.14.

Next, it can be suggested that adjacent teeth, surrounded by current-carrying conductors and connected by magnetic field lines, can be considered as two electromagnets attracting each other (Fig. 4.16).

So, the operation principle can be described as both "two-stage" and "mixed" since it contains two principles: "An electromagnet attracts/repels another electromagnet" and "Ampere's force acting on a current-carrying element." Another

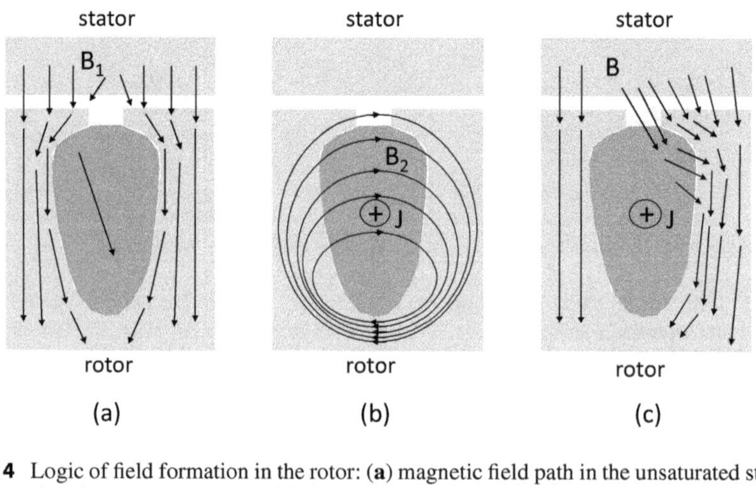

Fig. 4.14 Logic of field formation in the rotor: (**a**) magnetic field path in the unsaturated state. (**b**) Field distortion due to rotor currents. (**c**) Resulting combined field and torque production

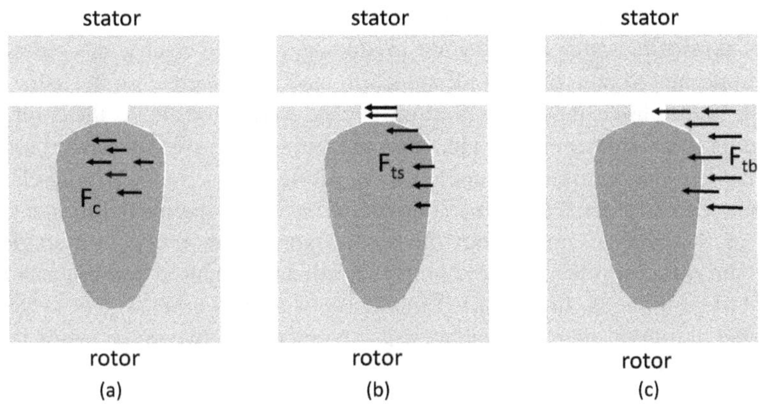

Fig. 4.15 Forces acting on: (**a**) the current-carrying conductor, (**b**) the slot walls, and (**c**) the tooth body

Fig. 4.16 The two "electromagnets"

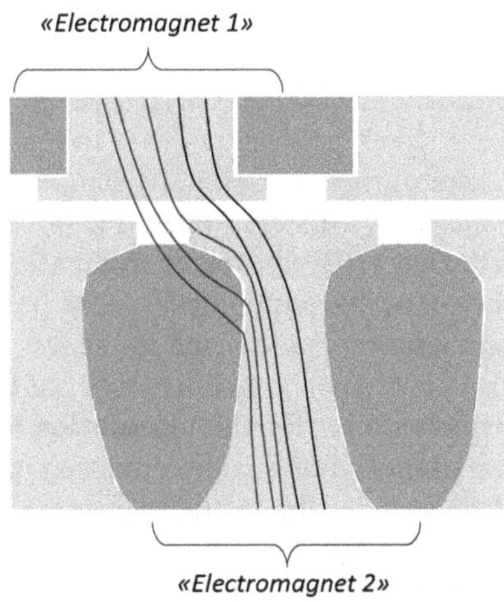

distinguishing feature of the operating principle of SCIM is that the machine transitions between the two ways of torque generation: during startup, high currents lead to saturation of the cores, and forces are mainly applied to the conductors, while in the steady-state operation, when saturation is not so pronounced the torque mostly comes from the forces acting on the teeth. Within this logic presented above, SCIM operating principle can be characterized as "two-stage," "mixed," and "floating"!

A proper description of the operating principle of the SCIM can be the topic for a separate discussion. The explanation proposed above is thought to be most suitable for training AI (hence the importance of the division into the two logical steps). The task for AI in designing a SCIM can be presented as selecting the correct combinations of the stator and rotor slots and optimizing the current distribution on the slots at each moment in time. Training AI to handle such cases will demonstrate that multistage operational principles are possible when producing torque in electric machines.

It is important to note that not all induction machines operate on the same principle: while in the classical SCIM with squirrel cage considered above, the combination of two principles is observed, in induction machines with solid/massive rotors, only one principle is taking place: "Ampere's force acting on a current-carrying element".

In some DC machines, the same combination of two principles of torque production as in SCIM takes place, but the torque generation process is single-stage because the rotor current is present being supplied through the commutator and do not need to be induced. In hysteresis motors, a two-stage process is observed: step 1—magnetization of the rotor; and step 2—interaction between the stator field and the magnetized rotor.

> The operation principles of various electric machines can be described as "single-stage" or "two-stage".

All the four principles of force/torque generation presented in Table 4.3 can be used for teaching the AI. As it was shown, in the real world these principles can appear in combinations, and transitions between principles or their combinations may occur during operation. AI should also be taught that machines can have "single-stage" or "two-stage" operating principles.

Criterion 4 "By spatial configuration of active parts"and **Criterion 5 "By the number of active part pairs"** can be presented together as in Table 4.4. The spatial configuration types are listed in the left column of the table. The number of active part pairs corresponds to the number of air gaps.

Criterion 6 is "By the orientation of the main magnetic flux relative to the air gap plane". The flux direction can be determined relative to the plane of the air gap that separates the interacting active parts. If the gap is uneven (e.g., due to the "toothed" design of the active parts), the plane considered is the one passing between the interacting parts at the level of the smallest gap. In rotating machines, the gap plane is curved—it may take the shape of a cylinder, disk, cone, or sphere, or even have a more complex shape. For linear and planar machines, the plane is flat. An additional spatial reference for defining the flux direction is the direction of movement (i.e., the relative motion of the active parts). To illustrate the location of the air gap plane for several typical machine configurations, we provide examples below. In all cases, the directions of motion (D) are indicated. The most common orientation of the flux relative to the air gap plane (P) is longitudinal (as in the machine shown in Fig. 4.17). In this configuration, the flux paths lie in the same plane as the direction of motion (D), which gives this configuration its name: "longitudinal".

It should be noted that the location of the flux path is generally independent of the spatial arrangement of the active parts. For instance, the active parts of the machine shown in Fig. 4.18 could belong to a linear machine or disk-shaped machine.

Table 4.4 Illustration of topologies on criteria 4 and 5

Types	With one airgap (per plane)	With two or more airgaps
Cylindrical (rotary and/or linear)	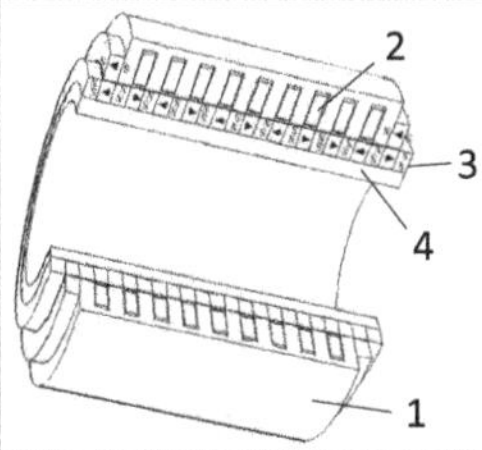	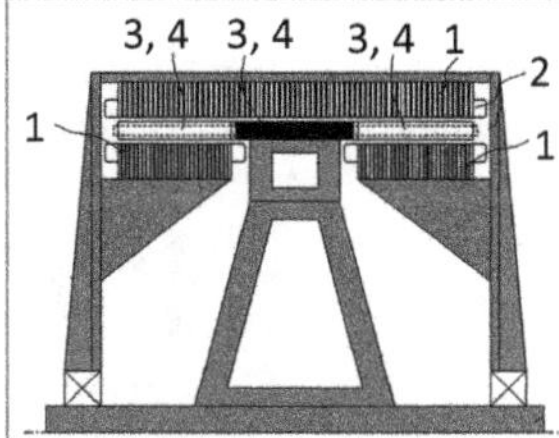
Disk-shaped	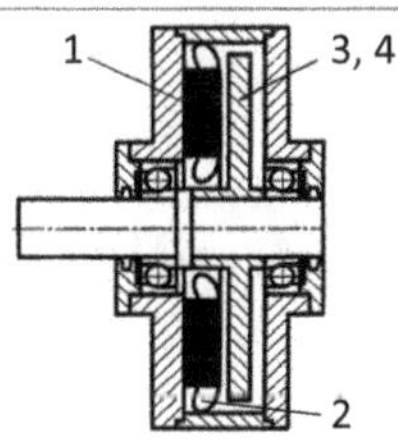	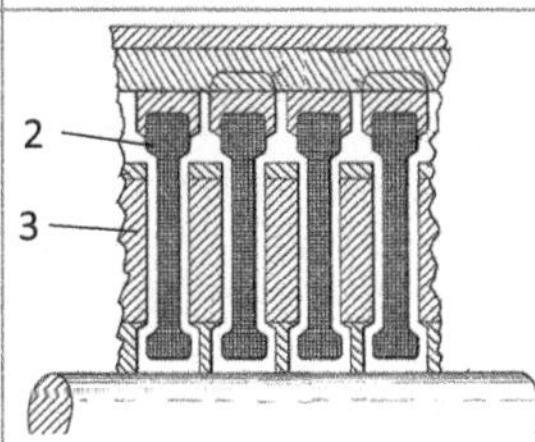
Linear or planar	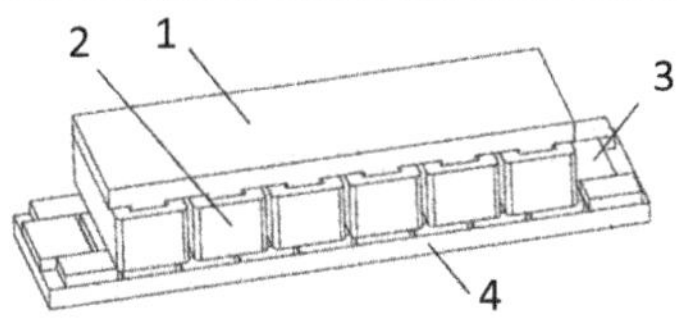	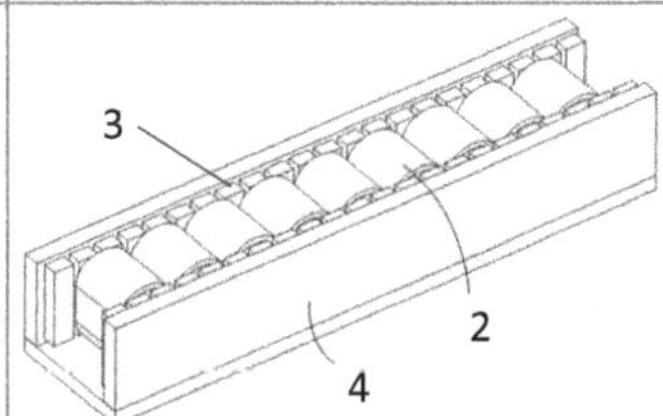
Conical	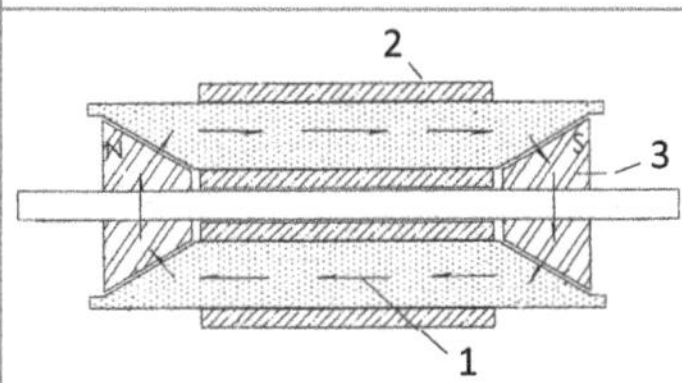	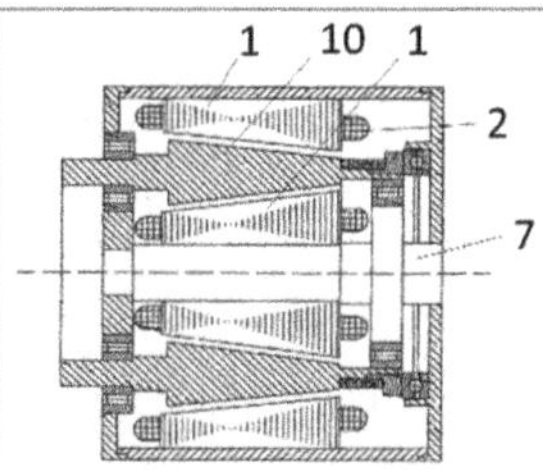
Spherical and elliptic	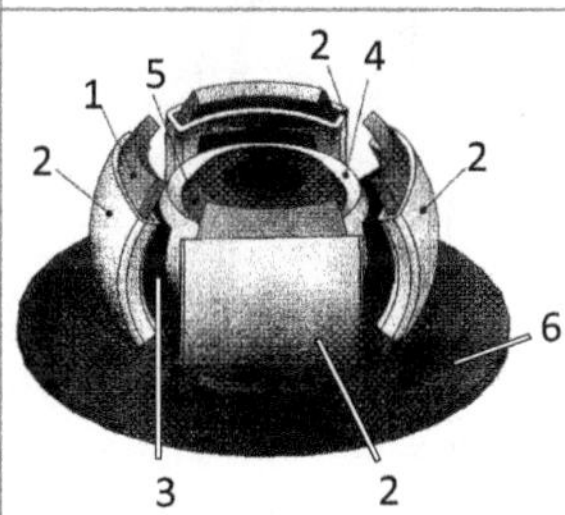	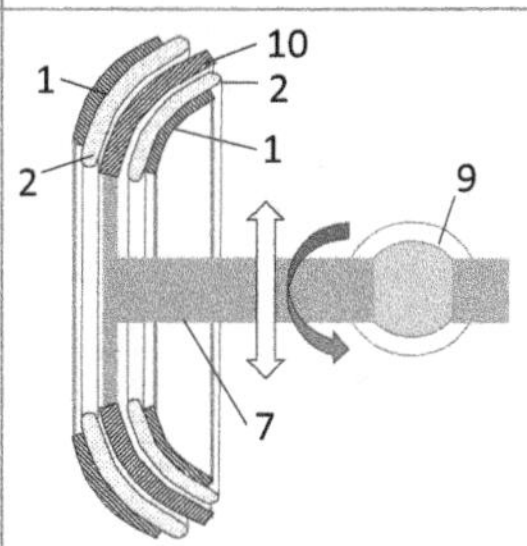

(continued)

Table 4.4 (continued)

Types	With one airgap (per plane)	With two or more airgaps
Combinations (e.g., mix of cylindrical and disc structures)	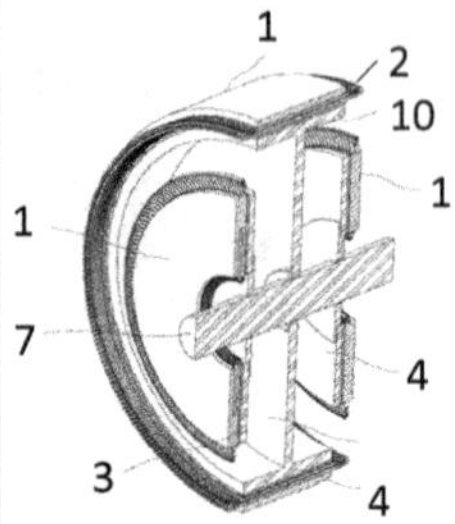	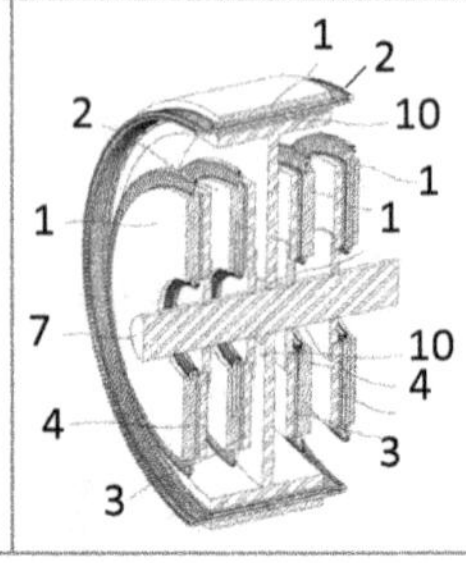

1—stationary core, 2—winding, 3—permanent magnets, 4—moving core, 5—spherical cup-shaped bearing, 6—base, 7—shaft, 8—magnetic flux path, 9—joint, 10—rotor
Pictures in the table are from patents CN101425732A, NO333170B1, CN107453578A, CN101604891A, CN106100284A, US4371801A, CN218183222U, US2006145555A1

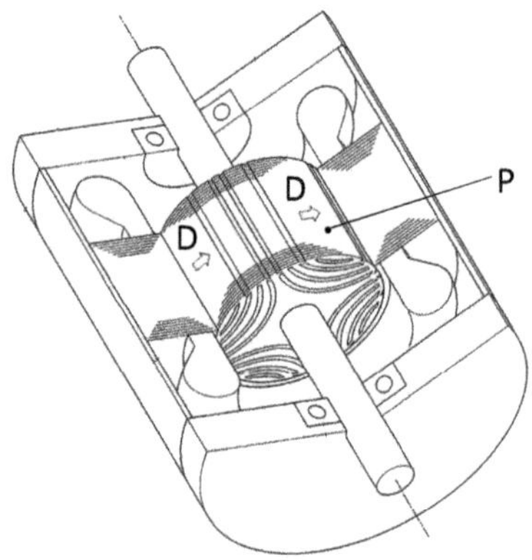

Fig. 4.17 Synchronous reluctance machine. (Picture from US10637310B2) (P—element of the gap plane, D—direction of motion)

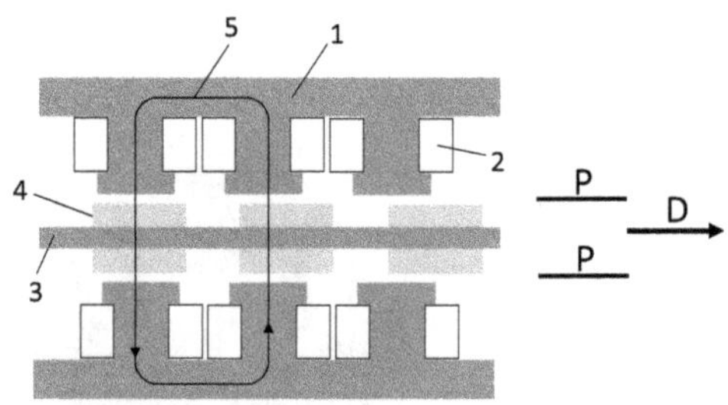

Fig. 4.18 Permanent magnet synchronous machine with two stators (1—stator core, 2—stator windings, 3—rotor core, 4—magnets, 5—magnetic flux paths)

To describe orientation of the main magnetic flux, it is necessary to define borders of the area of interaction of the active parts. For the most common topology with the airgap plane parallel to the axes of rotation, usually called radial-flux topology, the borders can be defined as in Fig. 4.19a. For the topology with the airgap plane perpendicular to the axes of rotation, usually called axial-flux topology, the borders can be defined as in Fig. 4.19b. A general interaction surface element can be simplified as in Fig. 4.19c.

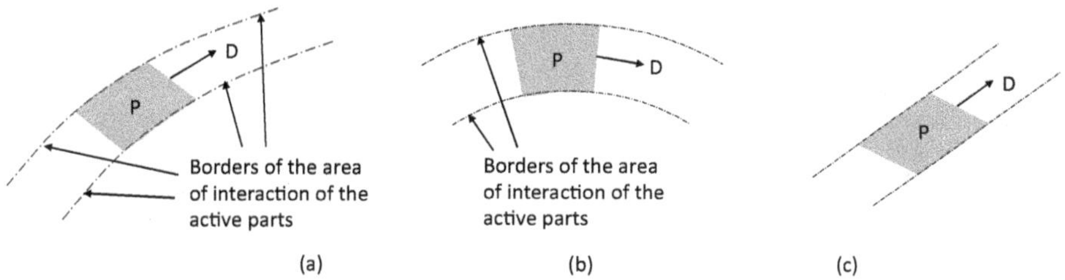

Fig. 4.19 Defining borders of the area of interaction of active parts: (**a**) "radial-flux" topology, (**b**) "axial-flux" topology and (**c**) general element

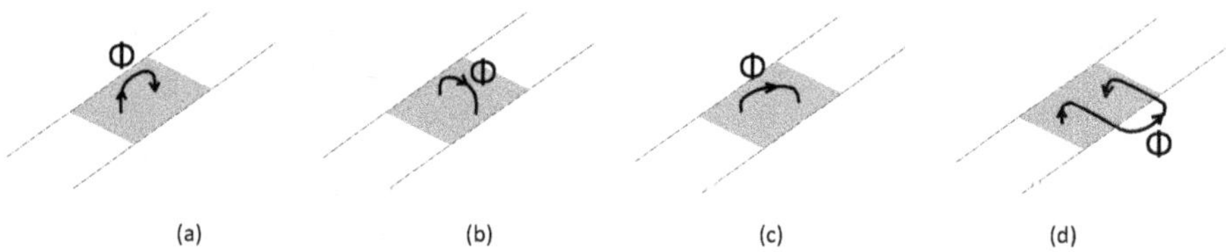

Fig. 4.20 Defining flux directions: "longitudinal" (**a**), "transverse" (**b**), "mixed" (**c**)

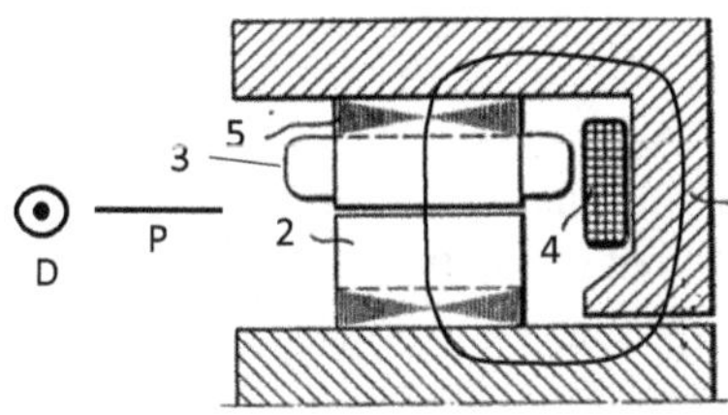

Fig. 4.21 Inductor-type machine from patent application US3851199A: 1—stator core, 2—rotor core, 3—armature windings, 4—excitation winding, 5—magnetic flux paths

Now we can define flux directions: longitudinal (Fig. 4.20a), transverse (Fig. 4.20b), and mixed longitudinal and transverse (Fig. 4.20c). If the flux path extends beyond the boundaries defined by the airgap planes, it is referred to as "lateral overshoot"—illustrated in Fig. 4.20d.

Let's illustrate these variants with real electric machines. For example, in the machine shown in Fig. 4.21, the flux paths lie predominantly in a plane perpendicular to both the airgap plane and the direction of motion. Additionally, the flux path extends laterally beyond the region of interaction between the active parts. This configuration can therefore be referred to as "longitudinal with lateral overshoot."

Figure 4.22 shows a machine with a mixed longitudinal and transverse flux orientation without lateral overshoot. In Fig. 4.23, a machine with a transverse flux orientation without lateral overshoot is depicted.

So, we have presented five examples corresponding to the main classification options under Criterion-6. The possible configurations are as follows: longitudinal,

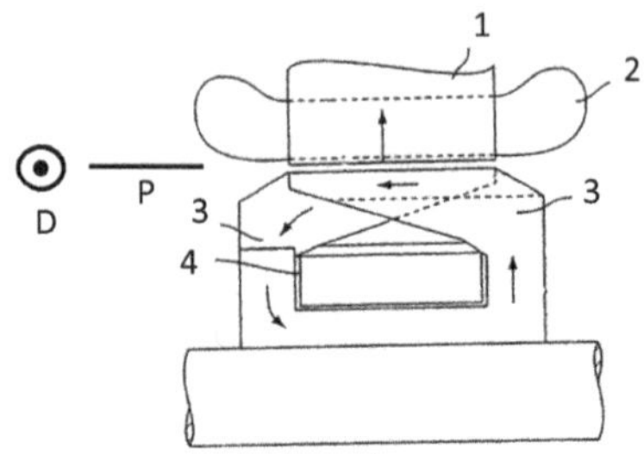

Fig. 4.22 Claw-pole machine from patent US6211596B1 (1—stator core, 2—armature winding, 3—rotor cores with "claws," 4—excitation winding)

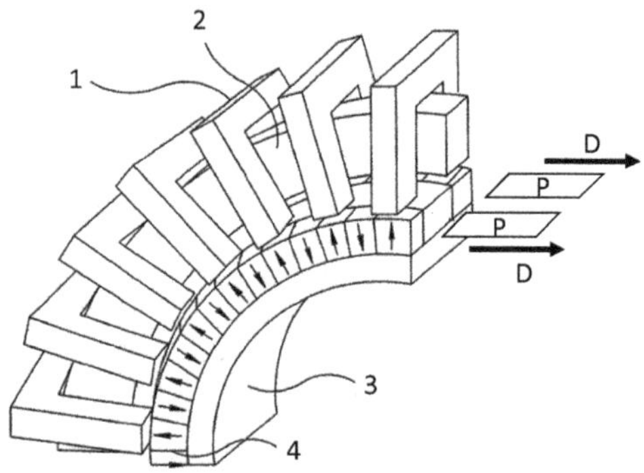

Fig. 4.23 Machine with transverse flux from patent application EP3058639A1 (1—stator core, 2—excitation winding, 3—rotor yoke, 4—magnets)

transverse, and mixed (longitudinal and transverse). Each of these configurations can either lack lateral overshoot or include it.

Integrated Solutions

Implementing any spatial configuration (e.g., from Table 4.4) usually requires specialized solutions for carrying structures, cooling systems, and other components. One promising direction of the electric machine evolution is integrated solutions. For example, a machine may lack its own shaft and bearings. Figure 4.24 illustrates the concept of integration of the active parts of a generator into the structure of a wind turbine. In this design, the generator utilizes the shaft and bearings of the wind turbine. Similarly, in some cases, neither the machine nor the entire system has a shaft at all. For instance, a rim-driven thruster with an integrated permanent magnet synchronous motor, shown in Fig. 4.25, does not include a shaft. Instead, special bearings are placed on the periphery of the rotor.

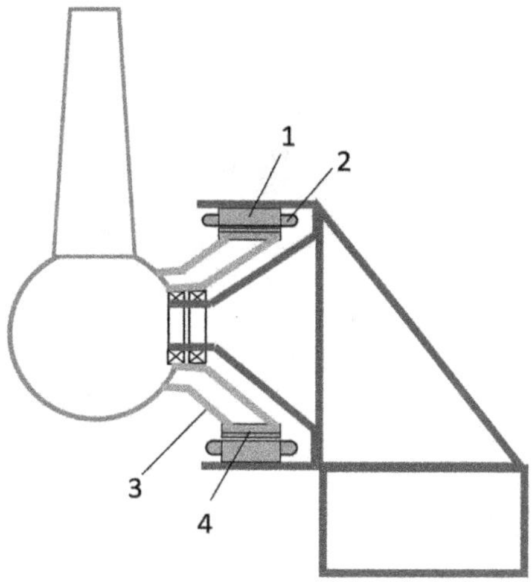

Fig. 4.24 Wind turbine with integrated generator (1—stator supporting structure, 2—stator active parts, 3—rotor supporting structure, 4—permanent magnets on the rotor)

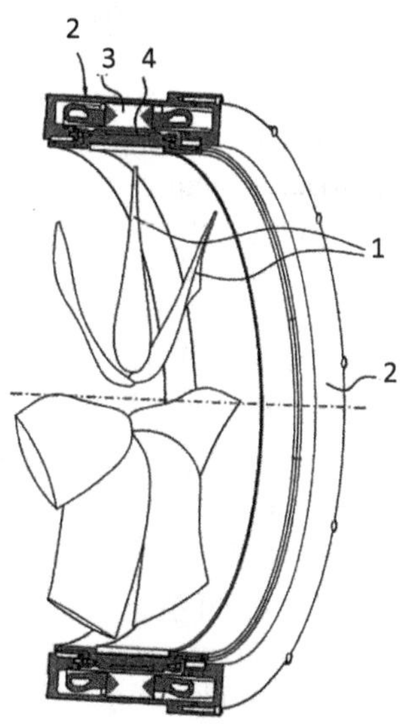

Fig. 4.25 Rim-driven thruster from patent application CN1897417A (1—blades, 2—rim, 3—stator, 4—rotor)

An even more radical example of integration presented in Fig. 4.26 involves configurations where the rotor teeth of a switched reluctance machines (SRM) serve as the working elements of a pump:

- Figure 4.26a—an axial pump integrated with a SRM from TWI305973B [8].
- Figure 4.26b—a gear pump integrated with a SRM from JPH11280664A [9].

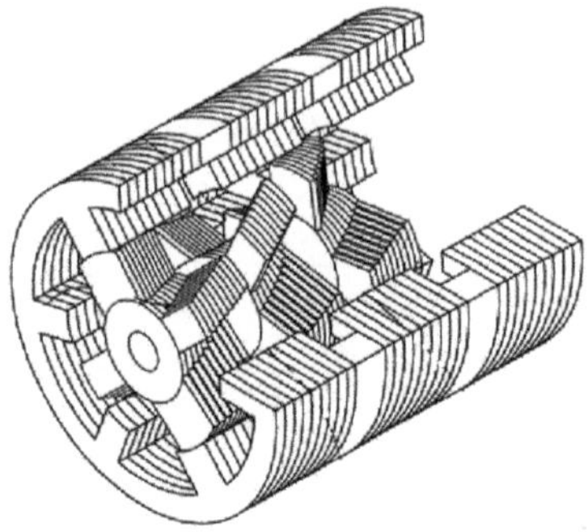

(a)

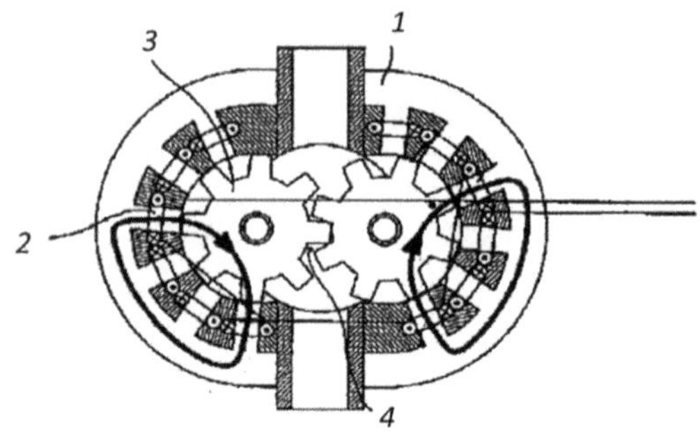

(b)

Fig. 4.26 Examples of radical integration: (**a**) an axial pump, (**b**) a gear pump

Knowledge Interconnectivity and Completeness of Understanding

Creating a new electrical machine requires comprehensive knowledge about all machine elements and all processes inside the machine. In [10], a structured approach to acquiring, organizing, and transferring knowledge was proposed, using a set of logical blocks represented by a *pentagram*, a well-known symbol and geometric figure. The Pentagram (Fig. 4.27) consists of nodes and triangles. A node can represent an object, process, or physical phenomenon. Nodes are labeled with letters, for example: "F"—Forces, "S"—Field sources, and "L"—Losses. There are two groups of triangles:

1. Central triangles, located at the center of the Pentagram, numbered 1, 2, 3, 4, and 5.
2. Pentagram rays, which are unnumbered.

The central triangles have nodes as their vertices, linking objects, processes, and phenomena into groups of three called "triads." The knowledge blocks corresponding to the nodes of the triads are described in Table 4.5.

The next, more complex level involves triads. In each triad, there are objects or processes united by a specific logic of interdependence. One could say that triads demonstrate objects or processes between which the connections are "strong," whereas the connections between any two nodes of the Pentagram that do not belong to the same triad are "weak" or even nonexistent.

The value of triads lies in enabling a more nuanced perspective on problems, moving from the logic of "element A positively influences element B" to "element A positively influences element B, but negatively influences element C." For example, instead of stating, "choosing a thinner stator yoke will reduce the machine's weight," one could say, "choosing a thinner stator yoke will reduce the machine's weight, but iron losses will increase." By considering multiple adjacent triads, one could further expand this logical chain to add, "… this will also decrease efficiency and increase vibrations." Thus, it is entirely feasible to simultaneously consider two triads to

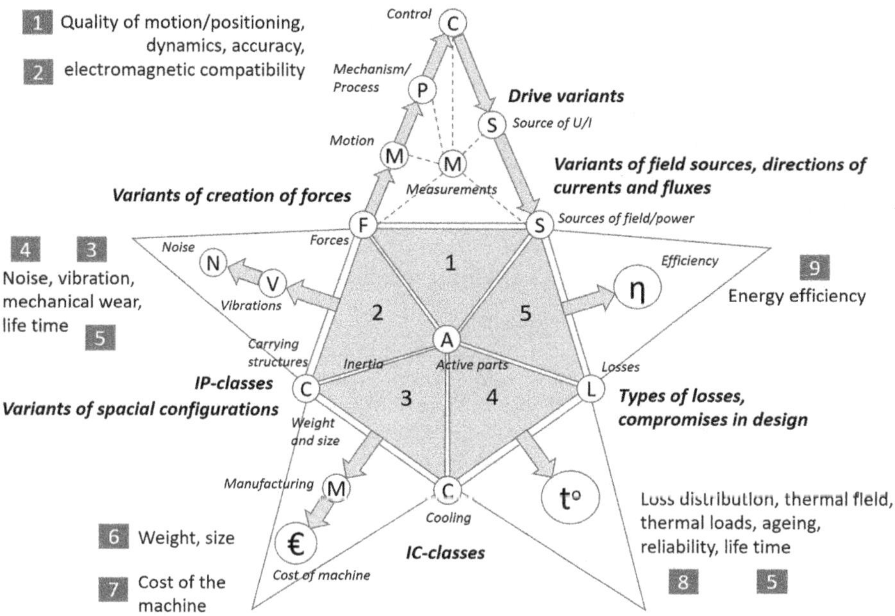

Fig. 4.27 Pentagram adapted for studying electric machines

Table 4.5 Main nodes of the Pentagram and knowledge blocks

Nodes	Knowledge blocks (example)
(A) Active parts	Ferromagnetic cores, conductors, insulation, permanent magnets. Winding topologies/schemes. Core shapes, e.g., like in Table 4.4
(S) Electromagnetic field sources	Variants of field sources: current-carrying conductors, permanent magnets. Magnet characteristics. Main configurations of field sources, current, and flux directions. How terminal voltages create currents in the windings
(F) Electromagnetic forces	Classification of force generation mechanisms. Classification by direction: tangential, radial, and axial forces. Points of force application. Calculation methods based on energy and co-energy, stress tensor. Field and force calculation methods, e.g., FEM
(L) Losses	Losses in ferromagnetic cores: hysteresis, eddy currents. Losses in conductors: "AC" and "DC" resistance. Losses in permanent magnets. Losses due to high-frequency effects coming from frequency converters. Mechanical losses
(C) Cooling	Water, air, hydrogen, and mixed cooling methods. Forced and natural cooling. Heat transfer mechanisms: radiation, convection, and conduction. Standards (IC classes). Cooling at high altitudes
(C) Carrying structures	Shaft, bearings, housing, and shields. Variants of spatial configurations of carrying structures—corresponding to active parts configurations. Dust and moisture protection (IP classes). Installation variants (B3, B5, etc.)

construct more complex logical chains. Understanding such interrelationships is characteristic of experienced designers. It should also be noted that understanding the connections between three elements is not cognitively challenging, which is why triads are likely the optimal logical unit for knowledge transfer. The possible material content corresponding to the triads of the Pentagram for an electric machine is presented in Table 4.6. This material is proposed for either self-study, discussions with colleagues, or for teaching others, and is grouped into "topics."

Table 4.6 Knowledge blocks corresponding to triads

Triads	Topics (example)
(1) Electromagnetic field sources—active parts—electromagnetic forces	*Topic: "Energy Conversion and Forces"* Variants of field sources. Variants of directions of currents and magnetic fluxes. Variants of active part structures Illustration of how field sources in active parts generate electromagnetic forces acting on the active parts Torque creation: field → forces → motion. Principles of operation of some machine (SCIM, DC motor, SM, etc.) Variants of force generation and application. Vectors and vector analysis. Distribution of forces applied to active parts Energy conversion. Active and reactive power, power factor, cosφ
(2) Active parts—electromagnetic forces—carrying structures	*Topic: "Forces, Vibrations, and Noise"* Variants of carrying structures. Protection classes Variants of active parts. Force application options How tangential and radial forces influence the selection of structural designs and dimensions of carrying structures How forces create vibrations and acoustic noise. How the structure and dimensions of carrying elements affect vibrations and acoustic noise Wear, service life, reliability, and mean time to failure
(3) Active parts—cooling systems—carrying structures	*Topic: "The Machine as a Physical Object: Mechanical Design"* Designing for minimum cost and automated production Integration of cooling systems and carrying structures Installation, adjustment, maintenance. Faults and repairs Variants of spatial configurations of active parts and supporting structures. Types of cooling systems. Protection classes Demonstration of how complexity and manufacturing cost depend on the spatial configuration of active parts, supporting structures, and cooling systems. Maintainability and reparability
(4) Losses—active parts—cooling systems	*Topic: "Heat Generation and Dissipation"* Losses and heat generation. Heat sources and their distribution in active parts. Spatial configuration options for active parts. Heat transfer mechanisms: radiation, convection, conduction. Thermal field. Thermal loads, insulation aging, reliability, and component service life. Repetitive short-term operating modes. Loss types. Cooling system types. Design trade-offs
(5) Electromagnetic field sources—active parts—losses	*Topic: "Energy Efficiency"* Variants of field sources. Variants of current directions and magnetic fluxes. Variants of active part structures Loss types. Field sources and losses in active parts. Power conversion efficiency. Efficiency at partial loads. Trade-offs: size/weight vs efficiency (as illustrated in Fig. 4.28)

In particular, when working with Triad (3), the goal is to explore how the interacting active parts (stator and rotor) can be cooled, focusing on more than just one or two primary methods.

Triad (1) is unique because it reflects the integration of the electric machine into the drive system (Fig. 4.29). The ray of this triad contains several relevant nodes required to describe the process of shaping the desired form of motion by generating voltages and currents and transmitting energy to the machine's active parts.

It is important to note that the triads and rays of the Pentagram correspond to the following key performance indicators:

1. Motion quality, dynamics, and accuracy.
2. Electromagnetic compatibility.
3. Vibrations.
4. Acoustic noise.
5. Machine service life.

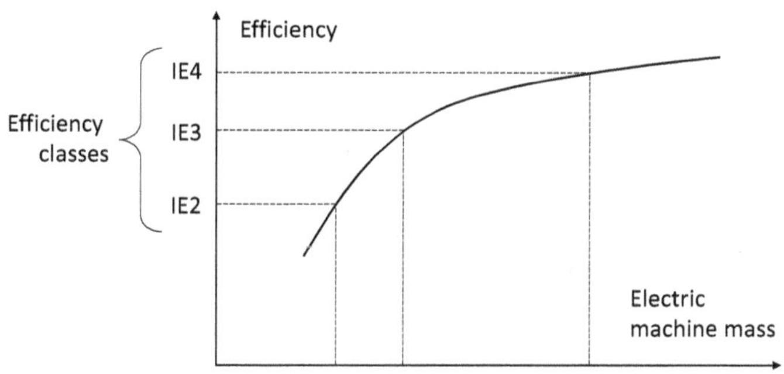

Fig. 4.28 Graph of machine mass dependency on efficiency requirements

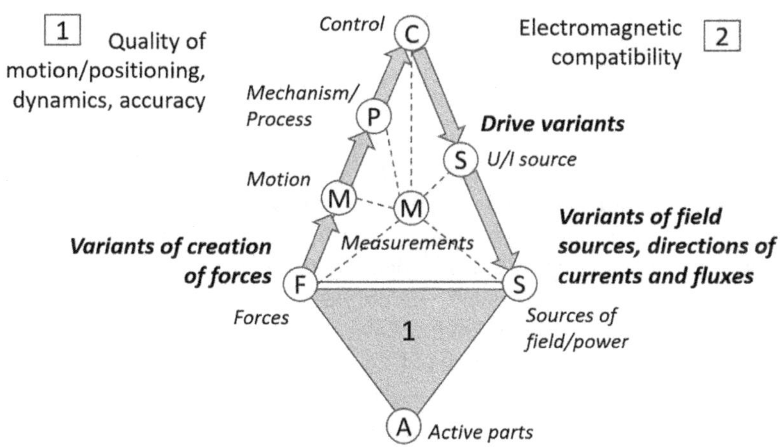

Fig. 4.29 Triad (1) and the inclusion of the electric machine into the drive system

6. Weight and dimensions.
7. Cost.
8. Reliability.
9. Energy efficiency.

The Pentagram, as a set of interconnected logical blocks, describes the entirety of physical processes, effects, their interrelations, and, to some extent, even the philosophy behind the design of electric machines. The Pentagram can also serve as a kind of navigator and "reminder" for the designer. In any new machine design, the rays of the Pentagram must be balanced.

Using the Pentagram in Teaching

The Pentagram can be applied not only in the design of electric machines but also as a tool in the educational process. Structuring the knowledge transfer according to the logical blocks of the Pentagram can make the learning process more organized and comprehensible for students. This approach can have a particularly positive impact when time is limited for conveying a specific amount of knowledge—for instance, within a university course limited to a single semester. The accelerating pace of technological innovation in the modern world demands changes in engineering education. A greater volume of knowledge must now be delivered in a shorter time. Consequently, the traditional curriculum on electric machines must be taught more quickly—within a limited number of lectures and seminars. If the Pentagram is adapted in the educational process, the knowledge blocks outlined in Tables 4.5 and 4.6 would naturally translate into topics for lectures and seminars. This methodology aims to consolidate all necessary information while setting an ambitious goal: to develop in students the broad professional perspective of an experienced designer, which is typically gained through years of practice. The approach leverages, among other things, the effectiveness of visual information, adhering to the principle that "a picture is worth a thousand words." From the outset, the knowledge that students are expected to acquire is clearly defined and visually presented.

It is worth noting that the Pentagram, drawn with a single continuous line (Fig. 4.30), is one of the oldest symbols, interpreted in various ways throughout history. One interpretation—symbolizing magical influence and the dominance of disciplined will over the phenomena of the world—can be applied to the realm of science and education. Rephrased, it reflects the influence of a scientist's will on processes within the Technosphere, particularly in creating technical systems. Another interpretation of the Pentagram—as a symbol of the "perfect human" (Fig. 4.31)—also aligns with the goals of the educational process.

Whether there is a connection between ancient meanings behind the pentagram symbol, like the one in Fig. 4.31 and the current use of the Pentagram proposed in this book, or if it is merely a coincidence, could serve as an engaging discussion topic between professors and students. Such discussions might spark new questions, such as:

Fig. 4.30 A pentagram drawn with a single line

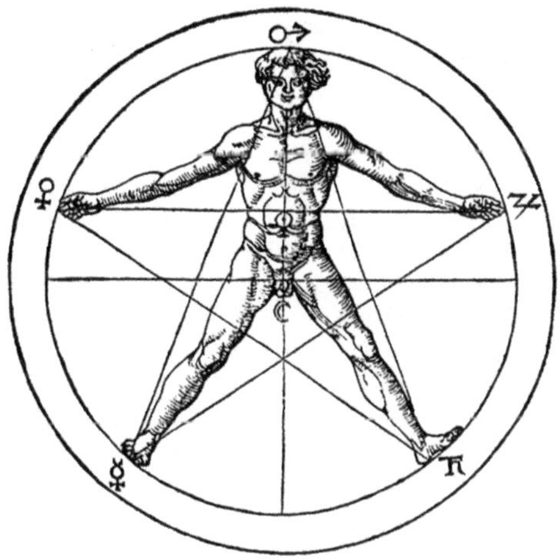

Fig. 4.31 A pentagram with elements and the "perfect human" (Agrippa's Man)

- "Does the possibility of using the Pentagram to structure logical relationships in electric machine point to an underlying interconnectedness of all things on Earth (in the noosphere)?"
- "Can other ancient symbols be adapted for educational or design purposes?"
- "How do the elements ('Earth,' 'Air,' etc.) in the classical pentagram of Fig. 4.31 correspond to the content of the Pentagram's rays in Fig. 4.27?"

Final Comments

This chapter essentially suggests taking a step back into the nineteenth century—into an era even before Tesla and Dolivo-Dobrovolsky—to "reassemble" Electromechanics. Pictograms could serve as new tools of language at the intersection of human and machine (AI) interaction. AI will need to be taught its first "words" and take its first "steps"—much like a child. After that, it will likely surpass its creators. AI must also be shown the space of possibilities, which it will then explore on its own. As for inventors, it is critical for AI to understand what constitutes the "basic building blocks" of electromechanical systems so it can assemble something truly novel from them.

The reader may not agree with some of the proposed classification criteria or even the overall approach. However, this chapter should be viewed as an invitation to engage in a broader discussion about shifting the paradigm in the classification of electric machines.

References

1. S. Barmada et al., Deep learning and reduced models for fast optimization in electromagnetics. IEEE Trans. Magn. **56**(3), 7513604 (2020)
2. H. Sasaki, H. Igarashi, Topology optimization accelerated by deep learning. IEEE Trans. Magn. **55**(6), 1–5 (2019)
3. N. Gabdullin, S. Madanzadeh, A. Vilkin, Towards end-to-end deep learning performance analysis of electric motors. Actuators **10**(2), 28 (2021)
4. I.P. Kopylov, *Electrical Machines* (Energoatomizdat, Moscow, 1986) 360 p
5. A.V. Ivanov-Smolensky, *Electrical Machines* (Mir Publishers, Moscow, 1982)
6. J. Pyrhoenen, T. Jokinen, V. Hrabovcova, *Design of Rotating Electrical Machines* (John Wiley & Sons, New York, 2008) 612 p
7. A.V. Ivanov-Smolensky, *Electromagnetic Forces and Energy Conversion in Electrical Machines* (Higher School, Moscow, 1989) 312 p
8. Y. Sheng-Ming et al., Switched reluctance motor with axial flow pump function, Patent TWI305973B
9. H. Tadayuki, Reluctance motor integrated type pump, Patent application JPH11280664A
10. A. Matveev, Applying the pentagram for structuring the process of studying electrical devices. Elektrischestvo **2**, 10–17 (2020)

Application of TRIZ in the Domain of Electric Machines

5

Theory of Inventive Problem Solving and Electromechanics

The value of innovations based on inventions is obvious. One of the goals of this book is to increase the number and quality of the inventions and thus to boost the innovative activities in the domain of electric machines. One of the popular methodologies for finding new solutions is the Theory of Inventive Problem Solving (TRIZ), created by G.S. Altshuller [1, 2]. Classical TRIZ is based on the following postulates [3]:

- Technology and technical systems develop in a generally consistent manner.
- The laws of technological development are knowable and can be used to find new technical solutions.
- The process of finding new solutions can be described as a sequence of intellectual, thought-based actions.

  TRIZ practitioners have created an information fund that includes, in particular:
- Methods for resolving contradictions and tables for their application.
- Databases of physical, chemical, and other effects, and ways to use them.
- Methods for using the resources of nature and technology.

Modern TRIZ includes a list of 40 basic methods for resolving contradictions. These methods only point to the direction and area where strong solutions may lie; however, they do not provide specific solutions—that work remains for the individual.

In this chapter, we will attempt to demonstrate the possibilities of TRIZ in the field of Electromechanics, highlight the accompanying challenges, and suggest ways to overcome them. The task is not trivial as TRIZ and Electromechanics are very different disciplines, as will be outlined below.

As defined in [4], TRIZ is an empirical, constructive, qualitative, universal methodology for generating effective ideas and resolving problems based on models of *contradictions* and methods for resolving them, extracted from known

A. Matveev, *Shaping the Future of Electric Machines*,
https://doi.org/10.1007/978-3-031-95248-7_5

examples of previously proposed solutions. Additional clarifications can be made: TRIZ is a *constructive* methodology because it contains practically reproducible models and methods that allow inventing new effective solutions and teaching people the processes, models, and methods of creating effective ideas. TRIZ is a *qualitative* methodology because its models and methods are recommendations based on the similarity of qualitative properties of models for transforming existing objects or creating new objects and are not based on strict mathematical axioms or mathematical rules for deriving solutions from some initial data. Thus, TRIZ is not a mathematical, quantitative theory, but a qualitative methodology. The formal concepts and notions of the theory take the form of categories, images, and metaphors. The multistep procedures of TRIZ used to solve problems are called algorithms.

Electromechanics is the branch of electrical engineering, a technical science that studies the application of electrical and magnetic phenomena for practical use. Electromechanics primarily relies on the concepts and tools of physics and mathematics. Like almost any subfield of engineering, Electromechanics uses specialized terminology. For example, the compactness of machines is described (formalized) by parameters such as "specific power," "specific torque," and quality of the energy conversion process—by "efficiency." Operational characteristics include parameters like "acoustic noise level" and "mean time to failure." These and many other terms of Electromechanics are not used in classical TRIZ. The language of Electromechanics requires high precision and clarity, while TRIZ benefits from flexibility, imagery, and metaphorical thinking. Therefore, the complexity and challenge of applying TRIZ in the field of Electromechanics arise from the differences in language and the way of thinking in these two disciplines.

The language of Electromechanics requires high precision and clarity, while TRIZ benefits from flexibility, imagery, and metaphorical thinking.

Types of Contradictions in TRIZ

TRIZ operates with contradictions: administrative, technical, and physical. An *Administrative Contradiction* (AC) arises when something needs to be done, but the method is unknown. An AC is formulated as follows: "*The system needs to be improved, but I don't know how to do it.*" For example: "*It is desirable to maintain high availability of the wind turbines while scaling them in size from 1 MW up to 10 MW, but it's unclear how, since repairing the large 10 MW generator, placed in the turbine's nacelle, in case of failure is not easy,*" or "*It's required to measure the vibrations of a machine's housing, but attaching the vibration sensor and connecting it to the monitoring system via signal cables is too complex*".

A *Technical Contradiction* (TC) is described as follows: improving one part (or one parameter) of a technical system by known means leads to an unacceptable degradation of another part (or another parameter). In other words, improving one parameter of the system results in the deterioration of another parameter. A more precise definition from [4] used in this chapter is as follows: a TC is a two-factor model, where the first factor corresponds to and supports the main useful function of the system ("plus-factor"), while the other factor does not correspond to or hinders this function ("minus-factor").

A *Compositional Contradiction* (CC) is a set of several TCs formulated for the same problem. The definition of a *Physical Contradiction* (PC) will be provided below in one of the examples.

Utilizing System Resources

In TRIZ, a *resource* refers to anything that can be used to solve a problem or advance systems development to a new level. Resources can include time, space, energy, substance, information, etc. Resources possess several properties, listed below, and for each of them the order from most to least favorable for problem-solving is given:

- Value: free → inexpensive → expensive.
- Quality: harmful → neutral → useful.
- Quantity: unlimited → sufficient → insufficient.
- Readiness for use: ready → modifiable → to be created.

It is worth noting that using a harmful resource, or getting some harmful effect eliminated, is considered an ideal solution.

Example of Mapping the System Resources

Let's demonstrate the approach to identifying available resources with an example involving an underwater propulsion unit—thruster, with rim-integrated motor (Fig. 5.1, source: patent application CN1897417A). The active parts of the electric motor in such application are usually protected from water penetration by encapsulation in epoxy or filling the rim with oil.

When the thruster diameter gets larger than 2–3 m, it becomes tempting to make the rim (and the stator inside the rim) segmented (like in Fig. 5.2) for simplicity of manufacturing and repair. The problem then can be formulated as follows: the windings of the stator segments of the thruster need to be connected into a single winding in such a way that the "connections" are reliable and long-lasting under water, while ensuring that the structure is possible to disassemble and that its production is automated with minimum manual labor. Mapping of the system resources for the task is presented in Table 5.1. The reader can try to come up with own solution to the problem.

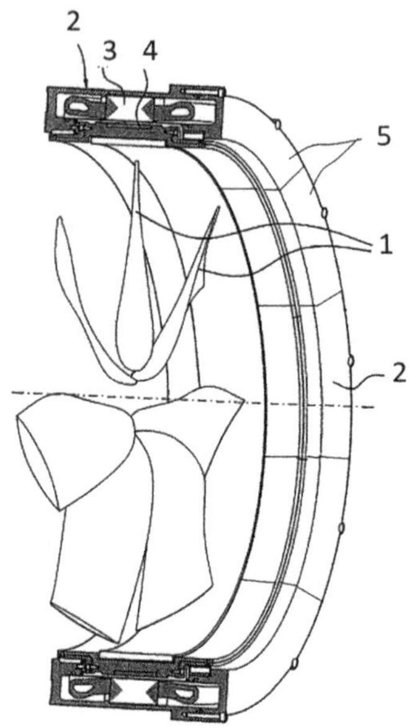

Fig. 5.1 Rim-driven thruster (1—blades, 2—rim, 3—stator, 4—rotor, 5—segments)

Example of Using One of the Resources: Spatial Resource

An example of working with *spatial* resources is discussed in [5], where the creation of an improved rotor for a synchronous motor with a starting cage winding is described. The rotor pole is shown in Fig. 5.3. The key aspect of the proposed solution is the transition from using varnish to using epoxy compound for impregnation.

Compared to varnish, epoxy compound has higher mechanical and electrical strength, as well as better adhesion. The wedge (7) and the rim's function of holding the coil becomes unnecessary under the new conditions because the coil (3) is firmly glued to the core (2) with the compound. Due to the impregnation, the coil becomes robust, and there is no need to support it. Therefore, the rim (1) on both sides can be trimmed, significantly reducing its weight and material costs. The wedge can also be eliminated. By removing the wedge and part of the rim, a new resource appears: "empty space" (Fig. 5.4).

The resulting through-channel exposes the inner surface of the working coil, increasing the area available for airflow. Most importantly, the short-circuiting elements of the starting cage winding can now be placed in this channel (Fig. 5.5). In the new pole design, these short-circuiting elements are positioned inside the working winding, allowing "deep" slots, which were previously impossible for machines of this type, to be made without obstacles. As a result, the starting torque should increase by at least 25%. This solution was disclosed in patent application SU1451802A.

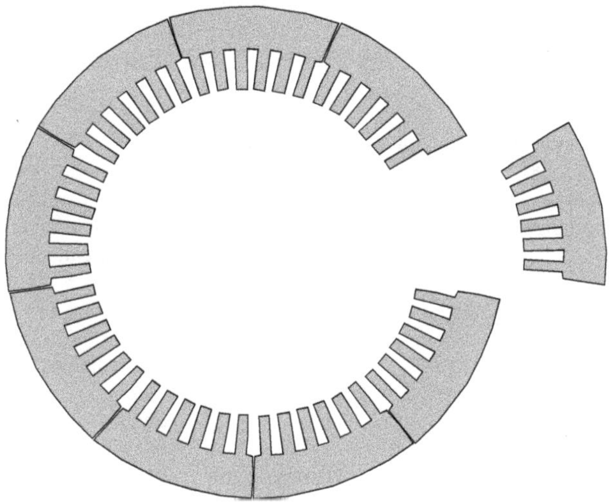

Fig. 5.2 Segmented stator core

Table 5.1 Resources of the underwater propulsion unit (thruster)

General description of all possible types of resources				Comments on resources available for use in the thruster
System-technical resources	System-related	Pertaining to general system properties	Efficiency, performance, reliability, safety, survivability, durability, etc.	None of these resources can be compromised to achieve the goal
	Informational	Pertaining to signal transmission	Accuracy, noise immunity, precision, completeness, coding methods and efficiency, data compression methods and parameters, etc.	Possibility to install and use temperature, magnetic field and humidity sensors
	Functional	Pertaining to function creation	Purpose (main useful function), auxiliary functions, negative functions, description of operating principle (functional model)	Positive function—thrust, motion in water, negative function—heating of the motor's active parts
	Structural	Pertaining to the composition of the object	List of components and connections between components, types of structures (linear, branching, parallel, closed, etc.)	Coils on teeth, connections of the winding sections, connectors

(continued)

Table 7.1 (continued)

General description of all possible types of resources				Comments on resources available for use in the thruster
Physical-technical resources	Temporal	Pertaining to time assessments	Event frequency, duration of time intervals, event ordering in time, delay/lead magnitude	Rim and stator assembly and disassembly sequences, service intervals
	Spatial	Pertaining to geometric properties	Object shape, dimensions—length, width, height, diameter, etc. Shape features—presence of voids, protrusions, etc.	Possible variation of shape and dimensions of the rim and the stator at design stage
	Material	Pertaining to material properties	Chemical composition, physical properties, special engineering properties	Oil or other encapsulation material and its composition
	Energetic	Pertaining to energy properties and manifestations	Types of applied and accounted energy, including gravitational, thermal, electromagnetic, mechanical forces, etc.	Magnetic field, electric field, motion of water and propeller, heat from active parts, electric current

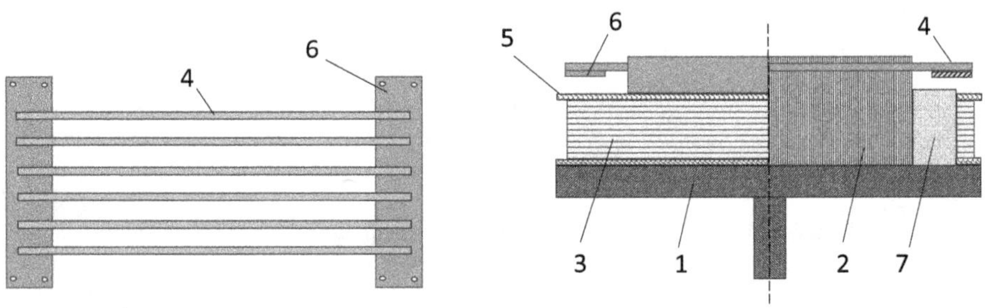

Fig. 5.3 Rotor pole (1—rotor rim, 2—pole core, 3—working coil, 4—starting cage, 5—insulating washer, 6—segment, 7—wedge)

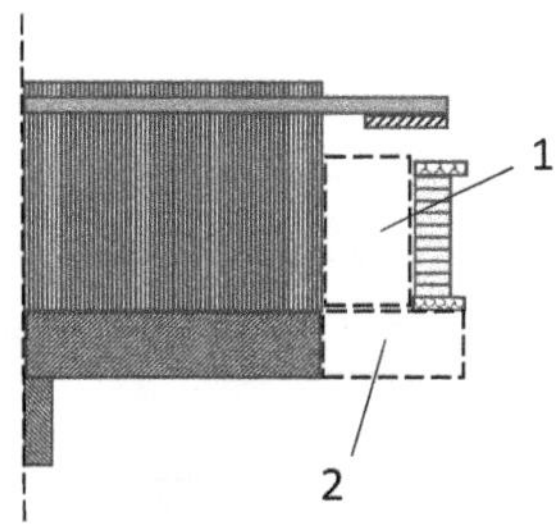

Fig. 5.4 Resource "empty space" (1—empty space instead of the wedge, 2—empty space instead of the rim)

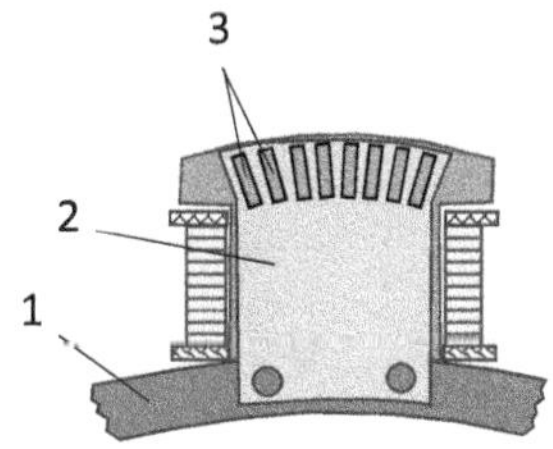

Fig. 5.5 Pole with new elements (1—rotor rim, 2—short-circuiting element, 3—deep slots)

Let's analyze the use of spatial resource in the example, based on the properties mentioned earlier:

- Value: free (material usage for the wedge and rim was even reduced).
- Quality: neutral.
- Quantity: sufficient.
- Readiness for use: created (the space was created by eliminating the wedge and part of the rim).

Physical and Chemical Effects in Electromechanics

In TRIZ, the use of effects and phenomena is given special importance. It is believed that their application when solving the problem allows for the strongest solutions. A solution to the problem can also be the elimination of a certain effect. The number of effects is calculated in the thousands. A physical effect is understood as a process or action that transforms an input physical quantity into an output physical quantity. Physical changes are reversible and do not alter the chemical properties of the object. Chemical effects in TRIZ are seen as involving changes of only the molecular structure of substances, and having the set of fields mainly limited to concentration, velocity, and heat fields. Chemical changes are irreversible and alter the chemical properties of the object.

Appendix G collects effects that are present (or may be present) in electric machines. The list is not complete, and it is recommended to study specialized

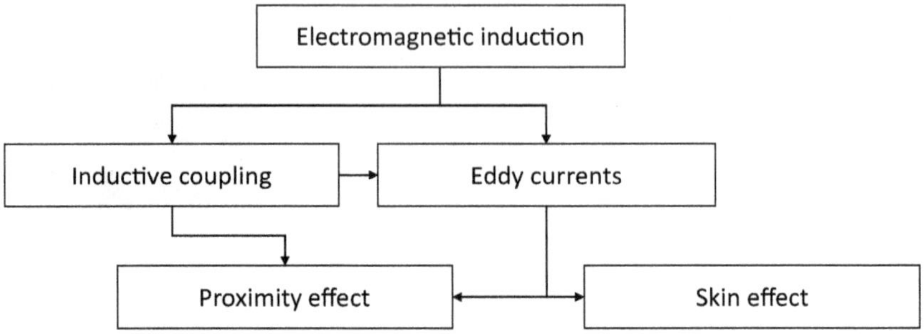

Fig. 5.6 Effects derived from electromagnetic induction

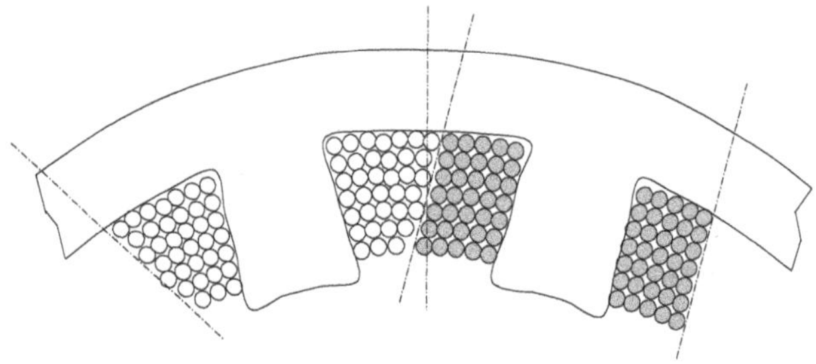

Fig. 5.7 Coils of different shapes in stator slots

reference books on physical and chemical effects. It is worth noting that effects can act together, and presence of some effects can lead to the appearance of other effects. An example with several well-known effects derived from electromagnetic induction is shown in Fig. 5.6.

It is notable that the role of physical and chemical effects in the life cycle of an electric machine manifests unevenly. During the production of the machine, both physical effects related mainly to the change in the shape of materials (stamping, cutting, casting) and chemical effects (polymerization of compounds during impregnation and consolidation of the winding), as well as physico-chemical effects (welding, soldering), are used. During the operation of the machine, mainly physical effects occur (electromagnetic induction and its derivatives, expansion when heated, heat transfer, etc.). Ideally, chemical effects should not be present at all during operation, but unfortunately, they occur, since the aging processes in electric machines are chemical in nature (oxidation, corrosion, etc.), and their manifestation increasing over time. At the end of the service life, if the machine goes for recycling, mainly physical effects will be used at this life cycle stage (cutting, melting); if the machine is simply discarded, then chemical effects (corrosion, decomposition) will take place.

Mathematical (primarily geometric) and biological effects, also used in TRIZ, are probably less applicable in the field of Electromechanics, although some examples can be found. Figure 5.7 shows a solution to increase the filling of the stator slot with copper. A special shape of the winding coils is used—the coils are divided into two groups with different shapes, while the cross-sectional area of the coils, regardless of shape, remains constant. The solution is based on the use of certain geometric proportions.

TRIZ Method of Compositional Contradictions

Now, when we have introduced resources and effects, it is also necessary to define the concepts of *Factors* and *Navigators*. Factors are properties or parameters of the system that can improve or deteriorate when making changes in the system. The list of 39 factors used in TRIZ is presented in Table 5.2. Navigators are indicators of solution techniques ("tricks") (listed in Table 5.3 and presented with examples in Appendix E). Factors and Navigators are linked by a "Matrix of Correspondence" (see Appendix F). Factors supporting the main useful function of the system are called "plus-factor", while Factors opposing this function are called "minus-factor".

The algorithm for applying the method of compositional contradictions [4] consists of eight steps (also illustrated in Fig. 5.8):

- Step 1: Formulate an administrative contradiction (this step can be skipped, although attempting to formulate the problem, especially in different ways, is very useful for deeper understanding of the problem).
- Step 2: Provide several alternative formulations of the technical contradiction (TC).
- Step 3: Determine pairs of Factors corresponding to the formulations in Step 2 (in Steps 2–3, the physical contradiction (PC) is also formulated).

Table 5.2 List of 39 TRIZ Factors

01. Productivity 02. Adaptability or versatility 03. Use of energy by moving object 04. Reliability 05. Manufacturing precision 06. Illumination intensity 07. Device complexity 08. Difficulty of control and measurement 09. Ease of manufacture 10. Stress or pressure 11. Ease of repair 12. Loss of substance 13. Area of stationary object	14. Object-generated harmful factors 15. Volume of stationary object 16. Length of stationary object 17. Area of moving object 18. Harmful factors affecting the object 19. Volume of moving object 20. Length of moving object 21. Shape 22. Speed 23. Moving object duration 24. Stationary object duration 25. Loss of time 26. Quantity of substance	27. Loss of information 28. Strength 29. Stability of the object's composition 30. Force 31. Ease of operation 32. Weight of moving object 33. Weight of stationary object 34. Temperature 35. Measurement accuracy 36. Power 37. Extent of automation 38. Use of energy by stationary object 39. Loss of energy

Table 5.3 List of 40 TRIZ Navigators

1. Parameter changes 2. Preliminary action 3. Fragmentation 4. Preliminary anti-action 5. Taking out 6. Homogeneity 7. Dynamization 8. Periodic action 9. Equipotentiality 10. Copying 11. Feedback 12. Local quality 13. Cheap short-lived instead of expensive long-lasting 14. Russian dolls	15. Discarding and recovering 16. Partial or redundant action 17. Application of composite materials 18. Intermediary 19. Another dimension 20. Universality 21. Thermal expansion 22. Spheroidality 23. Inert atmosphere 24. Asymmetry 25. Flexible shells and thin films 26. Phase transitions 27. Blessing in disguise	28. Beforehand cushioning 29. Self-service 30. Strong oxidants 31. Porous materials 32. Anti-weight 33. Skipping 34. Use of pneumatic and hydraulic structures 35. Merging 36. Vice versa 37. Color change 38. Mechanical oscillation 39. Mechanics substitution 40. Continuity of useful action

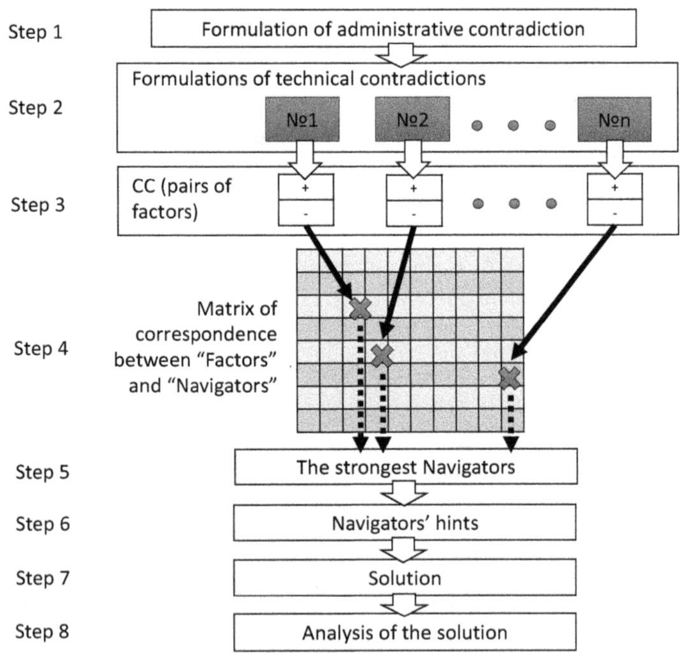

Fig. 5.8 Algorithm of the method of compositional contradictions

- Step 4: Find the numbers indicating the Navigators in the "Matrix of Correspondence" (Appendix F).
- Step 5: Select the most powerful Navigators from the found ones.
- Step 6: Write down the "Navigator Hints".
- Step 7: Find one or several solutions to the problem.
- Step 8: Analyze the solution from the perspective of TRIZ.

We will demonstrate the application of the method with three examples.

Example 1 Problem of electric insulation

Let us consider two problems related to electrical insulation that may have the same possible solution:

Problem 1A: The insulation of power cables can wear out due to mechanical impacts, which can lead to insulation breakdown and short circuits. If the insulation thickness was increased to reduce the negative effects of mechanical impacts, it would lead to worse heat dissipation from the copper conductors. In Fig. 5.9, the main structural elements of a cable with a voltage of 110–220 kV are shown. There is a need for a solution to the wear-out and heat removal problems.

Problem 1B: The insulation of windings of a high-voltage electric machine is often multilayered and quite thick, which leads to poor heat dissipation from the copper and, consequently, to the need to choose larger stator dimensions at the machine design stage to ensure lower copper losses and better heat dissipation. In Fig. 5.10, the cross-section of the slot with the bar winding of a 27 kV generator is presented. It should be noted that the hollow rectangular conductors in the example are cooled by a coolant circulating directly through them, i.e., the thickness of the outer insulation does not play a decisive role with this type of internal cooling. In contrast, for winding structures without inner coolant circulation, the insulation thickness will determine the thermal condition of the winding. There is needed a solution to this problem.

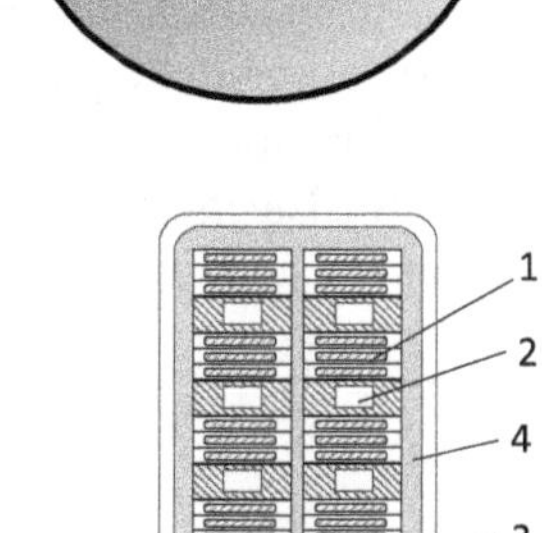

Fig. 5.9 Cable cross-section (1—cross-linked polyethylene insulation, 2—multilane copper conductor)

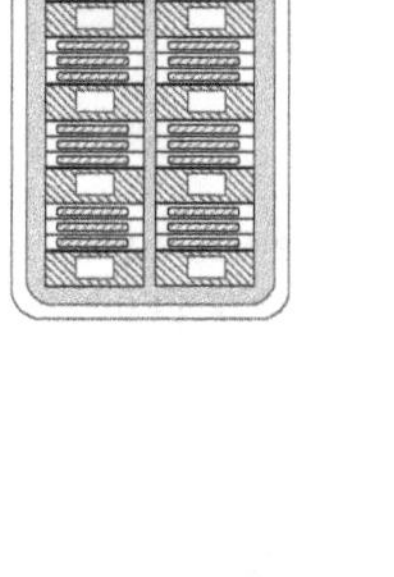

Fig. 5.10 Cross-section of electric machine winding (1, 2—insulated solid and hollow elementary conductors respectively, 3—insulation surfaces, 4—body insulation)

Problems 1A and 1B are similar, so we will try to solve them simultaneously.

According to the algorithm in Fig. 5.8, we start solving the problems 1A and 1B by formulating an administrative contradiction (AC) (*Step 1*):

Problem 1A. AC: The cable must last longer while also having lower losses.

Problem 1B. AC: The electric machine winding must last longer while also having lower losses.

In the following steps (2–4), we'll reformulate the problem as technical contradictions, selecting the Factors and Navigators. Steps 2–4 will be repeated for each problem formulation.

Formulation #1: (*Step 2*) The cable (or winding) subjected to mechanical (electrical) stress needs to last long but wear and emerging failures cannot be reliably detected during operation, which may lead to insulation breakdown.

Step 3—selecting Factor pairs:

- Factor supporting the main useful function of the system ("plus-factor"): cable (or winding) lifespan.
- Factor that opposes this function ("minus-factor"): inability to monitor the condition of the cable (or winding) insulation through measurements or other means to predict failures.

Step 4—selecting the most suitable TRIZ Factors and finding the Navigators using The Matrix of Correspondence from Appendix F:

Most suitable TRIZ Factors		Corresponding Navigators
Plus-factor	24. Stationary object duration	29 (Self-service), 15 (Discarding and recovering), 20 (Universality), 01 (Parameter changes)
Minus-factor	08. Difficulty of control and measurement	

Additionally, let's provide possible interpretations[1] of the TRIZ Factors for application in Electromechanics:

- Interpretations of Factor 24 (Stationary object duration) in the context of Electromechanics: service life, uninterrupted operation of an electric machine part/element/component.
- Interpretations of Factor 08 (Difficulty of control and measurement) in the context of Electromechanics: difficulty in detecting defects, including mechanical damage and insulation issues.

Let's carry out similar operations (steps 2–4) for two more formulations for technical contradiction (TC):

Formulation #2: We can attempt to increase insulation reliability by increasing its thickness; however, this impairs cooling, raises copper temperature, and increases electrical losses.

[1] In the future there can be formed some sort of a dictionary for electrical engineers using TRIZ, containing similar interpretations.

- Plus-factor: reliability of cable (or winding) insulation.
- Minus-factor: additional energy losses in the cable (or winding).

Most suitable TRIZ Factors		Corresponding Navigators
Plus-factor	4. Reliability	02 (Preliminary action), 28 (Beforehand cushioning), 01 (Parameter changes)
Minus-factor	39. Loss of energy	

- Interpretation of Factor 4 (Reliability) in the context of Electromechanics: reliability of operation for an electric machine part/element/component.
- Interpretation of Factor 39 (Loss of energy) in the context of Electromechanics: losses (electrical, etc.) in machine elements.

Formulation #3: Insulation strength can be increased by making it thicker; however, this reduces cooling, increases copper temperature, and raises electrical losses.

- Plus-factor: electrical and mechanical strength of the cable (or winding) insulation.
- Minus-factor: additional energy losses in the cable (or winding).

Most suitable TRIZ Factors		Corresponding Navigators
Plus-factor	28. Strength	01 (Parameter changes)
Minus-factor	38. Use of energy by stationary object	

- Interpretation of Factor 28 (Strength) in the context of Electromechanics: insulation strength.
- Interpretation of Factor 38 (Use of energy by stationary object) in the context of Electromechanics: efficiency reduction, increased losses (electrical, etc.) in machine components.

Based on the three alternative formulations above, we can combine the three Factor Pairs (Technical Contradictions, TC) into the *Compositional Contradiction* (CC): {(+24/−08); (+04/−39); (+28/−38)}. In this case, the CC consists of three pairs, but it can consist of 2, 4, or other numbers of pairs.

Step 5. Selecting the strongest Navigators using the following selection criteria: either the number of Navigator occurrences in the different formulations or the order within the relevant cells in the Correspondence Matrix (Appendix F). In our case, we obtained the following list of Navigators: 01 (three occurrences), 02, 29, 15, 28.

Step 6. Interpreting each of the Navigators (and the inventive tricks behind the Navigators) in the context of the given problem and thus forming a set of "hints" that together make a "story" about a possible solution. Let's take the Navigators one by one and for each of them give short possible interpretation:

- "29. Self-service"—The insulation itself prevents breakdowns or repairs defects.
- "28. Beforehand cushioning"—The insulation contains a feature that activates as issues escalate, preventing breakdowns.
- "01. Parameter changes"—There's an element within the insulation (possibly the "pre-placed cushion") that changes its state during a breakdown.

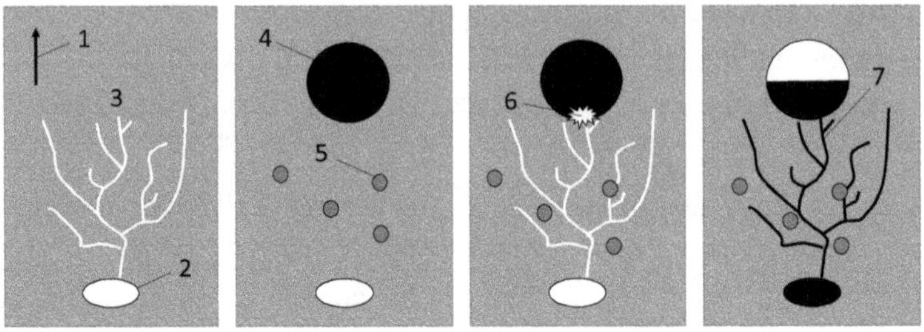

Fig. 5.11 The solution

- "02. Preliminary action"—This "cushion" should be set in place ahead of time.
- "15. Discarding and recovery"—Damaged parts undergo regeneration.

Having this "story", the inventor needs creativity, intuition, and insight to come to the solution.

Step 7. The solution:

The core idea is to initiate "self-healing" of insulation when electrical breakdowns or mechanical damage begin to develop. This solution is illustrated in Fig. 5.11. In the pre-breakdown stage, microchannels (3) appear in the polymer matrix of the insulation, which grow from micropores (2) in the direction of the electric field (1), often leading to rapid electrical breakdown.

In the proposed solution, microcapsules containing a liquid healing agent (4) and microcapsules with a catalyst (5) are pre-embedded in the matrix. Upon contact, the healing agent solidifies. The heat and mechanical forces acting at the tips of the growing channels are expected to be enough to break (6) the capsule with the healing agent that crosses the channel's path, releasing the agent, which flows into and fills the channels. Once inside, the agent begins to solidify upon contact with the catalyst, preventing further channel growth (7).

Implementing this solution will require research in polymer material chemistry. Siemens has a patent application [6] describing a similar approach applied for electric machine insulation.

Step 8. Solution analysis through the lens of TRIZ. In this step, we address a series of questions to understand how the solution was derived, as well as its adequacy and quality.

An essential question in TRIZ solution analysis is: *What Physical Contradiction (PC) was resolved by the solution?* Let's define PC as a set of oppositely directed and mutually exclusive system requirements for the same property of the system. It can be framed as follows: "*To improve the system, one part of it must simultaneously exist in different physical states, which is impossible.*" In our case, the PC is: "*The insulation must be both thick and thin at the same time*". This contradiction is precisely what our solution resolves.

Fundamental transformation (principle):[2] Separation in time and substance.

[2] One can find these and other TRIZ terms given below, in [2, 4] or other TRIZ literature.

Applied TRIZ standards: standard "Additives"—introduction of a highly active additive in small doses or introducing a substance in a chemical state from which it is released at the needed time.

Utilized effects: chemical (polymerization reaction, catalysis) and physical (electrical discharges, phase transitions, heating, shockwaves).

Utilized resources: material resources.

It is worth noting that other solutions exist for preventing mechanical wear of cable insulation used in harsh conditions—for instance, the use of armored cables. Thus, the solution we derived through the TRIZ algorithm is not the only possible one. It is recommended to check TRIZ literature [2, 4], to learn more about TRIZ terms and concepts.

Example 2 Problem of Extending the Operational Torque and Speed Ranges

In a hybrid or fully electric vehicle, the drive system needs to operate over a very wide range of rotational speeds, where high torque is required at low speeds. In such applications the permanent magnet synchronous motor (PMSM) is considered a good option due to its compactness, high efficiency, and relatively low rotor losses. However, at high speeds, the PMSM generates a high electromotive force (EMF), which requires implementing a field weakening control strategy. EMF is proportional to the number of winding turns and the rotational speed, so for high-speed operation it is preferable to design the winding with a small number of turns connected in series. Conversely, for generating high torque, a large number of turns connected in series is desirable. Thus, there is a conflict over the optimal number of winding turns for the machine to work across the different operating modes. This leads to either a machine optimized for high speeds (curve 1 in Fig. 5.12) or one optimized for high torque (curve 2).

One way to resolve this problem is by using a gearbox that could be engaged when high torque and low speed are required, allowing the PMSM to be designed with a smaller number of winding turns in series. However, the gearbox adds weight and complexity to the system. Another possible solution could be a motor with a

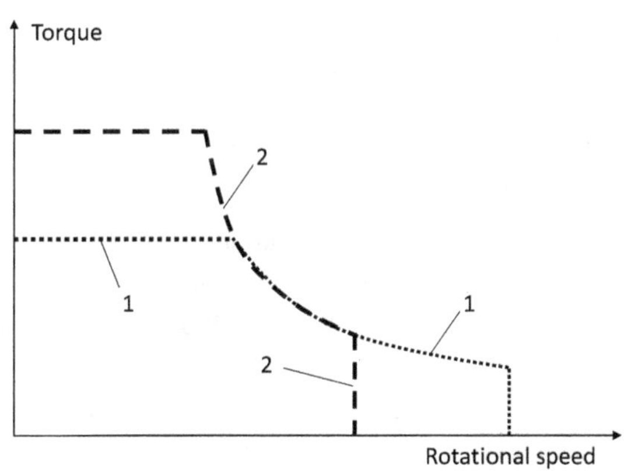

Fig. 5.12 Torque-speed characteristics

small number of winding turns but with increased conductor and core cross-sections—though this would result in a heavier and more expensive machine.

Looking for alternative solutions let's go through the same steps as in Example 1. We'll frame the problem as an administrative contradiction: *the system needs to operate over a wide range of speeds while having low system weight.* This time we will use six formulations of the problem. The formulations might sound similar to each other, but they will lead to different pairs of Factors.

Formulation #1: To increase power at low and high rotational speeds, it is necessary to design a larger and heavier machine or add a gearbox.

- "Plus-factor": Increased power
- "Minus-factor": Increased size and weight

Most suitable TRIZ Factors		Corresponding Navigators
Plus-factor	36. Power	8 (Periodic action), 10 (Copying), 19 (Another dimension), 13 (Cheap short-lived instead of expensive long-lasting)
Minus-factor	33. Weight of stationary object	

- Interpretation of Factor 36 (Power) in the context of Electromechanics: same as in TRIZ.
- Interpretation of Factor 33 (Weight of Stationary Object) in the context of Electromechanics: machine weight.

Formulation #2: To operate at high rotation speeds, the machine must have a low number of winding turns, but this limits the ability to generate the required torque at low speeds.

- "Plus-factor": Expand the range of operating rotation speeds
- "Minus-factor": Torque

Most suitable TRIZ Factors		Corresponding Navigators
Plus-factor	2. Adaptability or versatility	7 (Dynamization), 19 (Another dimension), 40 (Continuity of useful action)
Minus-factor	30. Force	

- Interpretation of Factor 2 (Adaptability or versatility) in the context of Electromechanics: operation over a very wide range of rotation speeds and/or torques.
- Interpretation of Factor 30 (Force) in the context of Electromechanics: torque on the machine's shaft.

Formulation #3: To enable operation at high rotational speeds, the machine must have a low number of winding turns, which limits the ability to generate the required torque at low speeds, necessitating the use of a gearbox, which may increase system weight.

- "Plus-factor": High power across all ranges
- "Minus-factor": Increased system complexity

Most suitable TRIZ Factors		Corresponding Navigators
Plus-factor	36. Power	40 (Continuity of useful action), 8 (Periodic action), 25 (Flexible shells and thin films), 15 (Discarding and recovering)
Minus-factor	7. Device complexity	

- Interpretation of Factor 36 (Power) in the context of Electromechanics: same as in TRIZ.
- Interpretation of Factor 7 (Device Complexity) in the context of Electromechanics: adding additional elements, such as a gearbox.

Formulation #4: To enable operation at high rotational speeds, the machine must have a low number of winding turns, which limits the ability to generate the required torque at low speeds, necessitating the use of a gearbox.

- "Plus-factor": Operation across all ranges
- "Minus-factor": Weight

Most suitable TRIZ Factors		Corresponding Navigators
Plus-factor	24. Stationary object duration	20 (Universality), 13 (Cheap short-lived instead of expensive long-lasting), 8 (Periodic action), 16 (Partial or redundant action)
Minus-factor	33: Weight of stationary object	

- Interpretation of Factor 24 (Stationary object duration) in the context of Electromechanics: continuous long-term operation in various modes.
- Interpretation of Factor 33 (Weight of stationary object) in the context of Electromechanics: weight of the entire machine or the machine's stator.

Formulation #5: To enable operation at high rotational speeds, the machine must have a low number of winding turns, which limits the ability to generate the required torque at low speeds. Consequently, a gearbox or an increase in the size and weight of the machine (stator and rotor) is necessary.

- "Plus-factor": Operation across all ranges
- "Minus-factor": Weight of rotor and gearbox

Most suitable TRIZ Factors		Corresponding Navigators
Plus-factor	2. Adaptability or versatility	3 (Fragmentation), 20 (Universality), 7 (Dynamization), 32 (Anti-Weight)
Minus-factor	32. Weight of moving object	

- Interpretation of Factor 2 (Adaptability or versatility) in the context of Electromechanics: operation over a very wide range of rotational speeds and/or torques.
- Interpretation of Factor 32 (Weight of moving object) in the context of Electromechanics: weight of the rotor or other rotor-attached elements (shaft, gearbox).

Formulation #6: To cover high rotational speeds, the machine requires a low number of winding turns, but this limits the required torque at low speeds, thus necessitating either a gearbox or an increase in the size and weight of the machine (stator and rotor).

- "Plus-factor": Operation across all ranges without adding a gearbox
- "Minus-factor": Weight of the machine (stator and rotor)

Most suitable TRIZ Factors		Corresponding Navigators
Plus-factor	29: Stability of the object's composition	33 (Skipping), 1 (Parameter changes), 5 (Taking out), 23 (Inert atmosphere)
Minus-factor	32: Weight of moving object	

- Interpretation of Factor 29 (Stability of the object's composition) in the context of Electromechanics: no need to add additional elements, such as a gearbox/multiplier.
- Interpretation of Factor 32 (Weight of moving object) in the context of Electromechanics: weight of the rotor or other rotor-attached elements.

Based on the six formulations we now can select the strongest Navigators: 8 (appears 3 times), 7, 40, 20, 19 (each appearing twice), as well as 3 and 33 (ranked first in the relevant cells in the Martix of Correspondence).

Then we interpret the Navigators to outline the "Navigator Hints":

- "8. Periodic action"—switch from continuous to periodic operation
- "20. Universality"—the object performs multiple different functions
- "7. Dynamization"—the object's characteristics should adapt to be optimal at each stage
- "40. Continuity of useful action"—maintain continuous operation
- "19. Transition to another dimension"—no concrete interpretation, but note the word "transition"
- "3. Fragmentation"—divide the object into independent parts
- "33. Skipping"—perform the process or individual stages at high speed.

The solution based on the "story" told by the Navigator hints can be as follows:

Divide the winding into sections ["Segmentation"], take the ends of the section windings out ["Transition to Another Dimension"], so that the connection scheme of the sections can be switched between two states ["Dynamization"]: (1) series connection of sections/turns (for low speeds) and (2) parallel connection of sections/turns (for high speeds) ["Versatility"]. Operate with the connection scheme optimal for the current rotation speed ["Periodic Action"]. Perform switching between states without stopping the machine ["Skipping"].

A similar solution is implemented by Yaskawa in their QMET-II system and their patent [7] (for the use in machine tools).

Now let's perform the solution analysis (like we did in Example 1):

- *Resolved physical contradiction*: many turns connected in series and few turns connected in series.
- *Fundamental transformation (principle)*: separation in time.

- *Standards used*:
  - Enable the object to be transformed into a controllable system.
  - Integrate the object into a more complex bi- or poly-system.
- *Effects used*: none.
- *Resources used*: structural, temporal.

Example 3 Problem of Cooling the Rotor of an Induction Motor in an Electric Vehicle

When using a squirrel cage induction motor (SCIM) in an electric vehicle (for instance, in the Tesla Model S), the SCIM rotor can overheat due to high losses in the squirrel cage under torque overloads. Around 20% of total losses occur in the rotor. Due to the rotor's rotation, it is difficult to dissipate this heat effectively. The SCIM rotor acts as a sort of "heat accumulator"—once it is overloaded, such as during rapid vehicle acceleration, it requires cooling, which may involve temporarily limiting the overloads. Factors affecting these losses include the cross-sectional area of the squirrel cage conductors/bars and the rotor temperature. If the rotor's size—its radius and surface area—is increased, the cross-sectional area of the conductors also increases, reducing their resistance and, consequently, reducing rotor cage losses and rotor temperature. Thus, lowering temperature requires increasing the rotor size and, consequently, its weight, which is not optimal for the overall size and weight of the electric machine. Rotor losses can be slightly reduced by cooling the rotor surface with a flow of a cooling agent (a substance such as gas). Increasing the amount of cooling agent will lower the temperature of the squirrel cage material, reducing its resistance and, consequently, the losses. However, calculations show that air circulation within the motor enclosure alone is insufficient. Additionally, the aluminum rotor winding is located in deep slots, so under intense, short-term overloads, the temperature deep within the slot can exceed critical levels despite the cooling effectiveness at the surface. It is important to note that high temperatures and significant temperature fluctuations can lead to reduced rotor lifespan and failures, decreasing machine reliability.

The problem can be framed as an administrative contradiction as follows: "*The rotor needs to be compact while not overheating*."

Let's formulate the technical contradictions:

Formulation #1: Increasing the rotor's size and weight reduces the current load, thereby reducing losses in the rotor cage and consequently lowering its temperature. Thus, to lower temperature, rotor weight has to be increased.

- "Plus-factor": lower temperature
- "Minus Factor": increased rotor weight

Most suitable TRIZ Factors		Corresponding Navigators[3]
Plus-factor	34. Temperature	26, 21, 20, 30
Minus-factor	32. Weight of moving object	

[3] In this example we are not indicating the names of Navigators and give just their numbers. The names can be found in Table 5.3.

- Interpretation of Factor 34 (Temperature) in the context of Electromechanics: rotor or squirrel cage temperature.
- Interpretation of Factor 32 (Weight of moving object) in the context of Electromechanics: rotor weight.

Formulation #2: To reduce thermal load and improve reliability, the surface area of the rotor needs to be increased.

- "Plus-factor": reduced thermal load and improved reliability
- "Minus-factor": increased rotor surface area

Most suitable TRIZ Factors		Corresponding Navigators
Plus-factor	04. Reliability	19, 2, 22, 16
Minus-factor	17. Area of moving object	

- Interpretation of Factor 04 (Reliability) in the context of Electromechanics: reliability.
- Interpretation of Factor 17 (Surface Area of Moving Object) in the context of Electromechanics: rotor surface area.

Formulation #3: Losses occur in the IM rotor, and they can be somewhat reduced by cooling the rotor surface with a flow of a cooling agent (a substance, like gas). Increasing the amount of coolant reduces the temperature of the squirrel cage material, lowering its resistance and thus the losses.

- "Plus-factor": reduced losses
- "Minus-factor": increase of the coolant flow.

Most suitable TRIZ Factors		Corresponding Navigators
Plus-factor	39. Loss of energy	14, 38, 29
Minus-factor	26. Quantity of substance	

- Interpretation of Factor 39 (Loss of energy) in the context of Electromechanics: losses.
- Interpretation of Factor 26 (Quantity of substance) in the context of Electromechanics: amount/flow of coolant.

Formulation #4: The rotor surface area can be reduced, but this will increase current loading, thermal load, and decrease reliability.

- "Plus-factor": Reduction of rotor surface area
- "Minus-factor": Decreased reliability.

Most suitable TRIZ Factors		Corresponding Navigators
Plus-factor	17. Area of the moving object	34, 04
Minus-factor	04. Reliability	

- Interpretations of Factor 17 (Area of the moving object) in the context of Electromechanics: Rotor surface area.
- Interpretations of Factor 04 (Reliability) in the context of Electromechanics: Reliability.

Based on the four formulation we now can select the strongest Navigators and interpret them to get the "Navigator Hints":

- "14. Russian doll"—One object is placed inside another, one object passes through a cavity in another object.
- "19. Transition to another dimension"—"Multi-story" configuration instead of "single-story," using the reverse side of the given area.
- "34. Use of pneumatic and hydraulic structures"—Instead of solid parts, use gaseous or liquid ones.
- "26. Phase transitions"—Use phenomena occurring during phase transitions, such as heat absorption.

This can bring us to the following solution:

The rotor shaft is made hollow, with cooling system elements placed inside the cavity, such as a heat pipe or concentric tubes that supply and remove the cooling agent. This way, heat is dissipated not only from the external but also from the internal surface of the rotor. When using a heat pipe, a phase transition (evaporation) occurs, accompanied by heat absorption.

A similar problem with heat dissipation from the rotor was addressed by Tesla company in their Model S and in the patent [8] obtained by Tesla, a solution with a heat pipe is described (Fig. 5.13). It should be noted that the solution in the patent is not necessarily the one implemented in the specific product. In particular, the relative drawback of the solution in [8] is the increased axial length due to the heat exchanger of the heat pipe.

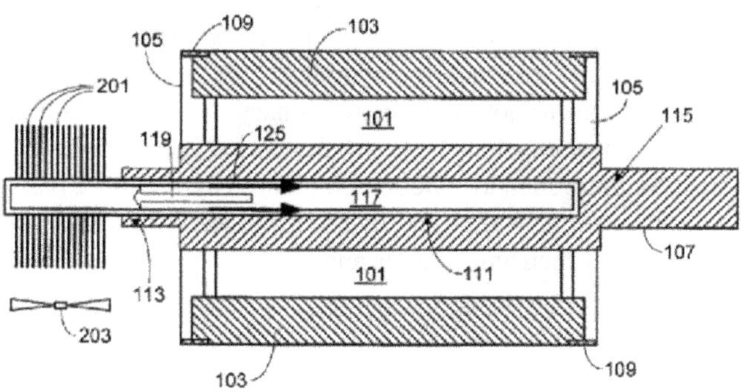

Fig. 5.13 Illustration from the patent US9331552B2 by Tesla [8]

Let's analyze the solution, like we did in Examples 1 and 2:

What physical contradiction was resolved: the rotor should be both large and small.

Fundamental transformation (principle) used: separation in space and substance.

Standards used: implementation of system functions at the micro level—substance and field level.

Effects used: convection and/or phase transition.

Resources used: space inside the rotor, rotor rotation, availability of the electric car's cooling system (antifreeze, radiator, hoses).

It should be noted that there exist other solutions to the problem. For example, one possible method of compensating for increased heat generation in the rotor during overload is the temporary intensification of forced cooling.

> TRIZ can guide the inventor to a "shot with a high probability of hitting the target".

This chapter and Appendices E and F contain all the necessary elements for the reader to try independently finding solutions to technology problems. Studying TRIZ literature is highly recommended but not mandatory for initial attempts. It is recommended to learn at least one algorithm (like the Compositional Contradiction algorithm explained above), develop practical skills in using it, and it is also helpful to study physical, chemical, and geometric effects, as well as learning to work with system resources.

Common Language for TRIZ and Electromechanics

It would be helpful to have some sort of a dictionary linking TRIZ and Electromechanics. Based on the three examples provided, an attempt can be made to create a fragment of a dictionary matching TRIZ terms with examples from the field of Electromechanics (Table 5.4). It is important to note that the dictionary can be practically unlimited in terms of additions from the field of electric machines and systems coexisting with electric machines. A comprehensive dictionary created in the future based on the analysis of numerous inventions in the field of Electromechanics could serve as an element of the synthesis of Electromechanics and TRIZ.

Table 5.4 Elements of the TRIZ-Electromechanics dictionary

Some TRIZ terms	Examples of interpretation of the terms in electromechanics
Stationary object duration	Service life, continuous operation time of a part/element/component of electric machine, continuous long-term operation in different modes
Difficulty of control and measurement	Difficulty in detecting defects, including mechanical damage and insulation breakdown
Reliability	Reliability of operation of a part/element/component of electric machine
Loss of energy	Losses (copper losses, etc.) in machine elements
Strength	Strength of insulation
Use of energy by a stationary object	Decrease in efficiency, increase in losses (copper losses, etc.) in machine elements
Weight of a stationary object	Weight of the machine
Adaptability or versatility	Operation over a very wide range of rotational speeds and/or torques
Force	Torque on the machine shaft
Device complexity	Addition of extra elements, e.g., gearbox, electronics
Weight of a moving object	Weight of the machine rotor or other elements attached to the rotor (shaft, gearbox)
Stability of the object composition	No need to add extra elements, e.g., gearbox
Quantity of substance	Amount/consumption of the cooling medium
Temperature	Temperature of the rotor, e.g., squirrel cage
Area of a moving object	Surface area of the rotor

TRIZ System Operator

The evolution of systems along the axes of "Supersystem-System-Subsystem" can be investigated using the framework of so-called System Operator (SO) with nine "screens" (Fig. 5.14). The use of SO makes possible describing and forecasting the development of technical systems. The attempt to apply SO for electrical machines is presented in Table 5.5. The cells of the table represent the "screens" from Fig. 5.14. The information in the cells of the table is just an example. Each electrical engineer can put into the table the elements of his or her own vision.

There is also a more complex framework—the 27-screen SO (Fig. 5.15), in which 2 additional planes are added to the 9 screens: the *anti-system* and the *competing system* planes. Thus, a forecaster or inventor can work with 27 screens simultaneously, which can be more productive than using 9 screens, and lead to new understanding of the problem domain.

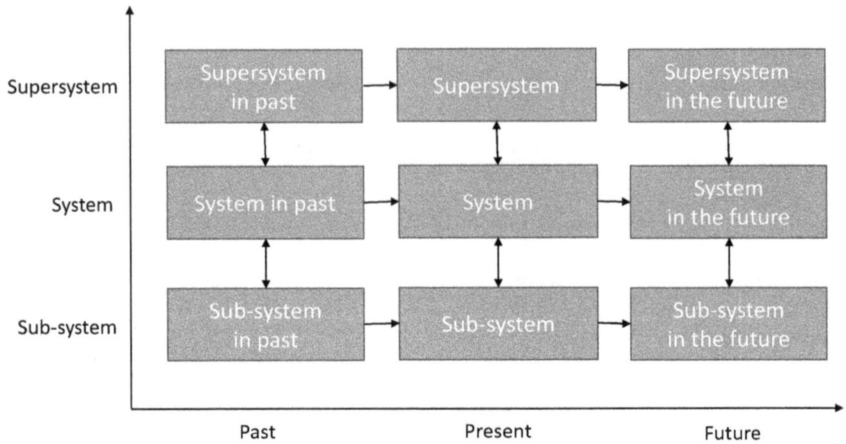

Fig. 5.14 System Operator with 9 screens

Anti-system

An Anti-System (AS) is a system that performs a function opposite to that of the system being considered. The AS is something that has external activity opposite in nature to the Main Useful Function (MUF) of the system being considered. For example, if the system is a pencil, and the anti-system is an eraser.

Let's try to find an AS for an electric motor. There is room for several alternative opinions here. The MUF of an electric motor is to create motion (e.g., rotational or linear) from the electric energy it consumes. One could assume that the AS of an electric motor might be an electric generator. The MUF of an electric motor can also be formulated as the most efficient conversion of electrical energy into mechanical energy with minimal heat loss. In this case, the AS could be an electric heater, whose task is to convert electrical energy into heat. One can try to work with each of these options or propose own variants. Any unconventional options can be fruitful and help to come to new ideas or new vision.

AS is a very effective means of analysis and synthesis, and it is most effective when "running" all nine screens of AS during analysis as we did above for the system—electric machines in Table 5.5. By analyzing an electric heater as an AS, we can point out, for example, the transition (at the "supersystem" level) from simple plug-in units from the recent past to digital control, the Internet of Things, "Smart Home" concepts, etc. in the present. At the "subsystem" level, we see a transition from lamps, nichrome, radiators in the past to infrared elements, and ceramics in the present. At the "system" level, we can observe the transition from lamp emitters to convectors.

Competing System

Competing systems (CS) are understood as all technical systems with the same MUF. If the MUF of an electric motor is to create motion, then an internal combustion engine (ICE) can be suggested as a CS. There can be other CS options as well.

Table 5.5 "9-screens" for electric machines

	Past (30 years ago)	Present	Future (assumptions)
Supersystem	Starters Sensors AC circuit breakers	Frequency converters Sensors High-temperature electronics GaN, SiC switches	DC circuit breakers Internet of Things Integrated sensors Ultra-high temperatures New applications
System	Direct start of induction machines	Integrated FC-EM Compactness	Multiphase SC windings Carbon nanotube cables High-frequency machines Digital twins
Subsystem	Solid conductors Permanent magnets: ferrites, SmCo Mechanical bearings	Litz wire Superconductors (SC) Permanent magnets NdFeB Magnetic bearings	New SC for room temperatures AC superconductors Carbon nanotubes Non-rare earth stronger magnets Additive technology-based components

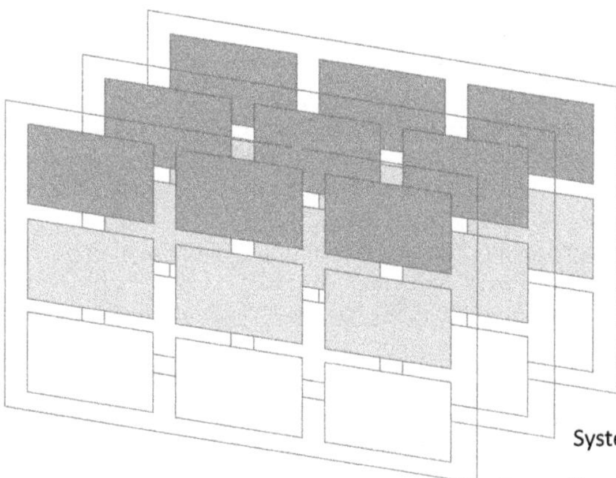

Fig. 5.15 System Operator with 27 screens

By analyzing ICE as a CS, we can point out, for example, the transition at the "supersystem" level from mechanical control systems in the past to electrical ones in the present. At the "subsystem" level, we see a transition from gasoline and diesel in the past to biofuels and hydrogen in the present. At the "system" level, we can notice the trend of hybridizing ICE with electric machines. It may be useful to follow news in the development of ICE at the "supersystem" and "subsystem" levels to fill the screens for the "future". For example, trends in the use of powder metallurgy and additive technologies are emerging.

It is possible to learn from anti-systems and competing systems and project the development in their domains to the own domain of technology.

Holding in mind the 27 "screens" simultaneously may give the strongest insights.

Concluding Comments

The TRIZ toolkit is not limited to the methods and concepts presented above. TRIZ includes other algorithms for finding solutions, as well as standards, Substance-Field models, and much more.

It is also important to remember that TRIZ is just one of the methodologies for creative thinking and can be regarded as sort of a trampoline for anyone learning to think "outside the box".

References

1. G.S. Altshuller, R.B. Shapiro, On the psychology of inventive creativity. Quest. Psychol. **6**, 37–49 (1956)
2. G.S. Altshuller, *The Algorithm of Invention*, 2nd edn. (Moskovsky Rabochy, Moscow, 1973)
3. https://en.wikipedia.org/wiki/TRIZ
4. M.A. Orlov, *The Fundamentals of Classical TRIZ* (Solon-Press, Moscow, 2006) 432 pages
5. V. M. Gerasimov, Rotor of a synchronous salient-pole electric machine: inventive tale, 25 pages, manuscript deposited at ChOUNB on 12.07.2005, no. 304, http://www.trizminsk.org/e/212008.htm
6. M. Brockschmidt, S. Kempen, F. Pohlmann, G. Schmidt, C. Staubach, Self-healing insulating layer for an electrical machine, Patent application WO2013/117373A1 by Siemens
7. A. Kumagai et al., Switching device of winding for motor in machine tool, Patent JP3596711B2 by Yaskawa
8. L. Fedoseyev, E.M. Pearce Jr, Rotor assembly with heat pipe cooling system, Patent US9331552B2 by Tesla

6 Philosophical Foundations of Evolution Management

In recent years, electrical engineers have gained new opportunities—greater computational power (including distributed), materials with enhanced properties, flexible manufacturing methods, compact power electronics, high-performance control electronics, and finally, software with artificial intelligence elements. Collectively, these tools enable the creation of new generations of electric machines with performances unseen before. At the same time, a notable issue is the lack of a widespread practice within the profession to discuss the future. This is a significant problem, as most electrical engineers do not have a clear understanding of the direction of the development of electric machines and Electromechanics. This can lead to an inability to meet the real challenges of the radically changing modern world and contribute to the necessary changes. As a result, motivation may decrease, frustration with the profession may set in, and some may even consider leaving the field altogether.

Attempts to rethink Electromechanics in the context of the new era raise a wide range of questions, some of which can be grouped as follows:

- *Questions about direction*: In which direction is the evolution of electric machines heading? Which direction should be considered the correct one? Where should the evolution be directed?
- *Questions about reference points and sources of knowledge*: From what philosophical perspectives (including ethical) should the evolution of electric machines and related systems be guided? On what basis should strategies for managing the evolution be chosen?
- *Questions about tools for managing the evolution*: Is it possible to manage the evolution of electric machines? Is anyone currently managing this process? What tools could be used for such management?
- *Questions about uniqueness*: Does Electromechanics deserve its own distinct branch of philosophy? Why can Electromechanics be singled out among other sciences? Is there any value in considering electric machines separately from other technologies?

A. Matveev, *Shaping the Future of Electric Machines*,
https://doi.org/10.1007/978-3-031-95248-7_6

- *Questions about essence*: Do electric machines have essential differences from other technical objects? What is the essence of electric machines?
- *Questions about neutrality*: Are electric machines neutral technology? Are they a "good," "evil," or "neutral" part of the Technosphere? Are there "good" and "bad" types of electric machines? Can anything be done to address this?
- *Questions about motivation*: Why engage in Electromechanics? If one chooses to engage, what questions should be addressed and on what grounds?

The answers to such questions may lie in the field of the philosophy of technology. The object and subject of the philosophy of technology are graphically described and defined in Fig. 6.1. In the case of Electromechanics, artifacts can be understood as electric machines, technical actions as the processes of developing and manufacturing the machines, and technical knowledge as the body of knowledge in Electromechanics as a science, supplemented by sections from other fields such as physics, chemistry, materials science, and others. Technical consciousness is what needs to be developed specifically in relation to Electromechanics. This is one of the topics of Chap. 6.

Seeking answers to the posed questions, we can turn to works of several prominent philosophers of technology. Studying the works, we can focus on the following aspects:

1. The worldview proposed by the philosopher.
2. The philosopher's answer to the question of what technology is and what its essence is.
3. How the philosopher views the relationship between humans and technology.
4. The philosopher's perspective on technical creativity and invention.
5. Examples relevant to Electromechanics.
6. Interesting, unconventional ideas related to the subject of our study.

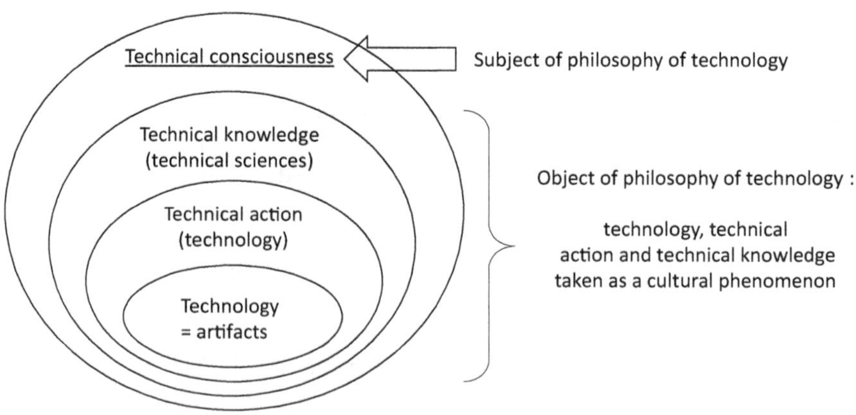

Fig. 6.1 Object and subject of the philosophy of technology

Through this review, we can systematize relevant information, synthesize some ideas, and attempt presenting elements of a worldview that may be of interest and benefit to electrical engineers.

It is notable that the emergence of the philosophy of technology occurred almost simultaneously with the rise of Electromechanics. The pioneering works of Kapp (1877), Engelmeyer (1898), and Bon (1898) in the philosophy of technology appeared around the same time as the invention of the dynamo machine (Siemens, 1872) and the induction machine (Tesla and Dolivo–Dobrovolsky, 1888–1889).

The Concept of Organ Projection by E. Kapp Ernst Kapp a German scholar, was the first to introduce the term "philosophy of technology." His contribution that may be of use to us is the theory of "organ projection." Kapp believed that humans, in all their creations, unconsciously reproduce their own organs and come to understand themselves through their inventions. For example, in his theory, the railroad network corresponds to the circulatory system, and the telegraph to the nervous system. Although Kapp's theory has been widely criticized [1, 2], some analogies in Electromechanics based on Kapp's logic may seem reasonable. For instance, the cross-sections of an electric cable and a nerve (Figs. 6.2 and 6.3) indeed show striking similarities. One could also draw comparisons between the heart's function (where tissues are excited by electrical impulses) and that of a magnetohydrodynamic engine (pump) or another electromechanical converter.

Although such analogics are relatively few, and they may not yield much in our exploration, the original idea of organ projection could be transformed into something different. P. Florensky, to some extent following Kapp's logic, argued that "as the center of the universe, its purpose and focal point, humans replicate its structure at the level of their own bodies" [3]. In this sense, the starting point for projection is not humanity but the universe itself. In I.P. Kopylov's developed theory [4], the organizing principle of the universe is electromagnetic. Planets behave like electric machines—for example, our planet functions as a unipolar machine. According to

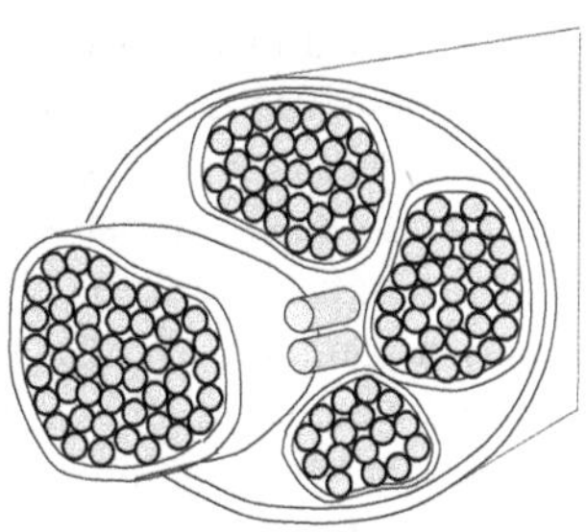

Fig. 6.2 Nerve

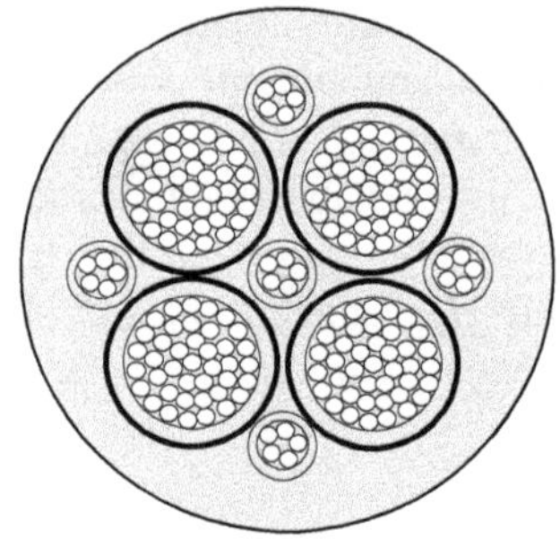

Fig. 6.3 Cable

this theory, one could say that electric machines are a direct projection of the electromagnetic universe!

Thus, the idea of "projections" could prove to be quite fruitful. Extending these analogies, we might add that when electric machines are used in exoskeletons or prosthetics for people with disabilities, a kind of *reverse organ projection* takes place. This trend of reverse organ projection could lead to the spread of cybernetic organisms, with Electromechanics playing a significant role in this development.

The Formula of *Technical Will* by P. Engelmeyer One of the earliest philosophers of technology was the Russian engineer, Peter Engelmeyer. Although he is rarely cited in philosophical literature [1], he was the first to provide a number of formulations and raise several questions that later became integral parts of the philosophy of technology. Specifically, Engelmeyer wrote: "Through technology, humans have freed themselves from the need to adapt their bodies to the surrounding natural conditions. On the contrary, humans adapt these conditions to their bodies, creating an artificial microcosm around them." Engelmeyer also provided a succinct formula: "Man is a technical being" (from his 1912 work *Philosophy of Technology*, as cited in [1]). It was only two decades later that similar concepts (such as "supernature," analogous to the "microcosm") and terms ("man-the-technician") were introduced by the notable philosopher José Ortega y Gasset.

Engelmeyer proposed investigating what technology represents, what goals it pursues, what methods it uses, and where its boundaries lie. He aimed to develop a comprehensive picture of technology. In [5], Engelmeyer addressed the relationship between technology and culture, art, and science. He wrote: "Whatever humans begin to do, in any transition from idea to object, from goal to achievement, we must pass through some form of specific technology. And all these technologies... have much in common. What is this commonality?" Thus, as early as 1898, Engelmeyer raised the question of the essence of technology, a question to which, as we will see below, the great philosopher Martin Heidegger provided a brilliant answer 50 years later.

Engelmeyer, like many engineers of his time, was interested in electrical engineering, though he was not an electrical engineer himself [1]. However, Electromechanics can certainly borrow one of his ideas. Engelmeyer believed that when creating technical objects, one should be guided not only by utility but also by truth, goodness, and beauty, and he proposed the so-called *formula of Technical Will* (W):

$$W = T^m B^n G^p U^o,$$

where, T—Truth, B—Beauty, G—Goodness, and U—Utility. The formula has two infinite extremes: “divine will” ($m = n = p = o = \infty$) and “devilish will” ($m = n = p = o = -\infty$). A special case would be purely utilitarian technology ($m = n = p = 0$).

In the context of modern Electromechanics, *Utility* is an unconditional criterion in the creation of electric machines. The concept of *Beauty* is relative; one could speak of the beauty of technical solutions in the design of the machines, but this beauty is often visible only to the designers. Beauty is also considered in fields like industrial design, where the goal is to improve the aesthetic qualities of objects produced in industry. *Goodness* and *Truth*, however, are not typically considered in classical Electromechanics, as is likely the case in other technical and engineering disciplines as well.

Goals on the Path to Happiness and Goodness According to F. Bon In the same year (1898) as Engelmeyer’s work, another philosopher of technology, Fred Bon, published his book *On Duty and Goodness*, in which he argued that the goal of technology is to strive for human happiness. However, even this goal is subordinate to the higher goal—the idea of Goodness. As we can see, the concept of goodness was important to thinkers of the late nineteenth century. Those who view happiness as the ultimate and highest goal of human aspirations must also investigate the means leading to this goal, as the most important factors in all technical tasks [1].

Bon’s worldview was religious, which was typical for his time. In such a view, *Goodness* can be defined, for instance, by sacred texts. Not all contemporary scholars will be willing to adopt this perspective, so we may need to consider a broader range of interpretations and worldviews. Bon’s work is notable for indirectly presenting a hierarchical structure, which we can try to illustrate as in Fig. 6.4, into which we could potentially integrate our work on electric machines. This hierarchical framework highlights how technological development, particularly in Electromechanics, could be aligned with higher ideals, suggesting that beyond utility, there might be a moral or ethical dimension worth considering in the design and use of technology.

Perception of Electric Machines During the Era of the First Philosophers of Technology It is noteworthy that the aesthetic and emotional perception of electric machines at the turn of the nineteenth and twentieth centuries differed significantly

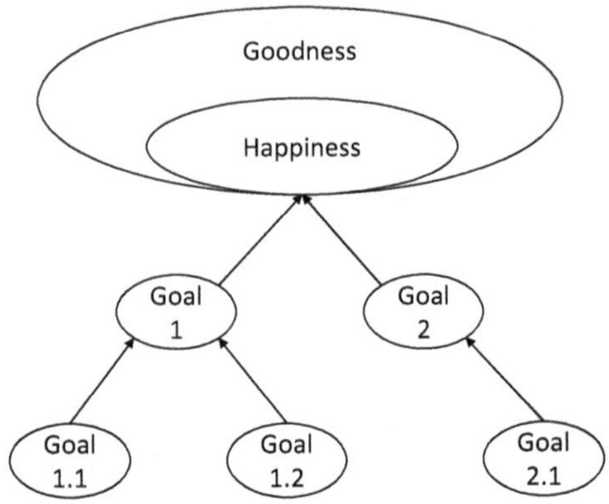

Fig. 6.4 An attempt to visualize Bon's concept

from the modern view. At that time, the *dynamo* became a symbol in the public consciousness of invisible, intangible, and incomprehensible electrical power, one of the manifestations of the distributed force within the universe. Electric machine was part of the "technological sublime," which replaced the "natural sublime" in culture [6]. The experience of the sublime involves glimpsing beyond the boundaries of the possible and briefly dwelling there. This feeling of the technological sublime can be seen as a transgressive experience—a crossing of boundaries. In the literature of that time, the dynamo was often compared to a temple or deity, and there were descriptions of reverence and worship toward the dynamo, as if serving it. This widespread perception likely influenced philosophers, especially those working in the early twentieth century.

The Hidden Mechanisms of Technological Development According to J. Ortega y Gasset Unlike the aforementioned predecessors, Spanish philosopher José Ortega y Gasset was a professional philosopher. In his well-known 1935 work [7], Ortega begins his discussion on technology by contrasting elementary needs—such as food, warmth, and movement—with modern phenomena like heating systems, agriculture, and car manufacturing. Ortega argues that these modern phenomena are not aimed at satisfying basic needs but, rather, at their opposite: the *cancellation* of these needs. His logic is as follows: if, whenever one feels cold, warmth is immediately provided, one would never feel the need for warmth, just as we do not typically feel the need to breathe—we simply breathe. Technology, by linking warmth to the sensation of cold, essentially eliminates warmth as a need, a deprivation, or a concern.

According to Ortega, technology is a response to nature or circumstances, resulting in the creation of an intermediary between nature and humanity—a "supernature" or a new nature built upon the primary one. By creating favorable conditions, humans generate this supernature, adapting raw nature to their own needs. Technology, thus, is the opposite of how living beings biologically adapt to their environment; instead, it represents the adaptation of the environment to the subject. In other words, technological action is the reverse of biological adaptation.

Ortega further claims that human desire for life and the need to exist in the world are inseparable from the passion for a good and comfortable life. He supports his hypothesis by stating, "Persons who become convinced that they cannot achieve a state of well-being, even a relative one, and that they must be content with mere existence in the world, often ends up committing suicide. Well-being, a comfortable life, and not just mere existence in the world, are the main human needs." He continues by explaining that the idea of a good life and well-being is a variable concept. Since human needs are a function of this concept, those needs are also subject to change. And because technology consists of acts born from and carried out within the system of these needs, it, too, becomes a constantly evolving reality.

Thus, according to Ortega, technology is a derivative of the human conception of well-being, which varies in every historical era. A more precise formulation is: "Technology is a function of the variable human program." The complete fulfillment of this program equates to achieving "well-being" and "happiness."

> "Technology is a function of the variable human program".

Essentially, Ortega provided a description of the driving forces behind the development of technology. We might attempt to illustrate this mechanism as shown in Fig. 6.5. This likely opens up opportunities for us to manage technological development by influencing human needs.

Regarding what technology offers humanity, Ortega writes: "Man is destined to create, to shape himself. To live means, first and foremost, to make maximum efforts to bring about what does not yet exist, to create man himself. Man must conserve his efforts to devote their excess to the incredible endeavor of realizing his existence in the world. The purpose and reason of technology lie in the use of the surplus strength man has thanks to technology itself. This is the mission of technology—to free man, granting him the opportunity to fully be himself."

Ortega foresaw the immense potential of technology: "A machine operating independently suggests intuitively that technology is a function separate from natural man, one that does not depend on him and does not consider human limitations at all. We know what a person can do with his immutable set of natural, biological actions. His horizon is limited. But the capabilities of machines, which people can invent, are, in principle, limitless." Ortega called engineers "the living embodiment of technology." He saw the risk that supernature might collapse and fail to support human existence.

Notably, on his path toward the humanization of the world, Ortega did not distinguish between material and immaterial (intellectual) technology. He believed that the world might become something like a materialized soul. It is difficult to say to what extent today's virtual reality spaces and artificial intelligence approximate Ortega's vision.

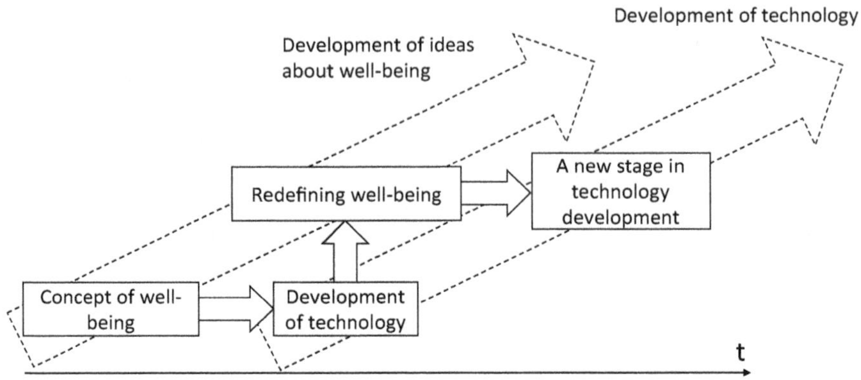

Fig. 6.5 A possible illustration of Ortega y Gasset's concept

Let us present one of Ortega's recommendations, which seems highly useful: "We must learn to step beyond our occupation, to look more closely at the appearance of life—it is always holistic. The highest ability in life is not conveyed by any profession or science, for this ability encompasses all professions and sciences, and much more."

Another of Ortega's thoughts was to mechanize as much as possible and to free human mental effort for managing our cultural treasures. Projecting this idea onto Electromechanics, one could imagine the future of Electromechanics where AI handles design, robots handle production, and electromechanical engineers focus on philosophical questions.

The Historical Context of Philosophers' Work is crucial for understanding the development of their ideas. Figure 6.6 illustrates the historical and technological context of the work of philosophers of technology, depicting key historical events and technological paradigms along a timeline. Dashed lines represent the philosophers' lifespans, while dots indicate the publication dates of their major works. This contextual visualization allows us to hypothesize which historical events and technological advancements may have influenced each philosopher's worldview. Lower part of Fig. 6.6 presents the sequence of technological paradigms [8]. The abbreviations used to describe the paradigms are as follows: MR—main resource, MI—main industries, KF—key factor, PI—paradigm achievements, and HA—humanitarian advantage.

Notably, the third paradigm, with Electromechanics as its key factor, was in a growth phase during the first two-thirds of the twentieth century. Currently, this paradigm is in its maturity phase.

Technopessimism of K. Jaspers German philosopher Karl Jaspers was one of the first and most influential technopessimists. In his book *The Spiritual Situation of the Age* (1932) [9], Jaspers writes that the gigantic technical apparatus, created to meet humanity's basic needs, reduces a person to the level of a simple function. People in this apparatus are shuffled like grains of sand, occupying one position or another.

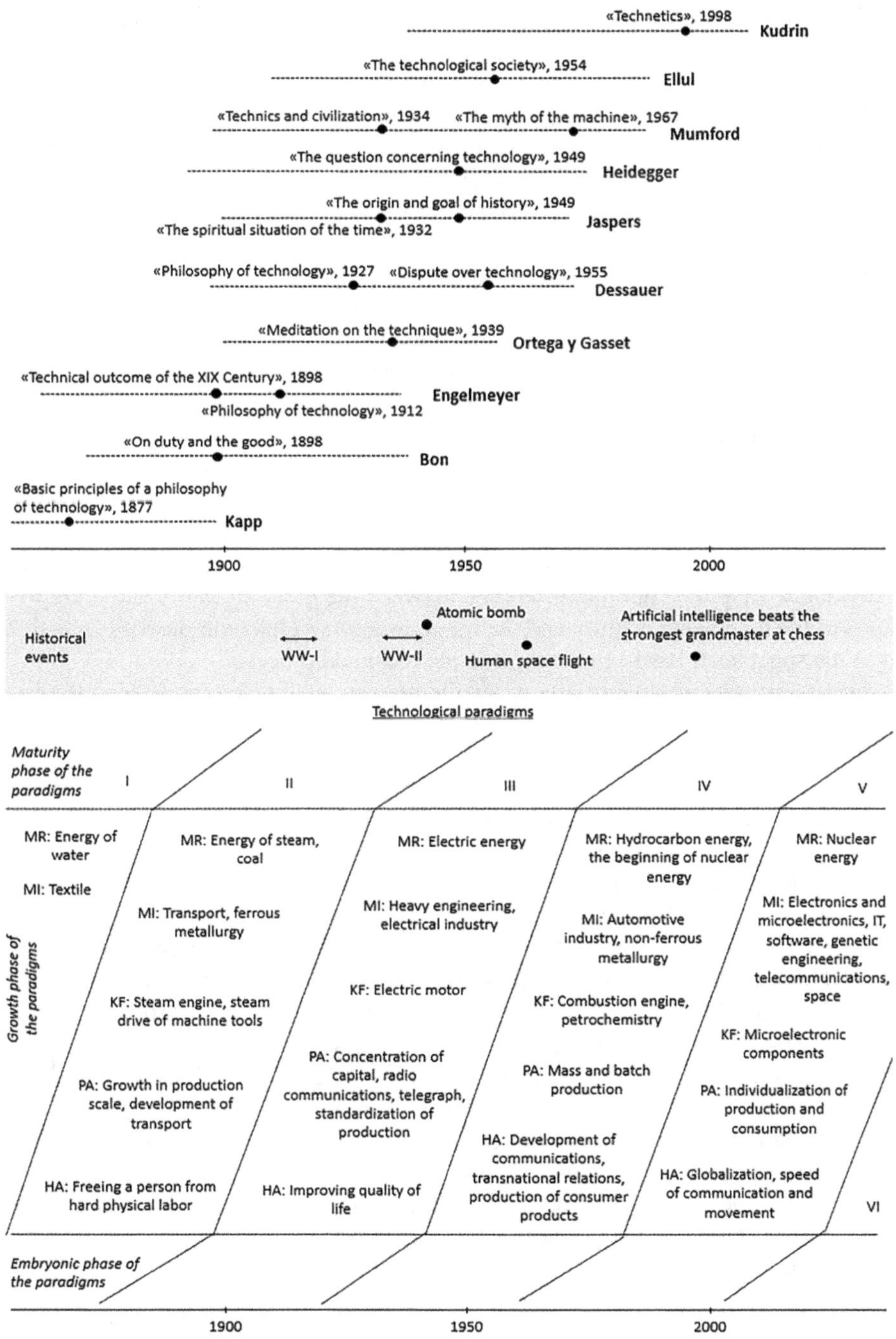

Fig. 6.6 The historical context of philosophers of technology

The human being is diminished to the level of a thing and loses their human essence. This apparatus is controlled by bureaucracy, which itself is an element of the system.

Jaspers refers to scientists and inventors, who push the boundaries of the technical world, as the "second great architects of the world," implying that the first architect is God. However, he sees technology as a threat overall. Jaspers warns that the world of technology is destructive to nature. Interestingly, even back then, Jaspers believed that science had already surpassed the point where humans could fully comprehend it.

After World War II, in his 1949 book *The Origin and Goal of History* [10], Jaspers takes an even more radical stance than in his 1932 work. He characterizes the postwar period as a time of catastrophic decline in spirit, humanity, love, and creative energy. The only thing that, in his view, retained greatness was science and technology. Technology had transformed the nature of work toward mass production, turned human existence into a technically perfect element of the machine, and the planet into a vast factory. Jaspers no longer describes inventors as the "second great architects," but rather as an anonymous mass of functional components in a process, where each link is connected to another, and participants no longer act as individuals. He points out the difference between the great inventors and entrepreneurs of the nineteenth century and the mass inventors of his time. Jaspers feels that even the spirit itself has been drawn into the technical process.

One aspect that could partially reconcile humans with technology, according to Jaspers, is the beauty of technical objects, which he understands as naturalness—in the sense that the bodies of animals and plants are natural to the biological world. This idea could be an interesting addition to Engelmeyer's formula of Technical Will.

Jaspers writes that one thing is certain: technology is in the process of transforming humanity itself. Humans can no longer extricate themselves from the technology they have created. Humans have become a kind of raw material destined for targeted processing. What was once the measure of all things—the human being—has become merely a means. Moreover, Jaspers predicts the emergence of a radically new environment, born from the spirit of technology, distinct from the "second nature" created by humans. Whether Jaspers foresaw the rise of artificial intelligence or something else in the 1940s, we can only speculate. Nonetheless, Jaspers raises the question: "Doesn't humanity have an origin that ultimately places all technology in dependence on it, instead of humanity becoming a slave to technology?" He still hopes that the source of control over technology will lie not within the technology itself, but in human consciousness and ethics.

Accepting the worldview described by Jaspers, one can pose the following questions: at what historical point did the Technosphere gain dominance over humans, and what role did Electromechanics play in this process?

F. Dessauer's Hypothesis on the Universal Force of the Technical The German scientist and philosopher of technology, Friedrich Dessauer, who was educated in electrical engineering and physics, presents a perspective that might resonate deeply with electrical engineers. Dessauer combined a fascination with technology and a strong faith in God. As a technooptimist, he believed that everything would ultimately be "good" because he trusted in the existence of an Almighty God who, by nature, is benevolent and would not abandon humanity [11]. In his 1927 work,

Philosophy of Technology, he argued that modern technology should not be seen merely as a means to ease human hardship; rather, the creation of technology is an act of participation in creation itself—one of the greatest experiences for mortals [2].

In [12], Dessauer offers three definitions of the essence of humanity:

- *Homo investigator*, the investigator, who studies causes and effects, weighing benefits and harms;
- *Homo inventor*, the inventive and organizing being, who not only has an innate need for knowledge (truth) but also a need for combinatorial creativity and organization, driven by environmental factors, nature, circumstances, fears, and desires;
- *Homo faber*, the fabricator, who can, primarily with hands (and tools), transform and transcend ideas from the intramental space—internal representations—into the external world, shaping the environment of perception.

Dessauer posits that within *techné* (from Greek τέχνη, meaning "skill" or "craft"), the acquisition of knowledge inherently involves mastering truth and, therefore, mastering skill. Like many philosophers before him, Dessauer speaks of an unseen world of ideas in relation to invention. According to Dessauer, an inventor may address their realized object by saying, "You could not have appeared earlier... until you became precisely this in my vision, because you could only be as you are now! And now you exist in the visible world. But I found you in another world, and you refused to enter the visible realm until I correctly discerned your true form in that other world."

Regarding motivation, Dessauer states that while economic factors drive technological creativity, an even more profound factor is humanity's fundamental desire to "shape the new." Dessauer provides a clear definition of technology: "Technology is the real existence of ideas through teleological (purpose-driven) formation and processing of nature's given inventory." For him, the essence of technology lay not in mass production and its products but in the act of technological creation itself.

Dessauer was likely one of the first to grasp the power of the realm of things, or the technosphere: "Technical objects and technologies originate from ideas, emerge historically, and, once they appear, persist with a force unique to each. Ultimately, the force of a technical object can combine with the forces of other technical objects. Thus, there exists a *common force of technology*, regardless of the differences in objects and their capabilities. This shared force is profoundly impactful; it alters the face of Earth".

> "There exists a common force of technology".

The Essence of Technology and the Concealment of Truth According to M. Heidegger In his 1949 work *The Question Concerning Technology* [13],

German philosopher Martin Heidegger stated that "the essence of technology is by no means something technological." He begins his argument by examining a well-known example from antiquity involving a silver chalice, exploring the "causality" of the chalice's creation. The first cause is *causa materialis*—the material from which the chalice is made; the second is *causa formalis*—the design or idea of the chalice's form and appearance; the third is *causa finalis*—the purpose of the chalice (a ritual). The silversmith, understanding these three types of causality, considers them carefully and gathers them together. Thus, the three types of causality manifest and interconnect, leading to the creation of the finished chalice. The craftsman himself is the fourth cause—*causa efficiens*.

As a thought experiment, one could imagine how this ancient philosophical example might be interpreted in Electromechanics, where the object is an electric machine: (1) *causa materialis*—metals (steel, copper, etc.); (2) *causa formalis*—the design (calculations, models, drawings); (3) *causa finalis*—the application (specification); (4) *causa efficiens*—electrical engineers and their tools.

But returning to Heidegger's argument: the four types of causality allow a thing to come into being. Through them, the thing becomes present. They bring it forth from non-existence leading to the fullness of its appearance. To describe the totality of causes behind the emergence of a thing, Heidegger introduces the term "*inducing to go forward*" (German *Ver-an-lassen*). The connection between the four types of cause (occasion) lies in their role in allowing something not yet revealed to come into presence. "Every occasion for the transition and emergence of anything from non-existence into presence is *ποίησις* (Greek *poiesis*), "*bringing-forth*" (German *Her-vor-bringen*)." As we can see, Heidegger has to employ new words because the language of his time was insufficient to convey these new meanings.

Thus, for Heidegger, technology is not merely a tool. "Technology is a way of revealing. It is the realm of bringing-forth out of concealment, the realization of truth. The essence of technology lies in the realm where revealing and unconcealment occur, where *ἀλήθεια* (Greek for "truth") comes to pass."

Heidegger distinguishes modern technology from the "poetic" technology of ancient and medieval craftsmen. He writes: "It too is a revealing of what is concealed. However, the revealing that modern technology is involved in is not bringing-forth in the sense of *poiesis*. The revealing that rules in modern technology is production, which places an unprecedented demand on nature to be a supplier of energy that can be extracted and stored as such. At the same time, such production always carries within itself the aim of reproduction, of increasing productivity in the sense of extracting the maximum benefit at the minimum cost."

Heidegger provides an example that might be appealing to electrical engineers: "The hydroelectric plant is set into the current of the Rhine. It sets the Rhine to supplying its hydraulic pressure, which then sets the turbines turning. This turning sets those machines in motion creating the electric current for which the long-distance power station and its network of cables are set up to dispatch electricity. In the context of the interlocking processes pertaining to the orderly disposition of electrical energy, even the Rhine itself appears as something at our command. The hydroelectric plant is not built into the Rhine river as was the old wooden bridge that joined bank with bank for hundreds of years. Rather the river is dammed up into the power plant. What the river is now, namely, a water power supplier, derives from out of the essence of the power station" [13]. He further adds: "Unlocking, transforming, storing, distributing, and switching... are ways of *revealing*. But the *revealing* never simply comes to an end. Neither does it run off into the indeterminate. The revealing reveals to itself its own manifoldly interlocking paths, through regulating their course. This regulating itself is, for its part, everywhere secured. Regulating and securing even become the chief characteristics of the *challenging revealing*".

Heidegger continues: "What kind of unconcealment is it, then, that is peculiar to that which comes to stand forth through this setting-upon that challenges? Everywhere everything is ordered to stand by, to be immediately at hand, indeed to stand there just so that it may be on call for a further ordering. Whatever is ordered about in this way has its own standing. We call it the "*standing-reserve*" (German *Bestand*). The word expresses here something more, and something more essential, than mere "stock." The name "standing reserve" assumes the rank of an inclusive rubric. It designates nothing less than the way in which everything presences that is wrought upon by the challenging revealing. Whatever stands by in the sense of standing-reserve no longer stands over against us as object."

Let us try to translate these thoughts into the context of Electromechanics. For example, an electric motor that has passed factory tests and is delivered to a pump station, where it is set to operate, is simply an object—we can perceive the motor this way. However, in doing so, its essence and its mode of existence will remain hidden from us. Once brought out of its concealment, it stands at the pump station solely as "standing-reserve" and only to the extent that it is placed there to ensure the possibility of supplying water. For this purpose, in its entire design and in every component, it must allow for such use, i.e., it must be ready for connection to the electrical grid and the pump. From the standpoint of standing-reserve, the motor is completely dependent; it exists only because it is set up to provide the torque that drives the pump's shaft at a certain rotational speed.

Heidegger poses the question: Who performs this productive supplying through which so-called reality emerges from concealment into standing-reserve? It seems obvious that it is humans. But to what extent are they capable of such revealing by their own efforts? Is there another, more powerful force? Heidegger introduces another concept—"*enframing*" (German *Ge-stell*)—which is the challenging summons that focus humans on bringing forth everything that emerge from concealment as standing-reserve, on extracting natural energies. Interestingly, the revealing of what is concealed does not happen solely within the human being or primarily through them. Engaged in this productive supplying, humans stand within the essential realm of enframing. *Ge-stell* is not an anthropological essence. Heidegger provides another clarification of enframing: it is the gathering force of the framework that places, i.e., compels humans to bring forth the real from its concealment by the process of setting it forth as standing-reserve. *Ge-stell* is the name for the essence of modern technology.

> "*Ge-stell* is the name for the essence of modern technology".

The mission of humans is the revealing of what is concealed, the uncovering of mystery, and the search for truth. Heidegger warns that when this mission is governed by enframing, it becomes an extreme danger. One of the risks is that humans may begin to perceive themselves as standing-reserve. The dominance of enframing also threatens with the danger that humans will no longer be able to return to a more fundamental revealing of what is concealed or to hear the voice of earlier truths. The more humans are consumed by what is revealed to their minds, the less time they have to reflect on the nature and origin of their insight: the more fully being is revealed, the more deeply it conceals itself, redirecting attention away from itself toward things.

At the same time, Heidegger also saw positive aspects of our existence within enframing—being caught up in the setting forth of the real as standing-reserve is, in the end, also a mission, sending humans on one of the paths to the revealing of what is concealed. As this mission, the essence of technology allows humans to enter into something they could neither invent nor arrange on their own. Moreover, enframing is fulfilled through the realization that allows humans to remain as beings required for safeguarding the essence of truth. Thus, humans are one of the guardians of the "essence of truth".

Heidegger pointed us in certain direction of thinking using only a few metaphors. Here we can try to express them in our own words, which might prove useful in our own reflections:

- Truth is the openness of the concealed.
- The *genuine truth* escapes among *the correct*; we must, by breaking through what is merely correct, seek the genuine. The merely correct is not yet truth. The genuine truth hides behind small "truths".
- Once, the bringing-forth of truth into beauty was called *techné*. The term *techné* also applied to *poiesis* in the fine arts. Every art, every bringing forth of the essential into the openness of beauty, is permeated by poetry.

The Sacralization of Technology and the Subordination to Efficiency According to J. Ellul In 1954, French philosopher Jacques Ellul argued in *The Technological Society* [14] that technology had transformed fundamentally: it was no longer merely a vast collection of means, each designated for a specific end, but had instead become "an environment in its own right." Furthermore, Ellul claimed that technology had subtly become "sacred" to humanity. Later, he explained this process in more detail: since humans inevitably sanctify their environment, they no longer sacralize nature but rather what has desanctified, desecrated, and even polluted it—namely, technology. According to Ellul, the consequences of this shift are not only environmental but also psychological, resulting in a form of dependency in humans' relationship with technology. By constantly improving, technology replaces values of the past, such as Christian values (like love for one's neighbor), humanistic values (morality), or republican values (freedom, equality, fraternity), with values such as work, utility, efficiency, economic growth, and progress. Ellul argued that it is not technology itself that enslaves us, but rather the sacrality we have projected onto it.

Ellul described technology not just as a collection of machines and innovations but as a set of methods being realized with absolute efficiency (for a given level of development) in all areas of human activity. He believed that in an all-encompassing pursuit of efficiency, technology had seized control over human activity far beyond the realms of production. Ellul outlined seven characteristics of modern technology that make efficiency an imperative: rationality, artificiality, technical automatism, self-augmentation (self-expansion or self-exaltation), monism (indivisibility), universality, and autonomy (self-direction). For instance the rationality of technology ensures logical and mechanical organization through labor division, production standards, and so forth. This creates an artificial system that "eliminates or subordinates the natural world." Technology's problems are always met not with "less technology" but with "more technology." Everything becomes subject to analysis through "input-output" and "cost-benefit" schemes in the search for the most efficient path to results—an approach and criteria familiar to electrical engineers.

Technology's problems are always met not with "less technology" but with "more technology".

Modern technology and technological systems have become universal phenomena of civilization, shaping a new social order where efficiency is no longer a choice but a necessity imposed on all human activity. According to Ellul, it is futile to think that we can separate technology from its uses, as the implementation of efficient methods inevitably leads to specific social and psychological outcomes, regardless of human intentions. This reality leaves no room for moral considerations: even a moral appeal to technicians could not alter the situation. At best, it would lead them to abandon their role as good technicians. Ultimately, technology recognizes only one principle—efficient order. Moreover, technology resists attempts to incorporate nontechnical relationships or ways of thinking into its framework or to subject it to their control.

Ellul proposed several actions and solutions to escape the dominance of technology:

- Examine and understand technology from within to realize that coexistence with it may be impossible.
- Perceive technology as merely tools—one of many resources—and accept it as it is, simply another thing among many, instead of believing in technology for its own sake or for the sake of society.
- Adopt an ethic of "non-use of power," where people agree not to do everything, they become capable of doing.

The last suggestion is perhaps the hardest to implement, given economic and military competition between countries.

L. Mumford's Megamachines American philosopher of technology Lewis Mumford viewed hierarchical civilizations as sociotechnical systems—Megamachines—in which people are reduced to standardized and interchangeable components. Mumford described Megamachine as the convergence of science, economics, technology, and political power into a unified framework, or, more simply put, a Megamachine is a social, economic, or political system where the gears are not made of iron and steel, but of people. Both in Mumford's time and today, Megamachines are created and sustained by large corporations and government entities.

Megamachine is a social, economic, or political system where the gears are not made of iron and steel, but of people.

In his work *The Myth of the Machine* [15], Mumford suggests that individuals can rebel against Megamachines. He argues that the most effective way to do this would be through the formation and support of smaller organizations—such as universities, societies, or labor unions—that could "revolt" against the vastness of the Megamachine through actions, ideas, and principles.

Like Ellul, Mumford believed that technology was no longer a simple tool but had become an active agent in reality, reshaping humanity in its own image. He distinguished between two types of technology: one that aligns with human nature and one that does not. Mumford advocated for the promotion of technologies that enhance the personal aspects of human existence, rather than those that restrict life solely for the sake of power.

Technetics by B. Kudrin The Russian electrical engineer and philosopher of technology, Boris Kudrin, proposed a new paradigm for studying technical reality. In this paradigm, technical products, technologies, equipment, materials, and waste are considered collectively as communities, or "technocenoses". Examples of large technocenoses include factories, cities, and industries. Examples of relatively smaller autonomous technocenoses (being autonomous over specific periods) include space stations, and submarines.

Typically, a technocenosis is defined as a set of technical objects (products) with relationships and connections among them, forming a hierarchically organized whole. A technocenosis, as a community of all products, is characterized by weak connections and minimal interactions between the elements that form the system. The number of elements can reach millions. For example, a large metallurgical plant might have around 10^5 electric machines, 10^6 low-voltage devices, and approximately 10^{10} different electrical products, assemblies, units, and components. The total number of products and parts that make up a modern large industrial enterprise can be estimated at 10^{11} [16].

The study of technocenoses is possible by distinguishing families of products—such as, e.g., electric machines, transport, boiler systems, etc.,—that can be categorized into a finite number of *types*. The method of analysis in this case can be similar to biological research methods.

In the 1970s, while examining the statistics on motors being repaired at several large enterprises, Kudrin discovered that the frequency of occurrences (ω) of machine types coming to repair follows the relationship: $\omega = 0.59a_i^{-1.84}$, where a_i is the machine type or "size" (defined by series name, such as "4A", and power rating, like 15 kW) [17]. In Fig. 6.7, depicting recurrence, the curve's right side shows many "rare" types with a small number of representatives (individual motors) within each type (i.e., many rare types), while the curve's left side represents a few types with a large number of representatives within each.

Analyzing these statistics over many years, Kudrin came to fascinating conclusions: although generations of engineers making decisions on motor purchases and generations of motor series changed at the examined enterprises, the distribution of motor types still followed the same curve (Fig. 6.7)—this dependence proved stable!

Subsequently, Kudrin concluded that, like natural systems—biological and informational—technocenoses have a common systemic structure that can be described by the distribution of types based on recurrence ("H-distribution"). As demonstrated above, the diversity of types in technocenoses is statistically characterized by a stable hyperbolic relationship of the form $= ka_i^{-b}$, where k and b are parameters, and a_i is the electric machine type.

> Technocenoses have a common systemic structure that can be described by the distribution of types based on recurrence ("H-distribution").

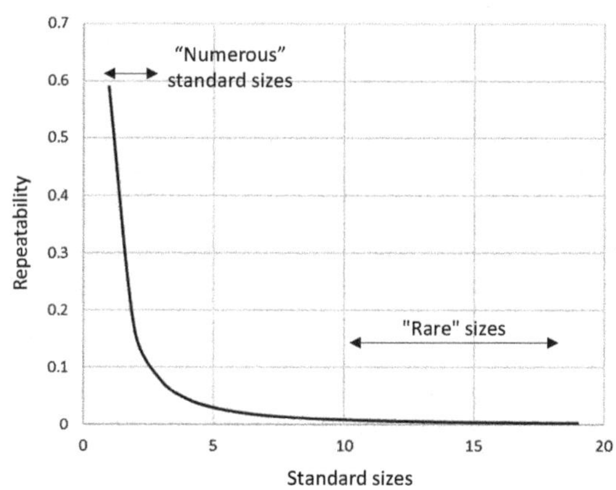

Fig. 6.7 Recurrence of electric motors undergoing repair

A technocenosis remains in a stable state when, within its type representation, 5–10% of individual products belong to unique, rare "castes," constituting 40–60% of the type diversity, while 40–60% of all individual products fall into mass "castes," covering 5–10% of the total number of types. Such proportions are also typical for biological systems.

Returning to the electric machines within technocenoses (at enterprises), regardless of the machine's parameters, it inevitably passes through a "rare caste". Its further position on the curve, including within the rare caste, is determined not by physical dimensions or power but by other factors. Analyzing the contents of rare castes at any enterprise consistently reveals unique small motors (low power) alongside large ones.

Kudrin defines technoevolution as the directed, gradual, and regular transformation of product types across generations. The fundamental difference between technoevolution and biological evolution is "informational selection" instead of the "natural selection" seen in living systems.[1] Informational selection involves the documented recording of information, with a spatial and temporal separation of the document itself, the reproduction method of the document, and the reproduction method of the product it describes.

The current generation of technology (and even more so for future generations) exists only as a small part of a broader technocenosis, most of which was created before any living person was born. The nested hierarchy of multiple technocenoses forms the planet's technosphere. Every new technical unit is introduced into a pre-existing technical environment—a specific technocenosis—thereby expanding it quantitatively and enriching its diversity in line with the patterns of hyperbolic H-distribution.

Kudrin named his theory "Technetics." Technetics is a fundamental science on the technical form of matter's existence, the general laws of technology, technical reality, and technoevolution, as well as on the construction and management of technocenoses. Technetics allows for establishing the laws of technoevolution, calculating the parameters of technical populations, and predicting the course of technoevolution.

Kudrin's worldview can be summarized as follows:

- The Technosphere has overlaid and transformed the biosphere. Technical reality has become universal, representing a natural process where "technical begets technical" beyond human will. The irreversible directed development of products and their communities (technocenoses) is governed by its own "technical" laws. Humanity serves only as a necessary subjective condition for the establishment of technical reality.
- The construction, maintenance, and development of technocenoses are grounded in their own technocenological postulates. The technical world is described by H-distribution models, where the concept of an "average" is inapplicable (lacking a mathematical expectation) and where infinite variance means any question may yield an arbitrarily large error (indicating an undefined solution).

[1] This was presented in Chap. 1.

- Global technical evolution dictates the emergence of new technical entities in such a way that each technetic unit modifies its surroundings to suit its own needs (as a type).

Kudrin raises the possibility of a shift from anthropocentrism to technocentrism as a worldview. He could be classified as a techno-pessimist. Since Kudrin is an electrical engineer with extensive experience who came to the philosophy of technology from his field and with a wealth of insight, his works can serve as valuable additions to the existing philosophical discussions on "technology as a whole."

It is important to note that the philosophers discussed above conducted their work prior to the emergence of the sixth technological paradigm. Capturing a snapshot of our current understanding (or its alternatives, or a synthesis of these) is essential as we approach the era of AI. The rapid advancements in AI, particularly from the early 2020s, appear poised to transform many aspects of our world.

Modern Philosophers of Technology Among contemporary philosophers of technology, Carl Mitcham and Andrew Feenberg stand out for their extensive publications.

Among Mitcham's numerous works, the book on ethics [18] (co-authored with A. Briggle) can be noted. In a noteworthy interview [19], Mitcham presents his critical view of the engineering community. According to him, the engineering community lacks the coherence found among scientists, whose work typically involves a shared pursuit of universal, transnational truths and the production of generalized knowledge. Engineers, by contrast, produce commercial products, competing not only across borders but also within their own nations and corporations. This fragmentation is further compounded by the frequent competition among engineers from different specializations, such as mechanical versus electrical engineering.

Mitcham even likens engineers to "guns for hire." He illustrates this with a story from 1945 told to him by one colleague: when Soviet forces occupied a German city with a large factory producing advanced electrical equipment, a delegation of engineers and workers from the factory approached them in just a few days after, asking, "when will you give us work?" These factory employees, Mitcham argues, were unconcerned with who employed them; they simply wanted to continue working. A similar ethical standpoint can be seen in the career of Wernher von Braun, who developed the V-2 missiles in Germany and later led the US lunar program, for which he received the National Medal of Science.

According to Mitcham, engineers often see themselves as "problem-solvers" without much concern for the origin of those problems. Engineers generally do not define their tasks—they are assigned to them. Moreover, Mitcham perceives that engineers do not merely represent technology but rather technology in service of capital, often forced to align with capital when conflicts of interest arise. He questions whether engineers can ever truly emancipate technology from the influence of capital, given the disproportionate power wielded by corporations.

Mitcham argues that the primary value of engineering lies in safety, and the ethical framework that might resonate most with engineers involves duty and cooperation. Interestingly, Mitcham identifies the rapid pace of change as one of the greatest challenges of the modern world—a pace he believes should ideally be slowed down.

Another modern philosopher Andrew Feenberg, author of the Critical Theory of Technology (CTT) [20, 21], provides a framework that aims to reconcile the positions of both technopessimists and technooptimists. Feenberg argues that philosophy of technology has evolved considerably since the time of Heidegger, and that contemporary philosophers must take an active stance in addressing the challenges posed by modern technology.

In CTT, Feenberg introduces two key concepts: the "Technical Code" and the "Democratization of Technology." The Technical Code (TC) is a set of rules that governs how technology is "realized" within a social context. The TC essentially translates social interests and ideologies into technical solutions. It acts as a filter, determining which of the technically feasible solutions will be implemented to achieve societal goals. Within the Code, goals are "encoded" and "ranked" based on criteria such as ethical acceptability, aesthetic appeal, social desirability, and so forth. Here, Feenberg draws upon the spirit of Engelmeyer's concept of Technical Will, which integrates values like *Goodness*, *Truth*, and *Beauty*.

> The Technical Code is a set of rules that governs how technology is realized within a social context.

Feenberg argues that the current realization of technology is distorted by the unequal distribution of power in society. Technology today often reinforces the power of "the few over the many." This creates an imbalance between what is determined by technical efficiency and what is influenced by external, non-technical values. Feenberg believes that access to influence is concentrated in the hands of technical experts and the corporate and political elites they serve. He proposes that democratizing technology can mitigate these distortions. Democratization would lead to more responsive feedback loops between disrupted ways of life, environmental sustainability, and the adoption of more socially and ethically appropriate technical solutions.

CTT asserts that the roots of the degradation of labor, education, and the environment do not lie in technology itself but rather in the cultural values embodied in its design. The technical systems, technologies, and associated standards that have entered our lives—and continue to be present—were once formulated based on certain values. The modern technical code has grown out of these values. Thus, Feenberg sees the potential and necessity for influencing technology through laws and standards.

As an example of philosophical thinking, Feenberg presents ten paradoxes in [22]. Table 6.1 provides author's attempt to interpret these paradoxes in the context of Electromechanics as an exercise. Such exercises can help develop the ability to think in complex and paradoxical ways.

Table 6.1 Feenberg's Ten Paradoxes with interpretation

	The paradox of …	Definition	Interpretation in Electromechanics, examples
1	the parts and the whole	The apparent origin of complex wholes lies in their parts but in reality the parts find their origin in the whole to which they belong	DC motors are used as alternatives to induction motors because the factory has available a DC grid. But the DC grid exists because there are many DC motors at the factory
2	the obvious	What is most obvious is most hidden	Cooling systems use the natural environment—air. But what to do in space or a vacuum train tunnel (Hyperloop), where there is no air?
3	origin	Behind everything rational there lies a forgotten history	Some machine types continue to be used even when surpassed performance-wise by newer types. The historical setup makes changing them costly and complex
4	the frame	Efficiency does not explain success, success explains efficiency	Early market entry, even with a suboptimal electric machine, can generate some profit which can be used for improvements and secure the market position
5	action	In acting we become the object of action	Engaging in electrical engineering, a person becomes an electrical engineer We live by the laws of the Technosphere, which we ourselves have created
6	the means	The means are the end	Someone becomes an electrical engineer to create electric machines, but actually, being an electrical engineer might be the ultimate goal
7	complexity	Simplification complicates	While electrification simplifies human life, it also complicates all systems and leads to unclear ecological consequences
8	value and fact	Values are the facts of the future	If environment protection is a value, the emergence of green technologies like wind turbines and generators for them is merely a matter of time
9	democracy	The public is constituted by the technologies that bind it together but in turn it transforms the technologies that constitute it	We live by the Technosphere's laws, but we also reshape it and can consciously influence it further
10	conquest	The victor belongs to the spoils	Humanity suffers the consequences of conquering nature. For example, waste from permanent magnet production often causes significant environmental damage

Concluding Reflections

By synthesizing the works of philosophers, we can derive the following elements of a worldview that have a direct relation to Electromechanics:

- The laws of mathematics, physics, biology, and chemistry act in nature. In the supernature (as defined by Ortega y Gasset), in addition to these laws, there exist some other laws that are either yet to be discovered or have been discovered but are not widely recognized.
- The Technosphere has an as-yet-unstudied field that affects the objects within the Technosphere and humanity. This field can be related to Heidegger's *Ge-stell.*
- Any electric machine is always part of some technocenosis, entering it, exiting it, and existing within it according to the laws of the supernature.
- The essence of our time is the interaction of technocenoses (as defined by Kudrin) with Megamachines (as defined by Mumford). Like everything else, electrical engineers and electric machines are objects of influence from the Technosphere (the embedded hierarchy of technocenoses), states, and corporations.

Also, based on the works of philosophers discussed above, we can attempt to answer the questions posed at the beginning of the chapter.

On Directions The main directions for the evolution of electric machines in the modern world are outlined, for example, in [23, 24]: these include increasing maximum achievable power and torque, improving controllability, raising maximum achievable rotational speed, enhancing energy conversion efficiency, increasing specific power and torque (increasing compactness, reducing mass), and enhancing the degree of mechanical integration with surrounding structures (integration into applications—supersystems). However, it should be noted that this pertains only to traditional electric machines created by humanity for its own needs. If we take size as a parameter characterizing objects in the physical world, we will see that traditional electric machines occupy a very small section on the logarithmic scale, which can represent the sizes of all known physical objects. The extreme sizes known to modern physics range from 10^{-18} m (the size of the Higgs boson) to 10^{26} m (the size of the observable universe—the largest known physical object). A fragment of the scale (from 10^{-8} to 10^{8} m), which has a more direct relation to electric machines, is represented in Fig. 1.4 (in the first chapter).

Clearly, one can point at two directions of evolution—to the right and to the left on the size scale. The movement to the right is associated with the exploration of outer space. As Dessauer wrote in 1955, "The cosmos contains much more than what man can dream of". The equally fascinating is the direction to the left on the scale. As the famous physicist R. Feynman stated [25], "There's Plenty of Room at the Bottom", referring to the field of technology that is now called nanotechnology.

Naturally, there are directions of evolution that are not related to electric machine size; several of them can be mentioned:

- Electric machines for operation in radically new conditions, such as in the atmospheres of other planets.
- The application of AI in the creation of electric machines and throughout their entire life cycle seems inevitable. Here, it should be understood that technocenoses, "enhanced" by robots and AI, will become the next superstructure over nature, in relation to which humanity will likely be even weaker than ancient humans were in relation to the "first nature".
- If a decision is made to limit the resource base of techno-ecosystems (which can be termed a strategy of "shrinking" technocenoses), the direction of recycling of electric machines that have reached their end of life will become especially important, along with the use of the materials obtained in new products, as well as the development of systems for automated recycling.

On Reference Points The foundation and tools for understanding the pressing issues and future of Electromechanics can primarily be sought in the philosophy of technology, as well as in classical philosophy. Important related disciplines are shown in Fig. 6.8. Within these disciplines, as well as in the field of the philosophy of technology, ethical reference points and sources of knowledge should be sought. Each electrical engineer can propose his or her own reference points, including even transcendental ones—which would be particularly interesting.

On the Tools In the context of the dominance of the Technosphere and Megamachines, the set of tools available to influence evolution is limited. Nevertheless, certain opportunities exist. At the core of everything is *thinking*. Thinking must be powerful, free, transcending all templates, and paradoxical. The entry points into the process of influencing evolution are ideas. One should learn to think and produce strong ideas. For example, one of the methodologies for normative (algorithmic) thinking—TRIZ—can be used at the initial stage as a precursor to free thinking.

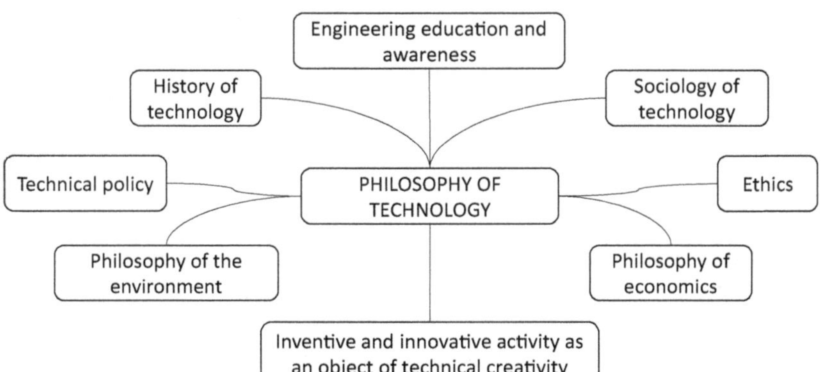

Fig. 6.8 Scientific disciplines related to the philosophy of technology

It is possible not only to discover the laws of the supernature (as defined by Ortega y Gasset) but also to write and implement one's own laws for the supernature. Influence on evolution will be achievable through the implementation of these new laws. This will require the creation of small expert groups built around new ideas and values. As an example of a new Technical Code element (likely feasible in the second half of the twenty-first century), the following rule can be proposed: "A new electric machine on planet Earth can only be produced from the materials of old electric machines."

In the future, it is conceivable to view AI as a tool for managing evolution. For instance, it will become possible to model directions of evolution: specifically, one could conduct experiments on the development of technocenoses consisting of digital twins of real objects.

On the Neutrality of Electric Machines According to Martin Heidegger, electric machines, like all other energy converters, inherently carry a negative connotation. However, from the author's perspective, electric machines are neutral entities. They can be viewed as "functions" or "potentials," embodying possibilities that are neither inherently positive nor negative. For example, consider a single electric motor: it could be installed on a combat drone or on a drone delivering food or medicine to remote areas. The electric motor and the drone construction could be identical in both cases, meaning that even the higher-level system, which is the drone in relation to the electric motor, does not inherently define the ethical value of the electric motor. This value is determined at an even higher level, suggesting that perhaps the ethical value of an electric machine is defined by "intention of the use". The question of neutrality is ultimately a significant and complex ethical issue.

On the Essence of Electric Machines One might speculate that the essential differences between electric machines and other technological objects could be tied to the electromagnetic nature of the universe itself. This thought might lead to the hypothesis that the essence of electric machines among technological objects is a quality of *naturalness* or *primordiality*. Electric machines could be seen as *archetypes of the universe*. Following Heidegger's insights, one might extend this idea to suggest that by continually producing electric machines, the universe is, in essence, reproducing itself. This metaphor suggests that electric machines serve as conduits through which the inherent electromagnetic structure of the universe manifests itself in technology.

On the Uniqueness of Electromechanics The uniqueness of Electromechanics lies not only in the essential differences of electric machines but also in two specific aspects: firstly, electric machines can be scaled across a uniquely broad range—from molecules to celestial bodies. Secondly, unlike other types of energy converters, such as internal combustion engines, steam engines, or jet engines, electric machines are reversible. This reversibility implies a potential freedom in energy flow, especially as future technological advancements may further minimize energy losses. This unique quality raises an intriguing question: does Electromechanics deserve its own distinct branch within the philosophy of technology? The author believes it does, though there is also merit in developing subfields of the philosophy of technology dedicated to other scientific and engineering disciplines.

On Motivation The uniqueness of Electromechanics and the distinct nature of the associated challenges may serve as a powerful motivation for engineers and scientists in the field. Moreover, electric machines are linked to a substantial portion of the technosphere through their applications and interconnections with various systems and subsystems. This makes any task within Electromechanics broader and more engaging, offering an expansive view of the field and its potential for meaningful contributions to both technology and society.

What, then, is the role or mission of the electrical engineer? Four tasks can be proposed:

1. Studying the laws of the supernature and influencing it.
   - It is necessary to study the laws by which the technosphere exists, and the forces acting within it—on electric machines and their supersystems and subsystems. This may be one of the most important scientific tasks.
   - Not only can we discover existing laws of the supernature, but we can also propose and establish our own rules for the supernature, which may balance the existing laws.
   - We should attempt to participate in managing the evolution and in developing the Technical Code (as defined by Feenberg).
   - It is important to keep in mind that not everything modern science and technology makes possible should necessarily be realized. We could attempt to incorporate this into the Technical Code.
2. Breaking free from the power of Megamachines.
   - Electrical engineers operate within multiple spheres of influence: the technosphere, states, and corporations. For a corporate electrical engineer, it is almost impossible to escape the Megamachine (at least during working hours and within job responsibilities). However, a scientist is even obligated to attempt this—there is more freedom in the academic environment.
   - The role of independent expert community is important, as one of the balancing tools against Megamachines. One tool could be the formation of small groups of philosophers of technology working in Electromechanics and related fields. History has shown the power of small groups.
3. Reflection on reality and the search for truth.
   - Electrical engineers must define for themselves what is good and what is true—these will be their reference points, their values. They must reflect on their activities and take a thoughtful position in their profession based on these values.

- The search for truth is one of the main tasks. Following Heidegger's abstractions, one must "swim against the current" and try to move "through" obstacles. Based on the essence of their science, electrical engineers are in a unique position to solve this task. On this path, a researcher may discover not just something new but something fundamentally different.

4. Self-realization.
   - Perhaps the best description of this mission comes from Ortega y Gasset: "Existence means, first and foremost, a state in which we are destined to realize the project that we are within the given circumstances. We are not allowed to choose in advance the world or the circumstances in which we live… I am not a thing but a drama, a struggle to become what I have to be. This aspiration, this program, which constitutes our 'I', bursts into the world, imprinting itself upon it, leaving its unique mark, while the world, in turn, responds to my impact by either accepting or rejecting it."

In conclusion, several theses are formulated for discussion, and are proposed for inclusion in a specialized section of the philosophy of technology—The Philosophy of Electromechanics:

- The technosphere has a certain, as yet unstudied, force field that can be related to Heidegger's *Ge-stell.*
- By creating an infinite number of electric machines through *Ge-stell*, the universe merely reproduces itself.
- The essence of electric machines as a family of technical objects is naturalness, primordiality. Electric machine is one of the archetypes of the universe.
- The interaction of technocenoses and Megamachines is a defining dynamic in the professional environment of electrical engineers.
- We can not only discover the laws of supernature but also propose and establish our own laws for it.
- Electromechanics must be future-oriented. The electrical engineer of the future is a "philosopher- engineer."

References

1. V.G. Gorokhov, V.M. Rozin, *Introduction into Philosophy of Technology: Teaching Aid* (INFRA-M, Moscow, 1998) 224 p
2. C. Mitcham, *Thinking Through Technology, the Path Between Engineering and Philosophy* (The University of Chicago Press, Chicago, 1994) 396 p
3. O.M. Sedykh, M.A. Khamenkov, Introduction into philosophy of technology: Russian context. Philos. Sci. Technol. **21**(1), 132–151 (2016) UDK 167
4. I.P. Kopylov, *Electromagnetic Universe*, 2nd edn. (MEI Publishing, Moscow, 1999)
5. P.K. Engelmeyer, *Technical Summary of the 19th Century* (Kaznacheev Publishing House, Moscow, 1898)
6. N.V. Nikiforova, The aesthetic dimension of technology: the dynamo as a technological sublime at the turn of the 19th and 20th centuries. Philos. Sci. Technol. **25**(2), 37–50 (2020)
7. J. Ortega y Gasset, Meditacion de la tecnica. La Naci6n de Buenos Aires, 1935; Espasa-Calne, S. A. Madrid, 1965

8. E.A. Sobolev, Innovation mechanisms of transition to the sixth wave of innovation in Russia. Rev. Econ. Law. Sociol. **4** (2015)
9. K. Jaspers, Die geistige Situation der Zeit. Berlin/Leipzig 1931; 5. Auflage 1932
10. K. Jaspers, Vom Ursprung und Ziel der Geschichte. München/Zürich (1949)
11. T.V. Filatov, Friedrich Dessauer and neo-Thomist interpretation of the philosophy of technology, Vestnik of Samara state technology university. Serija: Filosofija, No: 1(1) (2019), pp. 108–119, UDK: 167.7
12. F. Dessauer, *Streit um die Technik* (Herder, Frbg, 1959)., 206 p
13. M. Heidegger, *The Question Concerning Technology and Other Essays* (Garland Publishing, New York & London, 1977)
14. J. Ellul, *La Technique: L'enjeu du siècle* (Librairie Armand Colin, Paris, 1954)
15. L. Mumford, *The Myth of the Machine. Technics and Human Development*, vol. 1 (Harcourt Brace Jovanovich, San Diego, 1967) 1970 (Vol.2), 352 pages (Vol. 1)/495 pages (Vol. 2).
16. B.I. Kudrin, *Technetics; New Paradigm of Philosophy of Technology* (Tomsk University Publishing, Tomsk, 1998), 40 p
17. B.I. Kudrin, *Destribution of Electric Machines on Repeatability, Elektrifikatsija metalurgicheskih predprijatii Sibiri*, issue 2 (Tomsk University Publishing, Tomsk, 1974), 31–40 p
18. A. Briggle, C. Mitcham, *Ethics and Science: An Introduction* (Cambridge University Press, Cambridge, 2012)
19. C. Mitcham, A.A. Kazakova, Let us now think engineering: an interview with Carl Mitcham. Philos. Sci. Technol. **25**(2), 26–36 (2020)
20. T. Veak, *Democritizing technology: Andrew Feenberg's critical theory of technology* (State University of New York Press, Albany, 2006) 252 p
21. A. Feenberg, Critical theory of technology; an overview. Tailor. Biotechnol. **1**, 47–64 (2005)
22. A. Feenberg, Ten paradoxes of technology. Techné: Res. Philos. Technol. **14**(1), 3–15 (2010)
23. A.V. Matveev, Driving forces and lines of the evolution of rotating electrical machines. Part 1, Elektrichestvo, no. 1, 2021, pp. 44–54
24. A.V. Matveev, Driving forces and lines of the evolution of rotating electrical machines. Part 2, Elektrichestvo, no. 2, 2021, pp. 4–15
25. R. Feynman, There's plenty of room at the bottom: an invitation to enter a new field of physics, https://en.wikipedia.org/wiki/There%27s_Plenty_of_Room_at_the_Bottom

Conclusion

The electrical engineering industry and the corresponding scientific community have been shaped by history into a specific structure with distinct laws of existence and development, along with its own language. Recent advancements in information technologies, materials, components and manufacturing technologies open up possibilities for transforming this structure. The technical world is changing, and someone will inevitably make a breakthrough, essentially creating a new technical reality in fields adjacent to Electromechanics or even within Electromechanics itself. For companies, manufacturing clusters, and even countries, failing to keep pace with change leaders threatens their competitiveness. However, the ambition should not only be to follow others in order to avoid falling behind — it should also be to strive to become a driver of change oneself.

The book demonstrates that the evolution of electric machines adheres to specific known laws, and thus it is possible to forecast the future in the field of Electromechanics. Moreover, the main idea of the book is that one can even participate in shaping the future—meaning that, to a certain extent and within a very narrow corridor of possibilities, it is possible to influence the process of evolution. Here the key driving forces of evolution can be utilized—such as the emergence of new applications, the development of existing ones, appearance of new materials, new manufacturing technologies and the never-ending creativity of the inventors coming with new concepts and solutions. For example, one can search for emerging new materials and incorporate them in the development of new electric machines and electric drive systems. Another activity can be indicating the need for certain new materials, which could positively influcence the development. Another way of taking part in the evolution of electric machines is designing systems with mechanical integration of the machines into various applications. A particularly valuable strategy is to monitor events driven by major transformative forces — often difficult to predict, such as so-called 'wildcards' — and to seek ways to leverage them to influence the evolution of electric machines.

The foundation of conscious participation in evolution is free thinking, which is taught by philosophy. Through philosophy, one can attain an awareness of reality and one's place in it, and begin to employ specific tools for transforming reality.

A. Matveev, *Shaping the Future of Electric Machines*,
https://doi.org/10.1007/978-3-031-95248-7

Let's revisit the key tools for engaging the management of evolution, as outlined in the book:

Adopting philosophical thinking will help understand reality and define one's role in the evolution of electric machines.

Knowledge of trends, patterns, and lines of evolution will enable the forecasting of the future of electric machines.

The proposed new classification of electric machines may aid in exploring new concepts of machines. ***The pentagram*** serves as a guide in the study of electric machines, also acting as a navigator for the designers.

Working with the patent databases provides insight into contemporary Electromechanics, helping understand the state of the art in specific technological areas.

Methods of creative thinking, primarily the rich ***toolkit of TRIZ***, can be employed for solving technical problems in the field of electric machines and as a kind of a "simulator" for developing free engineering thinking.

Finally, one of the new tools—the ***constant*** C_M—can assist in quickly estimating the mass and dimensions of new electric machines, as well as assessing the quality of existing ones.

These tools can make it easier to form a conscious, active stance and to translate it into meaningful action. For an electrical engineer, motivations in this context may include personal fulfillment, contributing to the professional community, serving their country, and advancing the well-being of humanity as a whole.

Appendices

Appendix A: Comparison of the Constant C_M with Arnold's Machine Constant

C_M is a new machine constant, proposed in 2019 in [1]. However, the very first machine constant was introduced by E. Arnold back in 1896 [2]. Arnold's constant, along with various similar constants, continues to be used today—see, for example, the monographs [3–6].

The design of an electric machine usually begins with selecting the dimensions and materials of its active and structural parts, primarily the rotor's diameter and length [3]. The rotor dimensions are typically estimated using the machine constant or similar constants. For example, in [4], the question is posed: "What should the size of the motor be to produce the required torque?" and the solution path is provided by the expression $T = kD_\delta^2 l_\delta$, where T is the required torque, D_δ is the rotor diameter, l_δ is the magnetic core length, and k is a constant.

It is assumed that knowing the value of the constant k helps determine the rotor volume V_r, proportional to $D_\delta^2 l_\delta$, and subsequently the main dimensions of the machine. The physical meaning of the constant k is the specific torque per unit rotor volume:

$$k = \frac{T}{D_\delta^2 l_\delta} \sim \frac{T}{V_r} \tag{A.1}$$

The "mechanical power machine constant" C_{mec} in [5] corresponds to the constant k in expression (A.1):

$$C_{mec} = \frac{P_{mec}}{n_{syn} D_\delta^2 l_\delta}, \tag{A.2}$$

where n_{syn} is the synchronous rotational speed.

In some other sources, such as [6], alongside the constant representing specific torque per rotor volume, an alternative constant σ is proposed, representing specific torque per rotor surface area.

A. Matveev, *Shaping the Future of Electric Machines*,
https://doi.org/10.1007/978-3-031-95248-7

In [3], the machine constant is defined as:

$$C_A = \frac{D_\delta^2 l_\delta \Omega}{S_p} \sim \frac{1}{B_\delta A} \tag{A.3}$$

where C_A is Arnold's machine constant, Ω is the angular velocity, S_p is the total power, B_δ is the air gap flux density, and A is the linear load.

Comparing (A.2) and (A.3), we can conclude that $C_A \sim 1/C_{mec}$, meaning that both constants have the same physical meaning: they either represent the specific torque per rotor volume or its inverse.

Returning to expression (A.3), it is noted in [3] that Arnold's constant can only be used for scaling geometrically similar machines, i.e., machines with the same supply frequency and angular velocity. Additionally, to estimate C_A one must know not only the power and rotational speed but also the electromagnetic loads B_δ and A.

The selection of electromagnetic load values presents a separate challenge, as they vary widely. The choice of electromagnetic loads is influenced not only by the type of machine, its power, and rotational speed, but also by the properties of the active part materials and the cooling system design. In [3], it is noted that flux density typically ranges from 0.7 to 0.95 T, and linear load can vary from 20 to 200 kA/m.

In [5, 6], tables of machine constants are provided, with recommended value ranges based on machine type and cooling method. The values in such tables can differ by an order of magnitude, which is not surprising given the aforementioned variations in electromagnetic loads. Unfortunately, these tables do not cover all machine types, and the machine types are not always clearly defined, for example: "large water-cooled machines", "machines for aviation applications", or "enclosed machines".

Even for the same type of machine, the machine constant can vary significantly, e.g., in [5] mechanical power constant C_{mec} varies depending on power and the number of poles.

Thus, all the aforementioned authors use rotor volume as a key element in calculations. The machine constant represents either specific torque per rotor volume or its inverse. Two approaches are proposed for estimating machine dimensions:

- *The first approach*: Using the machine constant from some available sources, such as [5] or [6], estimate the rotor volume V_r, of the new machine, and then, based on V_r, estimate the main dimensions of the new machine.
- *The second approach*: Scaling from a known machine, i.e., using the rotor volume V_{r1} of the known machine 1 and the required torque T_2, calculate rotor volume V_{r2} of the new machine 2: $V_{r2} \sim V_{r1} \frac{T_2}{T_1}$.

In the first case, the problem is that not all machine types are represented in the available sources. In the second case, one must know the rotor dimensions of the existing machine, but this information is not always available—for example, catalogs typically provide only the overall dimensions. In both cases, it is necessary to know the relationship between rotor volume and the main machine dimensions.

Given the above, estimating the dimensions of a new machine is a labor-intensive task for a specialist in electrical machines and is almost impossible for a nonspecialist. Even the name of the highly variable coefficient C_A - "machine constant" - can be misleading.

Summarizing the state of the art, we can list the following drawbacks of existing approaches:

- Difficulty finding the machine constant for the required type of machine.
- Wide ranges of values for B_δ, A, C_{mec}, σ recommended in different sources.
- Undefined transition from rotor volume to its dimensions (D_δ, l_δ), including overall dimensions (D, L)—see Fig. A.1.
- Inability to estimate the mass of the new machine.
- Scaling is limited to geometrically similar machines.

An interesting point of research could be comparing the C_M with Arnold's classical constant C_A. To make such a comparison, expression like (A.3) from [3] should be derived for C_M as presented below.

In [1], it is shown that the machine's mass can be approximately represented as:

$$m = \frac{D^2 L \pi 1900 n^{0.09}}{4} \tag{A.4}$$

and using (A.4) and (1.7b), we can derive the expression:

$$C_M = \frac{D^2 L \pi 1900 n^{0.09}}{4} \frac{n^{0.5}}{P^{0.75}} \sim \frac{D^2 L n^{0.59}}{P^{0.75}}. \tag{A.5}$$

Thus, we see that the new constant differs significantly from the classical one:

- Arnold's constant:

$$C_A = \frac{D^2 l_\delta \Omega}{S_p}, \tag{A.6}$$

- The proposed new machine constant:

$$C_M \sim \frac{D^2 L n^{0.59}}{P^{0.75}}. \tag{A.7}$$

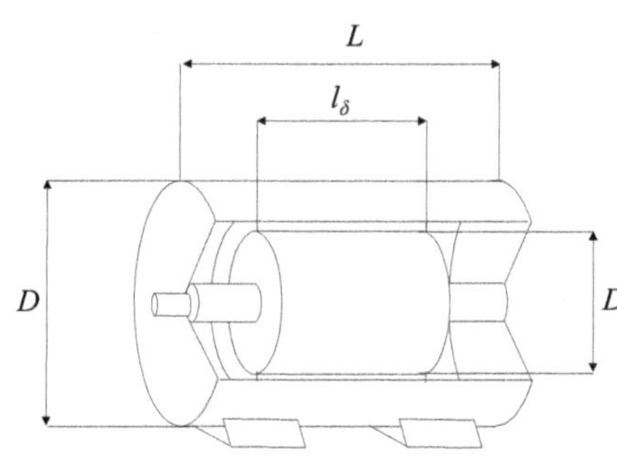

Fig. A.1 Overall dimensions of an electric machine and the dimensions of its rotor

The obvious difference between formulas (A.6) and (A.7) lies in the exponents for power (0.75) and rotational speed (0.59). The important distinction between C_A and C_M is that C_A has a physical meaning—it is the inverse of the specific torque per rotor volume, while C_M does not have such a meaning.

Appendix B: Examples of Size Estimation and Scaling

Universal Expressions for Size Estimation Based on Mass

In [1], formulas were proposed to estimate the overall dimensions of an electric machine, where the machine is simplified as a cylinder with a diameter D and length L, excluding elements like the shaft's output end, terminal box, feet, etc. The dimensions are estimated based on the mass, which in turn is evaluated through C_M:

$$\left\{ \begin{aligned} m &= C_M P^{0.75} n^{-0.5} \\ D &\approx \left(\frac{4m}{\pi \left(A n^a \right) C n^x} \right)^{1/3} \\ L &\approx \frac{4m}{\pi D^2 C n^x} \end{aligned} \right\} \tag{B.1}$$

where D is the outer diameter of the machine's housing in meters, L is the length of the machine's housing in meters, $A = 0.33$, $a = 0.22$, $C = 1900$, $x = 0.09$, and C_M is determined using (1.8).

It is important to note that C_M is not a highly accurate tool for predicting mass and the expressions (B.1) can be used only for rough estimates of the outer dimensions. The accuracy of the estimates may be influenced by the fact that electric machines can have different densities due to more or less of free space inside the machine not occupied by

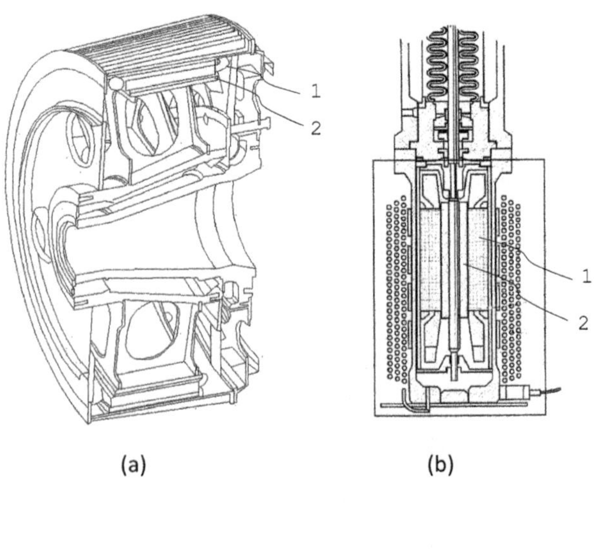

Fig. B.1 Examples of electric machines with different densities: low (**a**) and high (**b**)

the metal parts. In Fig. B.1a the picture from patent application CN101997370A shows wind turbine generator where the stator (1) and the rotor (2) form a thin ring leaving the space inside them almost empty. Figure B.1b shows opposite example borrowed from patent application US2014205475A1, where the stator (1) and the rotor (2) will occupy the entire space leaving no place for air. The difference in the density can be dependent on rotational speed (e.g., low speed machines have larger diameters and have more free space inside) or degree of integration with the drive mechanism.

Scaling

One of the potential primary applications of the machine constant C_M is scaling—estimating the mass (and dimensions) of a new machine based on the mass of a known machine.

To demonstrate the scaling process with an example, let's consider two machines of the same type, cooling method, voltage level, and efficiency but with different rated power and speed. We write the expressions for the mass of the known machine m_1 and the new machine m_2:

$$m_1 = C_M P_1^{0.75} n_1^{-0.5}, \tag{B.2}$$

$$m_2 = C_M P_2^{0.75} n_2^{-0.5}. \tag{B.3}$$

Let's express C_M from (B.2):

$$C_M = \frac{m_1}{P_1^{0.75} n_1^{-0.5}}. \tag{B.4}$$

Substituting (B.4) into (B.3), we get:

$$m_2 = m_1 \left(\frac{P_2}{P_1}\right)^{0.75} \sqrt{\frac{n_1}{n_2}} \tag{B.5}$$

Thus, for machines with the same type, voltage, cooling, and efficiency level, but with different rated power and speed, the mass of the new machine m_2 can be calculated based on the mass of the original machine m_1 using formula (B.5).

If the task involves scaling not only for new power and/or rotational speed but also for new voltage, cooling, and efficiency level, one should first calculate C_{M1} and at least approximately determine all the coefficients that define C_{M1} in expression (1.8) for the original machine. Then, calculate C_{M2} for the new machine, using the value of C_{M1} and adjusting for changes in coefficients such as K_e, K_c, or K_u. Finally, calculate the mass of the new machine.

Appendix C: Statistics on Patent Applications

This appendix organizes the results of a search for the number of patent families by specific types of electrical machines (EM). It also provides figures for the countries and companies with the highest number of patent applications for these types of EM. The data was collected using the open patent database www.espacenet.com.

For induction (asynchronous) machines, the search was conducted under class H02K17 “Asynchronous induction motors/generators”—the information is presented in Table C.1.

For synchronous machines with permanent magnets, the search was conducted using two methods:

1. A search under the class H02K21 “Synchronous motors/generators having permanent magnets”—the information is presented in Table C.2.
2. A broader search under the class H02K combined with the keywords “Permanent magnet”—the information is presented in Table C.3. Only data from the period starting in 1991 was analyzed.

For direct current machines, the search was conducted under the class H02K23 “DC commutator motors or generators having mechanical commutator; Universal AC/DC commutator motors”—the information is presented in Table C.4.

On switched reluctance machines the search was performed on a broader class H02K and key words “Switched (or variable) reluctance”—the information is given in Table C.5.

Table C.1 H02K17. Asynchronous induction motors/generators

	1961–1965	**1966–1970**	**1971–1975**	**1976–1980**	**1981–1985**	**1986–1990**
Number of families	814	1047	1469	1765	1575	1643
Leading countries	USA—262, Germany—250, France—220	Germany—419, USA—326, France—316	Japan—1005, Germany—285, USA—198	Japan—1345, Germany—216, USA—156	Japan—1132, USSR—209, Germany—161	Japan—929, USSR—360, Germany—189
Leading companies	Licentia—55, Siemens—48, GE—36	Licentia—65, Siemens—62, GE—33	Hitachi—55, Siemens—31, GE—21	Hitachi—162, Tokyo Shibaura—158, Mitsubishi—68	Hitachi—85, Tokyo Shibaura—71, Mitsubishi—62	Satake—104, Institut Elektrodinamiki—98, Toshiba—59
	1991–1995	**1996–2000**	**2001–2005**	**2006–2010**	**2011–2015**	**2016–2020**
Number of families	1247	1114	1258	1403	1935	1919
Leading countries	Japan—738, USA—231, Germany—220	Japan—542, USA—349, Germany—251	USA—511, Japan—442, China—430	China—819, USA—371, Japan—271	China—1290, USA—382, Japan—309	China—1386, USA—329, Japan—210
Leading companies	Fujitsu—78, Toshiba—68, Hitachi—49	Hitachi—53, Matsushita—33, Mitsubishi—31	LG Electronics—102, Mitsubishi—47, Siemens—26	Mitsubishi—34, LG Electronics—29, Siemens—23	Chongqing Leerja—89, Mitsubishi—42, Siemens—29	Siemens—39, Jiangsu Huanqiv—36, Wolong Elictric—34

Table C.2 H02K21. Synchronous motors/generators having permanent magnets

	1961–1965	**1966–1970**	**1971–1975**	**1976–1980**	**1981–1985**	**1986–1990**
Number of families	539	781	2197	2985	3018	3222
Leading countries	Germany—170, USA—169, France—145	Japan—336, USA—275, Germany—245	Japan—1881, Germany—306, USA—282	Japan—2611, Germany—306, USA—291	Japan—2630, USA—384, Germany—324	Japan—2694, USA—521, Germany—413
Leading companies	Siemens—36, Bosch—31, Philips—22	Siemens—33, Bosch—21, Licentia—14	Bosch—40, Hitachi—39, Matsishita—35	Hitachi—169, Matsushita—142, Citizen—59	Matsushita—176, Hitachi—131, Mitsubishi—64	Matsushita—86, Mitsubishi—71, Hitachi—52
	1991–1995	**1996–2000**	**2001–2005**	**2006–2010**	**2011–2015**	**2016–2020**
Number of families	2492	3735	4888	5623	5482	7262
Leading countries	Japan—1865, USA—646, Germany—502	Japan—2856, USA—1126, Germany—781	Japan—3487, USA—1974, China—1308	Japan—2758, China—2496, USA—2096	China—3050, Japan—2250, USA—2050	China—4925, Japan—2216, USA—1975
Leading companies	Matsushita—117, Sony—81, Nihon—73	Matsushita—200, Nihon—123, Hitachi—92	Matsushita—202, LG Electronics—117, Mitsubishi—116	Honda—198, Mitsubishi—141, Daikin—135	Mitsubishi—185, Denso—149, Daikin—64	Denso—185, Mitsubishi—163, Zhuhai Gree—126

Table C.3 "Permanent magnet" + H02K

	1991–1995	1996–2000	2001–2005	2006–2010	2011–2015	2016–2020
Number of families	2898	4068	5279	8121	11,254	18,473
Leading countries	Japan—1835, USA—894, Germany—717	Japan—2535, USA—1297, China—974	Japan—2875, USA—2013, China—1658	China—4263, Japan—2749, USA—2413	China—7817, USA—2681, Japan—2562	China—15,408, USA—2347, Japan—1999
Leading companies	Toshiba—78, Matsushita—71, Hitachi—57	Matsushita—129, Hitachi—106, Toshiba—84	Yaskawa—177, LG Electronics—111, Matsushita - 100	Mitsubishi—188, Honda—123, Siemens—120	Mitsubishi—234, Harbin Inst Technology—136, Univ Southeast—132	Zhuhai Gree—380, Mitsubishi—197, Harbin Inst Technology—162

Table C.4 H02K23. DC commutator motors or generators having mechanical commutator; Universal AC/DC commutator motors

	1961–1965	**1966–1970**	**1971–1975**	**1976–1980**	**1981–1985**	**1986–1990**
Number of families	643	859	2080	2578	2206	1570
Leading countries	Germany—212, France—177, USA—171	Japan—334, Germany—295, France—258	Japan—1742, Germany—268, France—204	Japan—2258, Germany—248, USA—233	Japan—1906, Germany—248, USA—231	Japan—1192, Germany—301, USA—413
Leading companies	Siemens—39, Licentia—32, Electronique & Automatisme—19	Siemens—38, GE—17, Bosch—15	Hitachi—73, Matsushita—52, Siemens—34	Matsushita—342, Hitachi—130, Canon—36	Matsushita—212, Hitachi—115, Mitsubishi—59	Hitachi—60, Matsushita—58, Mitsubishi—53
	1991–1995	**1996–2000**	**2001–2005**	**2006–2010**	**2011–2015**	**2016–2020**
Number of families	1339	1475	1738	1636	1167	1085
Leading countries	Japan—910, USA—3019, Germany—304	Japan—974, USA—527, Germany—384	Japan—975, USA—804, China—604	China—743, Japan—723, USA—619	China—728, Japan—419, USA—380	China—740, Japan—250, USA—223
Leading companies	Matsushita—48, Mabuchi—35, Tokyo Parts—32	Asmo—76, Denso—58, Matsushita—50	Asmo—115, Denso—98, Mitsuba—47	Mitsuba—106, Asmo—72, Denso—50	Mitsuba—59, Asmo—56, Johnson—42	Mitsuba—39, Denso—32, Johnson—11

Table C.5 "Switched (or variable) reluctance" + H02K

	1986–1990	1991–1995	1996–2000	2001–2005	2006–2010	2011–2015	2016–2020
Number of families	88	201	332	249	275	558	852
Leading countries	USA, Japan, Germany	USA, Germany, Japan	USA, Japan, Germany	USA—152, Japan—98	China—142, USA—113, Japan—71	China—352, USA—199, Japan—131	China—702, USA—133, Japan—69
Leading companies	Nidec SR Drives	Nidec SR Drives	Nidec SR Drives	LG Electronics	Mitsuba	Samsung	Shenzhen Huayin

Appendix D: Example of a Patent Search Methodology

There are numerous free and paid online tools for searching and organizing patent information. One of the best free resources is the European Patent Office's Espacenet patent database (https://worldwide.espacenet.com).

Patent searches can be conducted using keywords, patent classification codes, inventor names, patent holders, time periods, or combinations of these methods. It is advisable to familiarize yourself with the International Patent Classification (IPC) system, where inventions are divided into numbered classes and subclasses, each with its own code. Most patent search systems allow searches within specific classes, helping to narrow down search results.

When using keywords in the search, it's important to remember synonyms and alternative terms. For instance, the same type of machine may be referred to as "switched reluctance," "variable reluctance," or simply "reluctance".

Let's outline the search process using Espacenet as an example. Suppose our goal is to find patents on induction machines with solid rotors. We begin with a Boolean search, where multiple keywords or codes are connected by operators like AND, OR, or NOT. We can try one combination of keywords—"induction motor"—and one IPC code—"H02K", corresponding to "Dynamo-electric machines". The search fields in Espacenet's window are shown in Fig. D.1, representing the following query string: ***ftxt all "induction motor" AND cl any "h02k"***. Running this search returns 64,417 results (patent families), which is too many to review "manually." Therefore, it's necessary to narrow down the search scope. It's always a good idea to experiment with search terms and Boolean operators to find more relevant results.

Let's try to narrow down the search scope by selecting a more specific subclass (induction machines) within the "Dynamo-electric machines" class. Figure D.2 shows the corresponding code for induction machines—H02K17/00—in the IPC system. We'll also use a more specific keyword to describe the rotor type—"solid". The search fields for rotor type and the H02K17/00 code are shown in Fig. D.3, while Fig. D.4 shows the results window referring to 38 found entries.

Note that 38 entries are a manageable number that can be reviewed quickly to determine whether relevant information has been found or if the search needs to

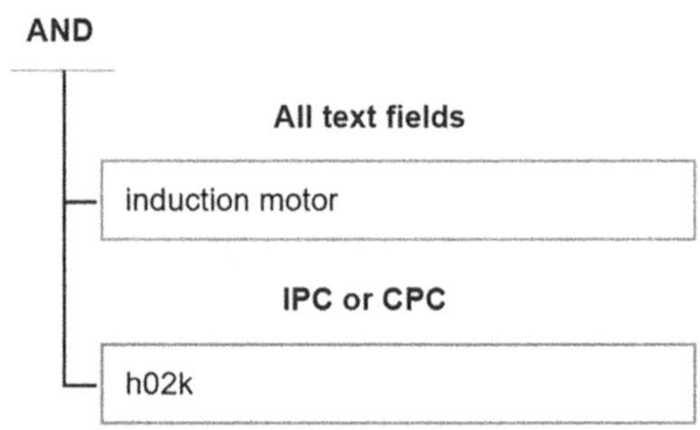

Fig. D.1 Search by machine name and code H02K

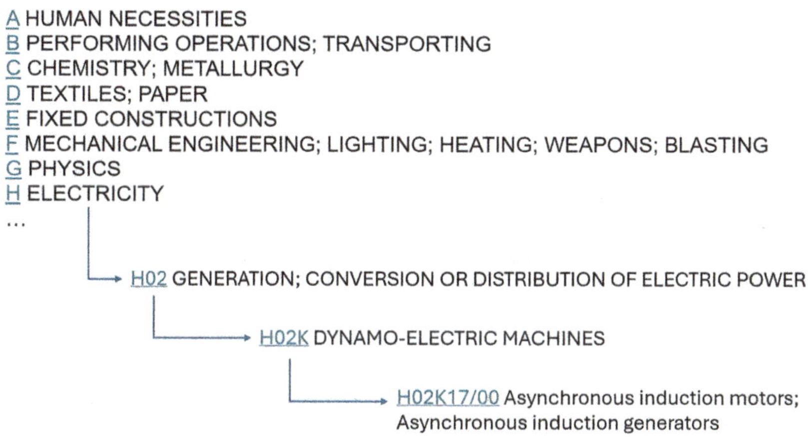

Fig. D.2 Patent classification (IPC) and selection of a specific code

Fig. D.3 Search by rotor type and code H02K17/00

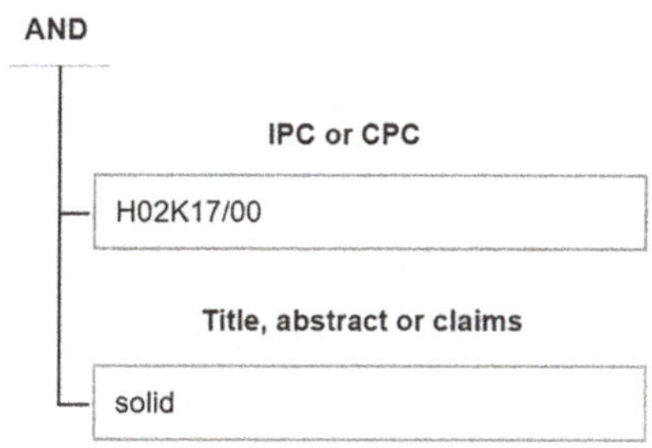

continue. In this example, the review shows little information on induction machines with solid rotors, meaning the search was unsuccessful.

Expanding the search scope, we move from the subclass code H02K17/00 to the broader class code H02K. This time, we specify that the term "solid" refers to the rotor ("solid rotor") as shown in Fig. D.5. Figure D.6 shows the results window with 350 found entries. A quick glance at the illustrations reveals relevant information - the search was a success.

Using a variety of keyword options is important because some patents deliberately contain vague wording—an attempt to "hide" the patent, which could later allow the patent holder to file a lawsuit against competitors who did not find the patent and infringed the owner's rights.

Another way to search for patents is through citations. Patents often cite other similar patents. When you find a patent of interest, look for references to other patents in it. This can lead to finding new inventions related to your search.

38 results in Espacenet:

Patent family 1: **Solid** rotor assembly
US5798559A

Patent family 2: **Solid** core angular position resolver
US6630763B1

Patent family 3: **Solid** body induction motor for vehicle ...
US2002149280A1

Patent family 4: Automatic transmission of automobile
CN103401397A(B)

Patent family 5: Method for operating an electrical ...
CA2980554A1

...

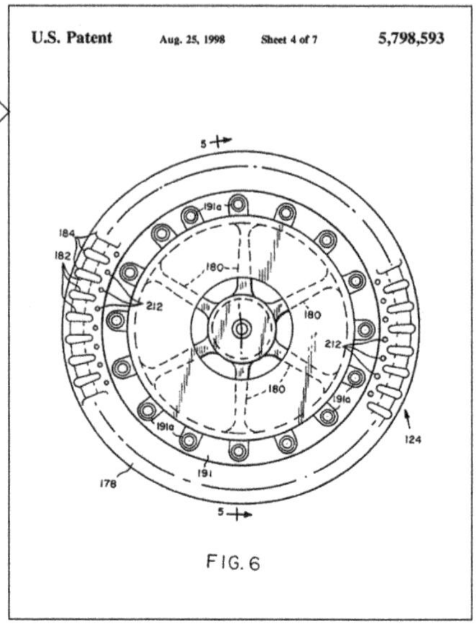

Fig. D.4 Search results window by rotor type and code H02K17/00

Fig. D.5 Search by rotor type and code H02K

AND

IPC or CPC

H02K

Title, abstract or claims

solid rotor

350 results in Espacenet:

Patent family 1: Slotted **solid rotor**
WO2005078895A1

Patent family 2: **Solid rotor** for motor
JPH10174391A

Patent family 3: **Solid rotor** type induction motor
JPS58163253A

Patent family 4: **Solid rotor**
SU163693A1

Patent family 5: **Solid rotor** iron core structure
CN203554101U

...

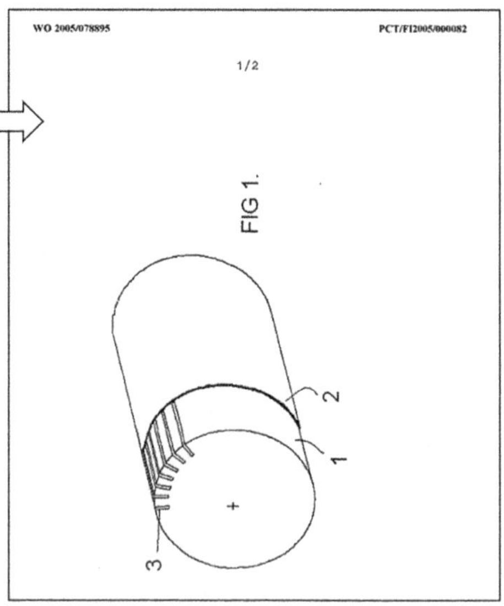

Fig. D.6 Search results window by rotor type and code H02K

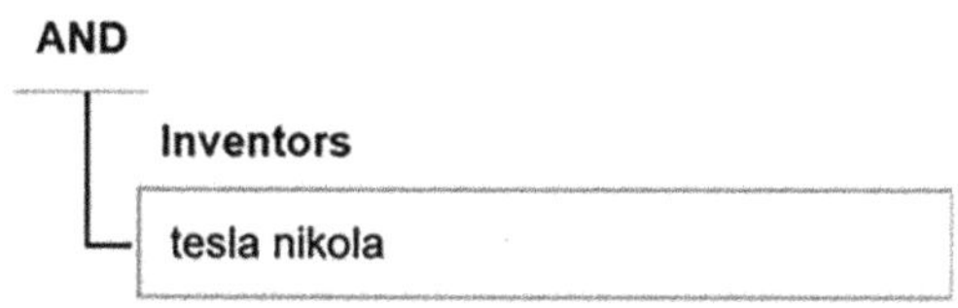

Fig. D.7 Search by inventor name

Another search method is by inventor's name. For example, Fig. D.7 shows the search fields for the name "Nikola Tesla." The results of this search will help you find most of Tesla's inventions that have been patented, with 88 found in the Espacenet database.

It is possible to search by patent owner. If you know who your main competitors are, in business or research, you can find the patents they hold. This allows you to see what they are working on and compare it to your own ideas. It can also help you predict whether your innovation might lead to a potential infringement lawsuit.

Lastly, a direct search by patent application or patent number is possible if you know the number.

A few practical tips:

- After finding several relevant patent results using one search method, try another method, expanding the search again. Looking at the search from a different angle can reveal valuable information you might have missed the first time.
- Use multiple databases if possible.
- Create a document to track your search history or at least save links to interesting patents.

In conclusion, after conducting a preliminary patent search, based on the results, you may consider hiring a patent attorney for a more in-depth search.

Appendix E: 40 Inventive Tricks

1. Parameter Changes

- *Change the state of matter. Modify the physical and chemical parameters. Change concentration or consistency. Change the degree of flexibility. Change the temperature, pressure.*

Example 1.1 The method of impregnating the winding in a vacuum and under pressure (VPI process) involves placing the stator with the winding into an impregnation vessel (vacuum chamber), creating a vacuum inside (to remove air), and then, without removing the vacuum, introducing impregnating varnish into the impregnation vessel. Once the varnish covers the entire stator, the vacuum is removed, and pressure is applied (to improve the varnish's penetration into the winding). After the pressure is released, the varnish supply is detached, a vacuum is created again, and only then the stator can be removed from the impregnation vessel. After impregnation, the winding is dried at an elevated temperature to remove the solvent, accelerate the varnish curing, and form stable varnish film. Thus, during the impregnation process, both pressure and temperature are adjusted.

Example 1.2 Figure E.1 shows the concept of a stepper motor using shape memory alloy (SMA) wires (patent CN104092349B). When current is passed alternately through the wires, they alternately heat up and cool down. Changes in temperature cause phase transitions in the SMA. As a result, the wires alternately elongate and contract, driving the gear plate.

Example 1.3 Electric centrifugal pumps used in deep-well and subsea applications, operating at high pressure, are typically filled with oil (a liquid dielectric). At high rotor speeds, significant friction losses occur due to the contact of the rotor surface with the oil. At rotational speeds above approximately 5000 rpm, the level of losses generally becomes unacceptably high. One possible solution is to use compressed gas instead of oil.

Example 1.4 Patent US8940825B2 describes a computer hard disk drive motor using a terminal sealing compound comprising urethane methacrylate, methacrylate, and acrylic acid in specific proportions. The proportions of the components are chosen so that the boiling point of the compound was above 150 °C to prevent the formation of gas.

Example 1.5 Patent US6664705B2 describes a generator with natural overheating protection, where a part of the stator magnetic core is made of a low Curie temperature material. When overheated, this part of the stator core loses its magnetic properties and the magnetic flux created by the rotor does not reach the stator windings, so the generator stops working and cools down.

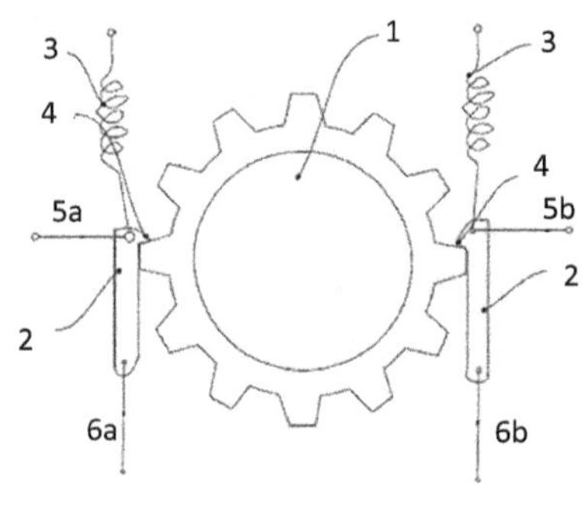

Fig. E.1 SMA stepper motor (1—gear plate, 2—driving arms, 3—elastic connecting pieces, 4—convex hooks, 5a—reset SMA wire, 5b—sub-reset SMA line, 6a—driving SMA wire, 6b—sub-driven SMA line)

2. Preliminary Action

- *Perform, in advance, the required change of an object (either fully or partially). Pre-arrange objects such that they can come into action from the most convenient place and without losing time for their delivery.*

Example 2.1 To eliminate condensation after downtime or storage at subzero temperatures, the motor is warmed up before starting. During the warm-up, only two phases are used, while the third phase is disconnected (using electronic switches). As a result, current flows only to one part of the winding, heating it up but preventing the rotor from rotating. To ensure the motor warms up evenly, the part of the winding to which the current is applied is changed every few seconds.

Example 2.2 Before connecting a high-power generator to a load, it is allowed to run idle to ensure that the oil circulation system in the bearings and the cooling system reach normal operating conditions.

Example 2.3 In an induction motor drive, if the motor start needs to be synchronized with the release of the mechanical brake or if the maximum possible starting torque is required, it is possible to use the machine's pre-magnetization mode with a selectable magnetization time. This feature is implemented in modern frequency converters.

Example 2.4 To achieve more precise execution of the setting (in terms of timing and position), the electric drive can take up the slack in the gearbox before starting or reversing. This function is also implemented in modern frequency converters.

Example 2.5 To ensure the stability of the utility grid in case of an emergency shutdown of power units, a so-called "spinning reserve" is created—power units where the turbine and generator are spinning but are not fully loaded.

Example 2.6 Patent US8664817B2 describes an oil-filled submersible motor designed for operation in deep wells at temperatures above 180 °C. The motor's insulation uses polyamide film and epoxy resin. At high temperatures, the epoxy resin releases an amino acid that can damage the polyamide film. To prevent this issue, the oil filling the motor includes special additives that break down the amino acid.

3. Fragmentation

- *Divide an object into independent parts. Make an object easy to assemble or disassemble. Increase the degree of fragmentation.*

Example 3.1 When designing a stator core with minimal slot opening and concentrated winding, placing the winding into the slots can be very labor-intensive, mak-

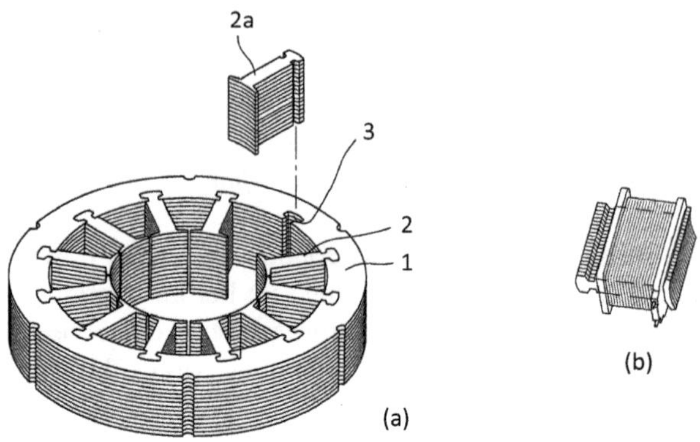

Fig. E.2 Segmented stator concept from patent application US2005218749A1: Segmented stator (**a**) and one separated stator segment with a coil (**b**)

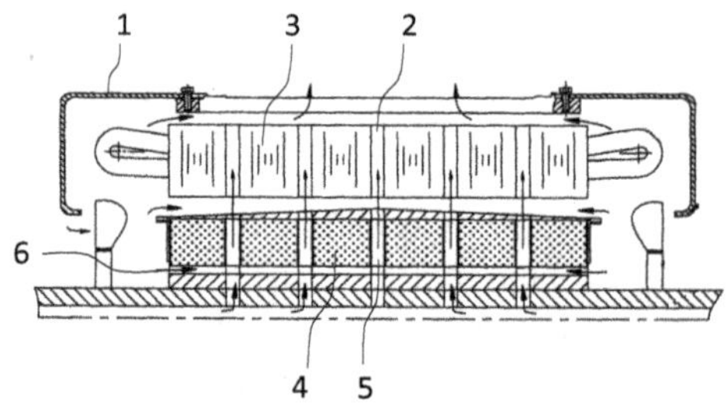

Fig. E.3 Radial ventilation system (1—housing, 2—radial stator channels, 3—stator core packages, 4—rotor packages, 5—radial rotor channels, 6—axial rotor channels)

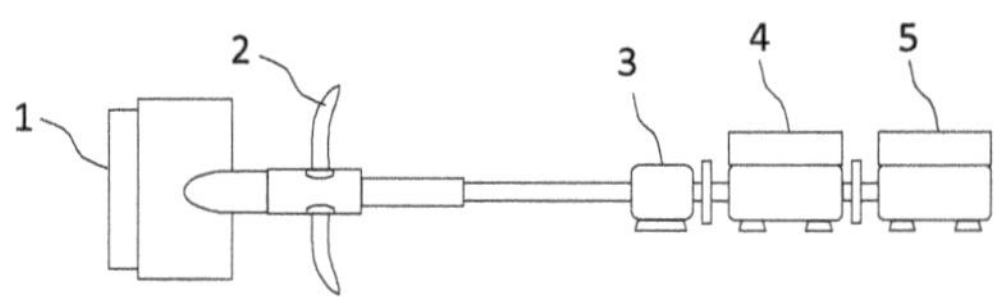

Fig. E.4 Propulsion system (1—rudder, 2—propeller, 3—thrust bearing, 4, 5—motors)

ing it hard to achieve a high copper fill factor. A possible solution is to manufacture the stator core from segments, as shown in Fig. E.2. The coils are wound on individual segments, which are then assembled into a ring, and the coils are connected to form a single winding.

Example 3.2 In high-power machines with air cooling, the stator core is divided into axial packages, with small gaps left between them. These gaps are used as channels for the flow of cooling air. In this way, the so-called radial ventilation system is implemented Fig. E.3 (example from CN203086307U).

Example 3.3 In the hull of a marine vessel below the waterline, the space near the keel is restricted (vertically and to the sides of the propeller shaft), so a single large electric motor will not fit. The solution is a propulsion system with two electric motors arranged in tandem (Fig. E.4). An additional advantage is that this configuration provides a redundant system.

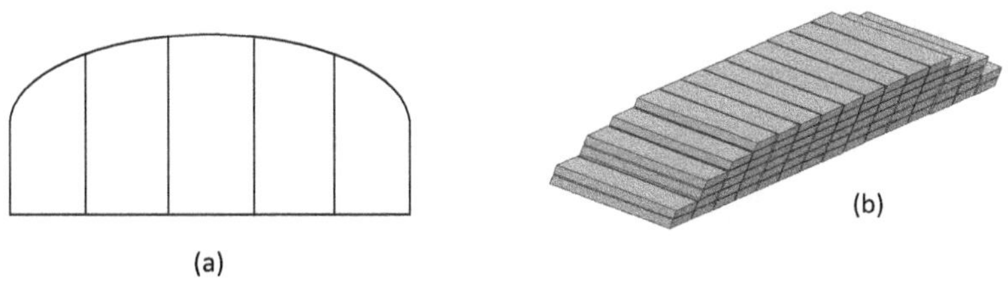

Fig. E.5 Some options for magnet segmentation: (**a**) A magnet divided into five segments while maintaining its arc-shaped form. (**b**) A magnet composed of multiple pieces, segmented in both tangential and radial directions

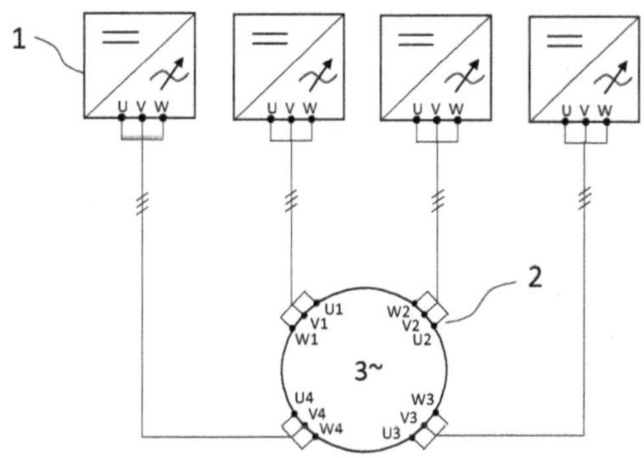

Fig. E.6 Redundant system (1—inverter, 2—separate three-phase winding in the motor)

Example 3.4 To reduce losses in magnets due to eddy currents, the magnets are divided into segments (Fig. E.5). The number of segments and the direction of segmentation depend on the supply frequency, the presence of higher harmonics, and the machine's rotation speed.

Example 3.5 In some cases, low voltage and low power frequency converters (FCs), which can be completely implemented on one printed circuit board, are made of two connected boards—separate power and control parts. This is done for ease of testing and repair.

Example 3.6 To ensure the drive system remains operational in the event of a fault in one part of the winding, the winding is divided into electrically independent sections, each of which is a three-phase winding powered by a separate inverter. This creates a redundant system (Fig. E.6).

4. Preliminary Anti-action

- *Preemptively apply stress to the object that is opposite to the unacceptable or undesirable operating stresses.*

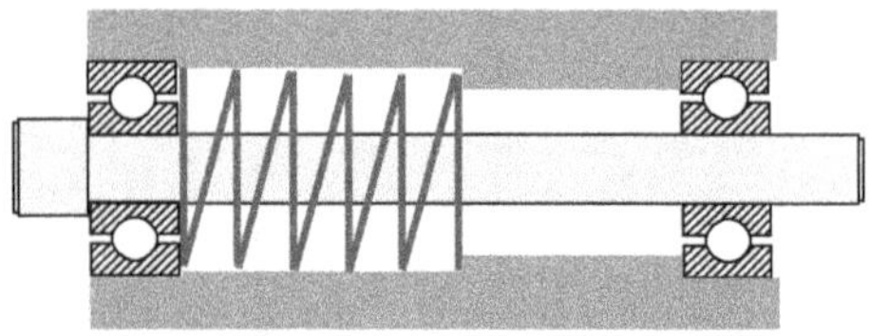

Fig. E.7 A preloaded bearing

Example 4.1 In the rotors of high-speed machines with permanent magnets, a carbon fiber thread is wound around the outer surface of the magnets (forming a bandage) with a certain tension to create a pre-stress, which ensures that the magnets are securely held at high rotational speeds.

Example 4.2 Ball bearing systems can be preloaded using a spring (Fig. E.7). This is done to eliminate radial and axial play, reduce operational noise by stabilizing the rotating mass, minimize both repeatable and nonrepeatable runout of the rotational axis, decrease the likelihood of damage from vibrational loads, and increase stiffness.

Example 4.3 Before using the machine for an unusually long period of time under high load, the machine can be cooled to a lower than normal temperature to prevent overheating.

Example 4.4 According to the principle of pre-stressing, the machine's compound shaft tubes can be pre-twisted in the direction opposite to the operating torque.

5. Taking Out

- *Separate an interfering part or property from an object, or single out the only necessary part (or property) of an object.*

Example 5.1 In the tunnel thruster with an integrated permanent magnet synchronous motor (Fig. E.8), only the active parts of the motor are present: The stator is integrated into the annular wing, and the rotor is mounted on the periphery of the screw, with several blades. An integrated electric machine does not have a shaft, mechanical bearings, end supporting structures, or a housing. It contains only the essential active components.

Example 5.2 In an electric machine with a rotor winding made of high-temperature superconductors (HTS), a localized low-temperature zone is created on the rotor, surrounded/isolated by a vacuum (Fig. E.9, concept from patent application US2005160744A1).

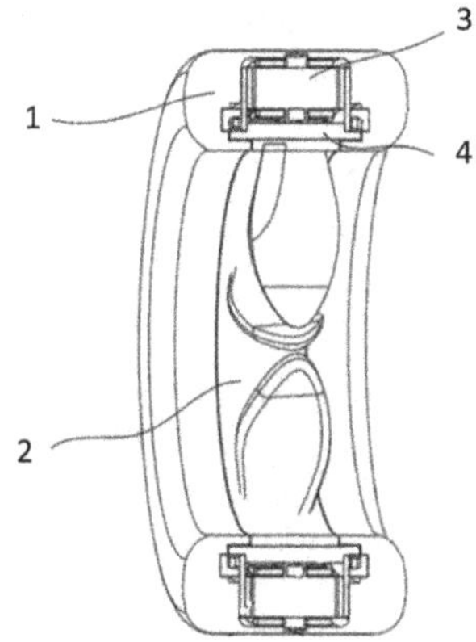

Fig. E.8 Tunnel thruster from patent application CN119010411A (1—annular wing, 2—rotor rim, 3—stator, 4—rotor with magnets)

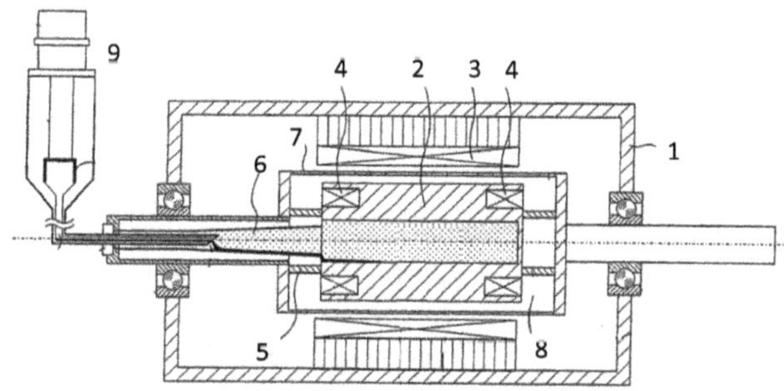

Fig. E.9 HTS machine and cryogenic system (1—housing, 2—HTS rotor, 3—stator winding, 4—HTS winding, 5—hollow cylindrical suspension elements, 6—lateral cavity, 7—gas-cooled chamber, 8—vacuum-filled space, 9—cryostat)

Example 5.3 Typically, in a motor drive system, both the electric machine (EM) and the control electronics (frequency converter, FC) have terminals for connecting the power cable that links these devices. In low-power systems with a short distance between the EM and FC, where disconnection of the cable is not anticipated, it is possible to forgo the terminals and solder the winding ends directly onto the FC's printed circuit board.

Example 5.4 When using underwater drive systems at great depths (1–5 km), where pressures reach 100–500 atm, the motor control electronics is separated from the electric machine and placed in a sealed capsule, often spherical, where normal atmospheric pressure is maintained.

6. Homogeneity

- *Make objects interacting with a given object of the same material (or material with identical properties).*

Example 6.1 If the rotor is made of a material with a higher coefficient of thermal expansion than the stator material, the expansion can reduce the air gap to an unacceptable value during operation when the parts are heated by losses. To avoid this, the coefficient of thermal expansion of the rotor should not significantly exceed that of the stator.

Example 6.2 In electrical connections, it is desirable to use the same materials or compatible materials. For example, connecting copper and aluminum conductors or terminals can lead to oxidation of the aluminum part of the contact over time.

Example 6.3 In generators of large wind turbines, uneven distribution of the coolant speeds in the stator cooling channels can lead to significant temperature gradients in the winding or magnetic core of the stator along the axial direction. There are methods to improve the uniformity of the axial temperature distribution in the stator, such as those proposed in patent application WO2021037193A1.

7. Dynamization

- *The characteristics of the object (or external environment) should change to be optimal at each stage of operation. Divide the object into parts that can move relative to each other.*

Example 7.1 A conventional electric motor has two active parts: a stationary one—the stator; and a rotating one—the rotor. In a birotative machine (Fig. E.10), both active parts rotate, i.e., there is no stator, but two rotors. It can be said that the stator starts rotating and becomes one of the rotors. Energy is transferred to the rotor windings through slip rings.

Example 7.2 The control electronics is traditionally stationary. However, in some synchronous machines, diode rectifiers that supply the field winding are located on the rotor, receiving power from a rotating transformer.

Example 7.3 In the US6492753B3 patent, the rotor can be shifted in the axial direction, reducing the EMF induced in the stator windings. The axial shift is used in field weakening mode at high speeds (Fig. E.11).

Example 7.4 When controlling a PMSM over a wide range of rotational frequencies, adaptive algorithms can be used (patent EP2497266B1). For example, sinusoi-

Fig. E.10 Birotative machine (1—first rotor, 2—second rotor, 3—slip rings, 4—brushes)

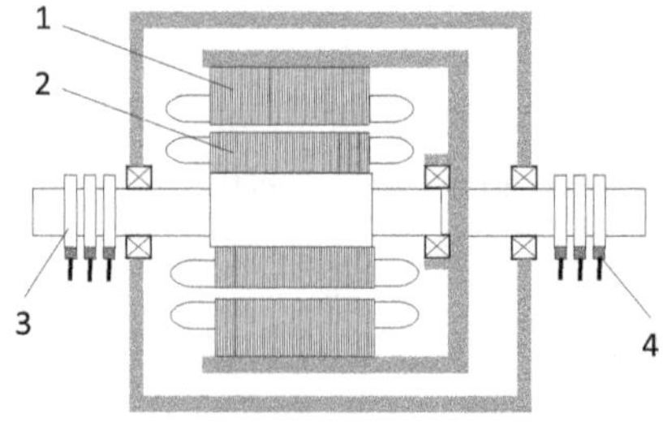

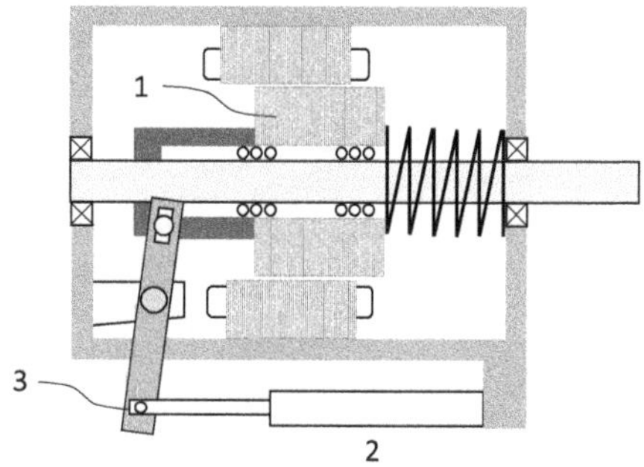

Fig. E.11 Electric machine with mechanical field weakening (1—rotor, 2—actuator, 3—lever)

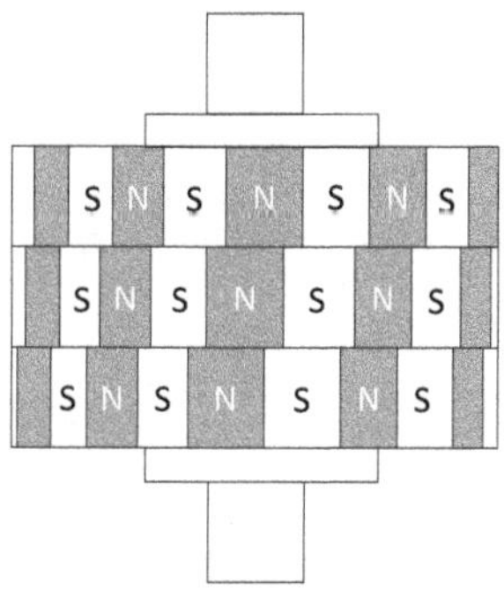

Fig. E.12 Skewed magnets on the rotor

dal waveform can be applied at low speeds and trapezoidal at high speeds. In another example, the carrier frequency of the PWM can be adjusted depending on the rotational speed.

Example 7.5 To reduce torque ripple, the so-called "pole skewing" is used—this involves axially dividing the rotor into several segments (Fig. E.12).

8. Periodic Action

- *Instead of continuous action, use periodic or pulsating actions. If an action is already periodic, change its magnitude or frequency. Use pauses between actions to perform a different action.*

Example 8.1 In a switched reluctance machine, at each moment of time, part of the winding coils is not used in the process of torque production. Low current pulses can be applied to these inactive coils to determine the angular position of the rotor (using nonuniform magnetic conductivity). Measurements are made periodically, with the frequency of switching of the coils.

Example 8.2 In PMSM, it is possible to switch to a design with a multiple of the number of poles and increase the fundamental frequency by the corresponding number of times.

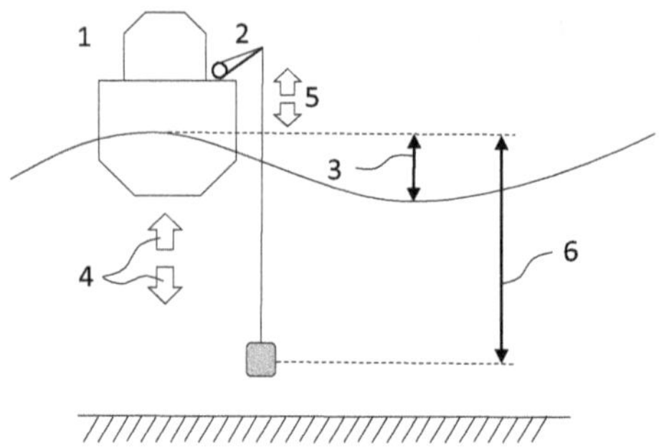

Fig. E.13 Illustration of the AHC system (1—offshore vessel, 2—winch, 3—wave height, 4—vertical movements of the vessel relative to the seabed, 5—vertical movements of the wire relative to the vessel, 6—distance of the payload from the vessel)

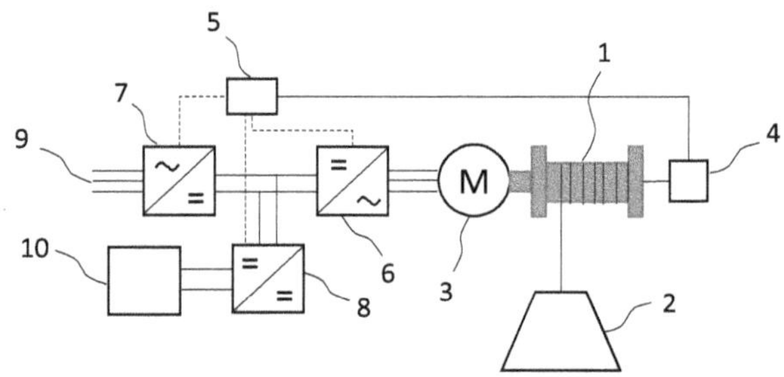

Fig. E.14 Winch drive system (1—drum, 2—payload, 3—electric machine, 4—rotor position sensor, 5—control system, 6—motor control inverter, 7—grid-side inverter, 8—DC/DC converter, 9—vessel's electrical grid, 10—energy storage)

Example 8.3 On offshore marine vessels (Fig. E.13), winch drive systems with active heave compensation (AHC) can be used. In such a system, wave motion is compensated by automatically coiling or uncoiling the winch wire in the opposite direction at the same linear speed (accordingly, the direction of rotation of the winch drum and the driving motor changes periodically). This way, the payload maintains its position relative to the seabed. AHC winches can be used in systems with underwater vehicles and for lifting equipment that needs to operate near or on the seabed. AHC system functionality may include wire tension control. The electrical part of the AHC system (Fig. E.14) implies the accumulation of energy in the battery when unwinding the wire when the vessel moves up and the use of the accumulated energy when winding the wire onto the winch drum when the vessel moves down. A group of inverters with a common DC link is used to control the energy flows.

Example 8.4 In electric drive systems, the microprocessor can periodically switch between different tasks.

9. Equipotentiality

- *If an object has to be lifted or lowered, change the object's environment so the need to lift or lower is eliminated or performed by the environment.*

Example 9.1 The design of the wind turbine generator from segments allows the wind generator to be repaired directly in the nacelle, without lowering it to the ground, by replacing a failed segment.

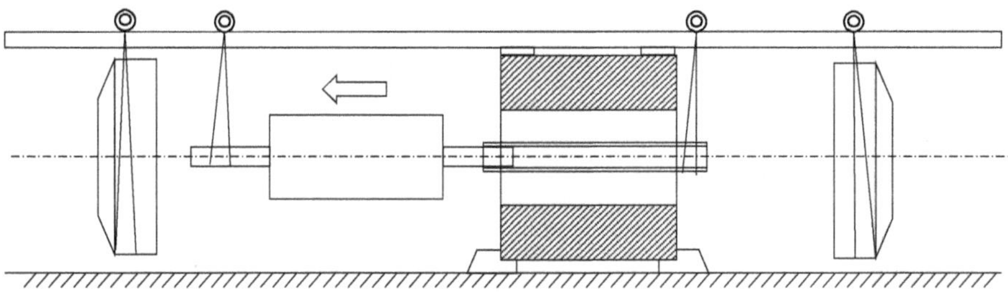

Fig. E.15 Mechanism for removing the rotor from the machine

Example 9.2 The method of removing the rotors from the machine so as not to damage the cores is shown in Fig. E.15. This method does not involve movement in the vertical plane.

Example 9.3 Giant hydrogenerators are assembled from segments directly at the hydropower plant.

10. Copying

- *Instead of an inaccessible, complex, expensive, inconvenient, or fragile object, use simplified and inexpensive copies of it. Replace the object or system of objects with their optical copies (images). Use scaling (enlarge or reduce the copies). If visible optical copies are used already, switch to infrared or ultraviolet copies.*

Example 10.1 When designing and commissioning an electric drive, testing is usually required, which involves significant time and material costs. To address this issue, equipment simulators based on field-programmable gate arrays (FPGAs) can be used. FPGAs are semiconductor devices that can be configured by the manufacturer or developer after production. The purpose of FPGAs is to simulate in real time the power part of an electric drive, which includes a frequency converter, a motor, and a drive mechanism.

Example 10.2 For exhibitions, "empty" motors—without active parts—are used. This reduces their weight and simplifies transportation and installation on the exhibition stand. Visually, such a motor, of course, does not differ from a motor with active parts.

Example 10.3 With the spread of additive technologies, particularly 3D printing, it has become possible to "print" structural elements of electric machine from relatively inexpensive materials, such as plastic. These "printed" elements can be useful, for example, in testing the assembly process of the machine. It is also possible to "print" cooling system channels and test their efficiency.

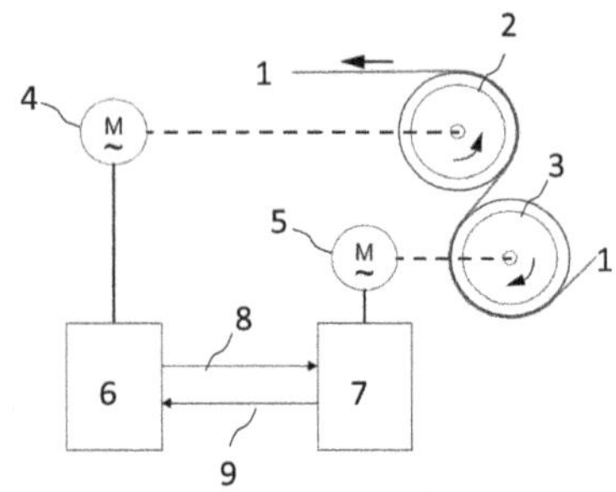

Fig. E.16 Structure of the master-follower system (1—sheet/web, 2—roller No. 1, 3—roller No. 2, 4—motor No. 1, 5—motor No. 2, 6—inverter of motor No. 1, 7—inverter of motor No. 2, 8—torque setting, 9—feedback)

Example 10.4 Using a digital twin instead of physical testing. Thanks to increasingly advanced multidisciplinary models and the growth of computational power, modeling is increasingly applicable across multiple stages of electric machine development. This includes not only design but also manufacturing, assembly, and testing, including testing within a higher-level system.

Example 10.5 To maintain uniform tension on a sheet of metal in a rolling mill or a sheet of paper in a pulp and paper production, a master-follower system is used (Fig. E.16), in which, depending on a number of production parameters, the same torque on the rolls can be maintained. The torque of one motor is "copied" into the other by means of a setpoint transmitted between the motor inverters, for example, via fiber optic communication.

Example 10.6 For measuring and controlling temperature, a thermal imaging camera is used instead of integrated sensors.

11. Feedback

- *Introduce feedback (referring back, cross-checking) to improve a process or action. If feedback is already used, change its magnitude or influence in accordance with operating conditions.*

Example 11.1 The patent application TW201815025A proposes a quality control system for coil winding with automatic feedback regulation, which includes an automatic winding machine (AWM), a measuring system (MS), and a control platform (CP). Based on the results of electrical testing performed by the MS, the data is transmitted to the CP for analysis. The CP can adjust the control parameters in the AWM based on the analysis results to improve the quality of the coils.

Example 11.2 In some cases, to improve the quality of motion in the electric drive system, it is possible to transition from simpler sensors to more complex ones, such as from Hall sensors to incremental or absolute encoders.

Example 11.3 The "master-follower" system from Example 10.5.

Example 11.4 Use of failure prediction techniques as part of the company's business model.

12. Local Quality

- *Transition from a uniform structure of the object (or the external environment, external influence) to a nonuniform one. Different parts of the object should have (or perform) different functions. Each part of the object should be in conditions that are most favorable for its operation. Make each part of an object fulfil a different and/or complementary useful function.*

Example 12.1 In the pump station control system (Fig. E.17), a single frequency converter (FC) is used to sequentially start several motors, with each motor being switched directly to the grid once it reaches 50 Hz. See also Trick 20, "Universality."

Example 12.2 In machines with superconducting windings, only the superconducting windings themselves are cooled to the critical temperature, not the entire machine. See also Example 5.2.

Example 12.3 Patent SU1575058 proposes to use integrated heat pipes with an evaporation zone, a condensation zone and a condensate pipeline to cool the rotor; all elements of the system are located on the rotor. See also Example 26.1.

Example 12.4 In high-voltage machines with conductors with a rectangular cross-section, the electric field intensity is higher at the corners of the conductors than

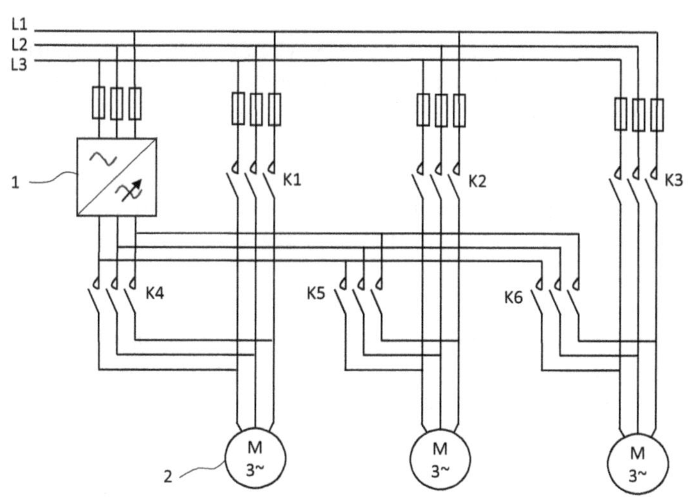

Fig. E.17 Pump station control system: 1—FC, 2—electric motor

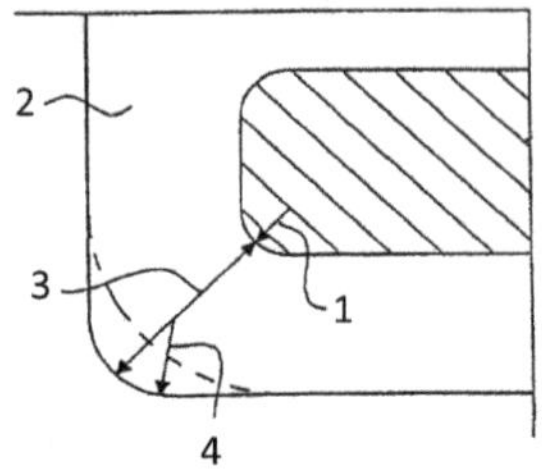

Fig. E.18 Insulation structure (patent US6404092B1): 1—conductor corner radius, 2—insulation, 3—insulation thickness, 4—insulation corner radius

along their sides. To ensure reliable operation of the insulation, its thickness in the area of the corners of the conductors is made greater than in the area of their sides (Fig. E.18).

13. Cheap Short-Lived Instead of Expensive Long-Lasting

- *Replace an expensive object with a set of cheaper objects, sacrificing some qualities (e.g., durability).*

Example 13.1 In applications with low rotational speeds, a direct drive with a low-speed electric motor is often used. Alternatively, a system consisting of a gearbox and a high-speed motor can be employed. This system will generally have lower reliability and a shorter service life but will be more cost-effective.

Example 13.2 In applications where a smooth start, typically implemented through frequency converters or soft starters, is usually justified (e.g., for pumps), a direct-on-line start of the motor still can be used. This approach will lead to increased wear of the motor and the mechanism but will help reduce capital costs associated with the power electronics equipment.

Example 13.3 To reduce total system costs when installing an electric drive system, "single-use" fuses can be used as protection devices instead of more expensive automatic circuit breakers.

Example 13.4 It is possible to save costs by not using filters between the frequency converter and the electric machine (EM) in systems with long cables, although this is likely to lead to premature degradation of the EM insulation due to travelling wave effect.

Example 13.5 Insulation systems vary in structure, type of materials, additives, and temperature tolerance. Insulation rated to last 100,000 hours is different from insulation rated to last 10,000 hours.

14. Russian Dolls

- *Place one object inside another. Place multiple objects inside others. Make one part pass (dynamically) through a cavity in the other.*

Example 14.1 The counter-rotating propulsion system of a flying apparatus (Fig. E.19) can use concentric shafts (one shaft passes through the cavity of the other).

Example 14.2 Patent US11211854B2 proposes combining a conventional active zone with a winding in slots and a slotless winding, as shown in Fig. E.20.

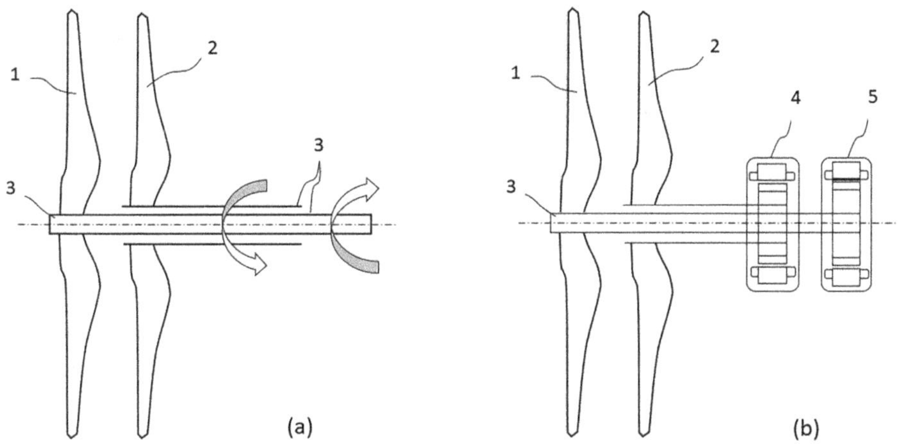

Fig. E.19 Concentric shafts (1—first propeller, 2—second propeller, 3—shafts, 4—first motor, 5—second motor)

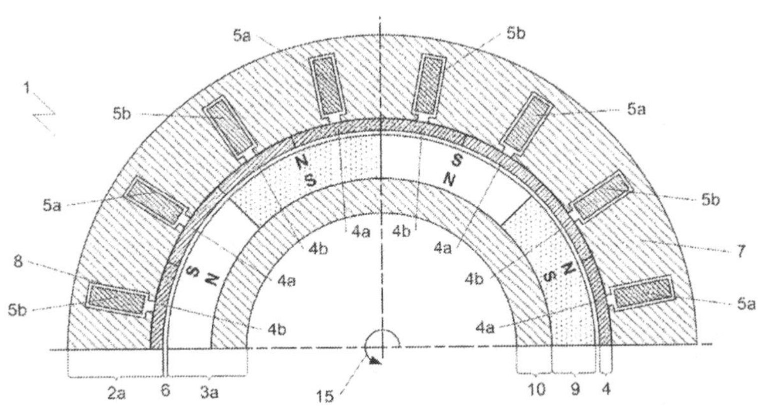

Fig. E.20 Hybrid machine (patent application US2019115814A1): 4a, b—airgap winding, 5a, b—slot winding

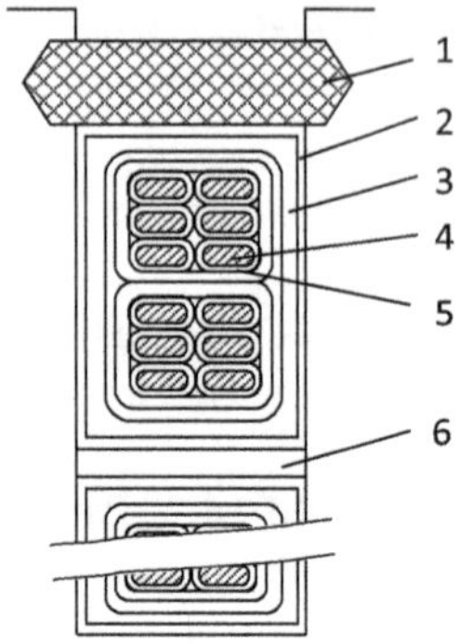

Fig. E.21 Insulation system (1—wedge, 2—insulating spacers, 3—body insulation, 4—conductor, 5—turn insulation, 6—interlayer insulation)

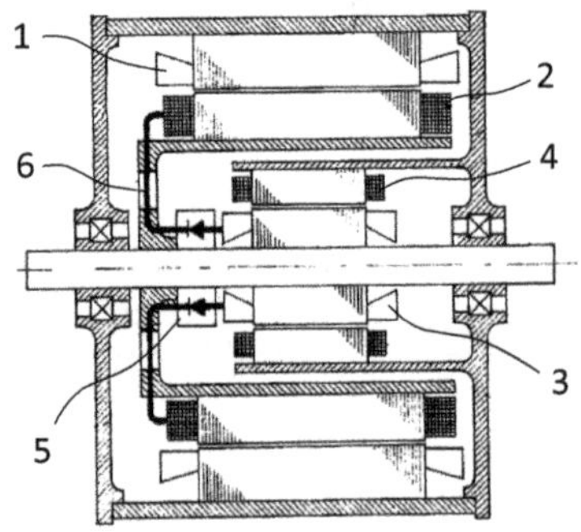

Fig. E.22 Machines with an integrated exciter (1—stator winding, 2—exciter winding, 3—exciter rotor winding, 4—exciter stator winding, 5—diode rectifier, 6—power cable from the rectifier)

Example 14.3 In a synchronous machine with an exciter on the shaft, to save axial space, the exciter is located inside the rotor, as shown in Fig. E.22 (example from FR2327666A1).

Example 14.4 In high-voltage machines, such as turbogenerators and hydrogenerators, the insulation is multilayer/multilevel (Fig. E.21). High-voltage cables are also multilayered (Fig. E.22).

Example 14.5 Machines with two or more air gaps, as shown in Example 19.1.

15. Discarding and Recovering

- *Make portions of an object that have fulfilled their functions go away (discard by dissolving, evaporating, etc.) or modify these directly during operation. Conversely, restore consumable parts of an object directly during operation.*

Example 15.1 To start a SCIM, an inverter or soft starter can be used, which are switched off after starting, and the SCIM is switched directly to the grid using a bypass contactor (Fig. E.23). See also Example 12.1.

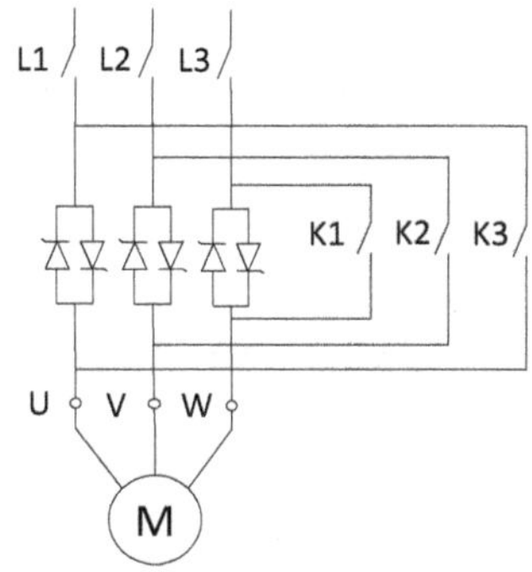

Fig. E.23 Soft starter with bypass contactor

Example 15.2 Energy regeneration/recovery: Braking energy, instead of being converted into heat, can be stored in accumulators (batteries, capacitors, flywheels) and reused. See also Example 8.3.

Example 15.3 See the example "Self-healing Insulation" in Chap. 5.

Example 15.4 Permanent magnets that have lost some of their magnetic properties, for example, due to high temperatures, can be remagnetized in special pulse magnetizing machines.

Example 15.5 In a machine where the stator is made up of segments with independent winding sections, if a winding fault occurs in one of the segments, that segment can be disconnected, allowing the machine to continue operating at a reduced load.

16. Partial or Redundant Action

- *If achieving 100% of the desired effect is difficult, aim for "slightly less" or "slightly more" instead. This approach can significantly simplify the task.*

Example 16.1 If the application requires certain torque that would necessitate the development of a new specialized electric motor, which is unacceptable due to time or cost constraints, a readily available oversized standard motor with higher torque can be used instead.

Example 16.2 When producing a winding with compound filling, it is difficult to measure the exact amount of compound needed for the mold in which the winding is placed, so it is filled with a surplus. After hardening, the excess compound is simply removed from the mold.

Example 16.3 Coreless machines have low inductance, and for optimal performance, it is desirable to use a high PWM frequency to reduce current ripple. In some applications, a reduced PWM frequency can be used—resulting in the motor operating with lower efficiency, while the inverter operates with higher efficiency, meaning the overall system efficiency will not change significantly.

Example 16.4 If cooling large parts of the machine is problematic, one can cool some elements only locally.

17. Application of Composite Materials

- *Transition from homogeneous materials to composite materials.*

Example 17.1 The motor housing can be made of fiber reinforced polymer material with integrated cooling ducts.

Example 17.2 In high-speed machines with permanent magnets fastened on the rotor surface, carbon fiber retaining screens/sleeves are used to secure the magnets in place.

Example 17.3 The windings of coreless machines may be composed of copper conductors, enamel (insulation on the conductors), glass fibers (woven with the conductors to provide mechanical strength), and cured epoxy resin, thus being a composite material.

Example 17.4 In machines with high supply frequency and high rotational speed, litz wire is used instead of solid conductors to reduce copper losses. Litz wire consists of multiple strands, each individually coated with insulating varnish.

Example 17.5 The thin steel sheets used in machine cores (laminations), separated by thin layers of insulation, form a composite material.

Example 17.6 To reduce the weight of electric machines, the supporting structures can be made from composite materials.

18. Intermediary

- *Use an intermediary carrier article or intermediary process. Use an intermediate object that transfers or transmits the action. Merge one object temporarily with another (which can be easily removed).*

Example 18.1 Transformers are used to match the voltage level of the grid and the electric machine.

Example 18.2 Gearboxes are used to match the rotational speed of the available electric machine with the required speed of the drive mechanism.

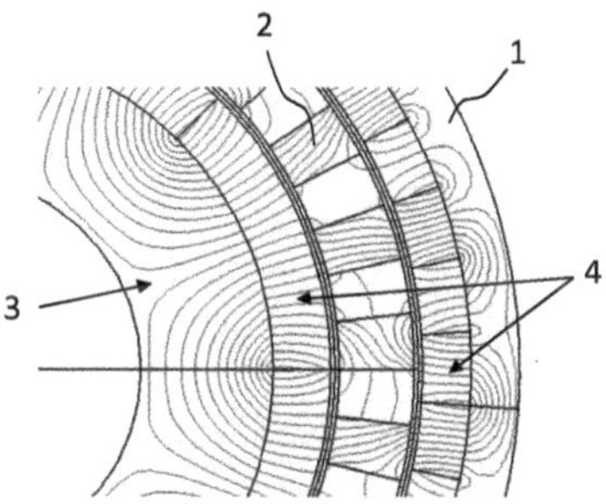

Fig. E.24 Main elements of the magnetic gearbox and magnetic field: 1—yoke of the outer ring, 2—"fingers," 3—yoke of the inner ring, 4—permanent magnets

Example 18.3 For convenience in connecting a multiphase cable to a low-power electric machine, the ends of its winding are soldered to a small printed circuit board, which has traces connecting the winding in the required configuration and serves as an interface for the cable.

Example 18.4 Flexible couplings are used to connect the shafts of the electric machine and the mechanism, and to protect against shocks and vibrations.

Example 18.5 In magnetic gearboxes (Fig. E.24), "fingers"—ferromagnetic elements—are located between the outer and inner rings, transferring and configuring the magnetic field.

19. Another Dimension

- *Difficulties related to the movement (or placement) of an object along a line are eliminated when the object gains the ability to move in two dimensions (i.e., on a plane). Accordingly, tasks related to the movement (or placement) of objects within a single plane are resolved by transitioning to three-dimensional space. Use multi-story arrangement of objects instead of a single-story one. Use the reverse side of the given surface. Utilize optical flows that fall on an adjacent surface or on the reverse side of the existing surface. If an object contains or moves in a straight line, consider use of dimensions or movement outside the line. If an object contains or moves in a plane, consider use of dimensions or movement outside the current plane. Tilt or reorient the object, lay it on its side.*

Example 19.1 Multilayer machines (Fig. E.25) have multiple rotors and stators (example from WO2007139299A1).

Example 19.2 In systems for machining and measuring three-dimensional geometries, a system of rotational motion can be used in conjunction with a linear motion mechanism driven by a magnetostrictive drive (Fig. E.26).

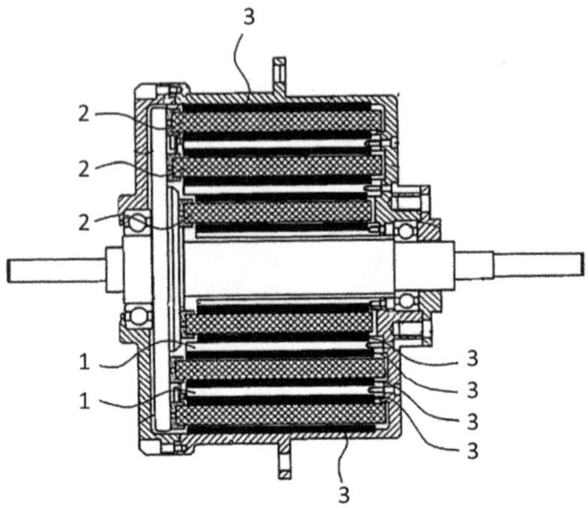

Fig. E.25 Multilayer electric machine (1—rotors, 2—stators, 3—magnets on the rotor)

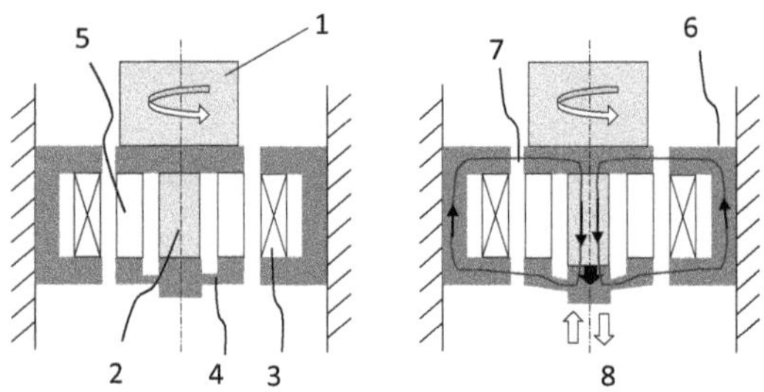

Fig. E.26 Rotary-linear actuator (1—spindle, 2—magnetostrictive actuator, 3—coil, 4—preloaded diaphragm, 5—nonmagnetic material, 6—magnetic material, 7—magnetic flux, 8—linear displacements)

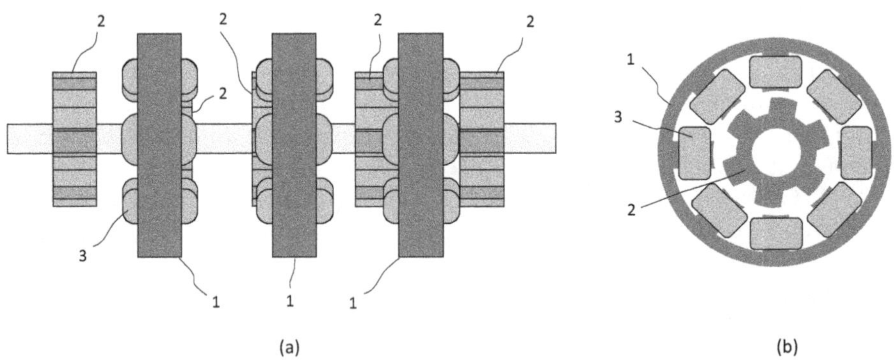

Fig. E.27 Rotary-linear actuator of the reluctance type with 3 stators and 5 runners (1—stators, 2—runners, 3—stator coils): Side view (**a**), shaft-end view (**b**)

Example 19.3 In a rotary-linear actuator of the reluctance type (Fig. E.27), rotation is created based on the principle of a switched reluctance motor, while linear motion is achieved by activating specific pairs of stators and runners.

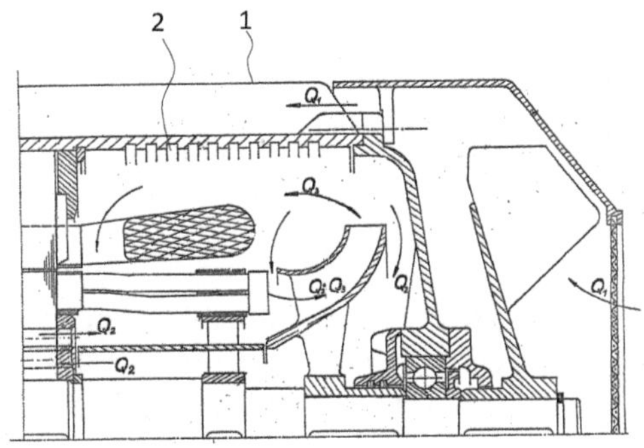

Fig. E.28 Cooling system of the electric machine (1—fins on the outer side of the housing, 2—fins inside the housing)

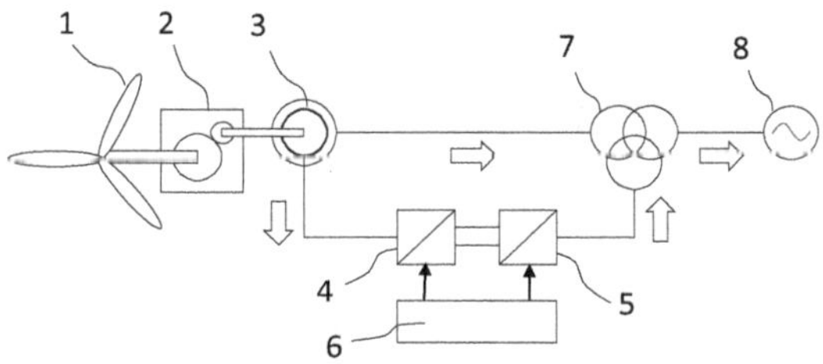

Fig. E.29 Diagram of the electric power take-off of a wind turbine (1—blades, 2—gearbox, 3—double-fed machine, 4, 5—inventors, 6—control system, 7—transformer, 8—grid)

Example 19.4 In printed circuit boards for motor control, both surfaces can be used for component mounting.

Example 19.5 In patent DE2558405B2, cooling ribs are present both outside and inside the housing of the electric machine (Fig. E.28).

20. Universality

- *Make a part or object perform multiple functions; eliminate the need for other parts.*

Example 20.1 The car's starter-generator performs two functions: It accelerates the internal combustion engine (ICE) and then generates energy for the car's electrical system.

Example 20.2 In a wind turbine's power conversion system with a double-fed machine, a single three-winding transformer is used to transfer energy to the grid from both the stator and rotor windings (Fig. E.29).

Example 20.3 A cable sheath made from magnetic material increases the inductance of the cable, thereby acting as a filter for high-frequency current oscillations and interference in the cable.

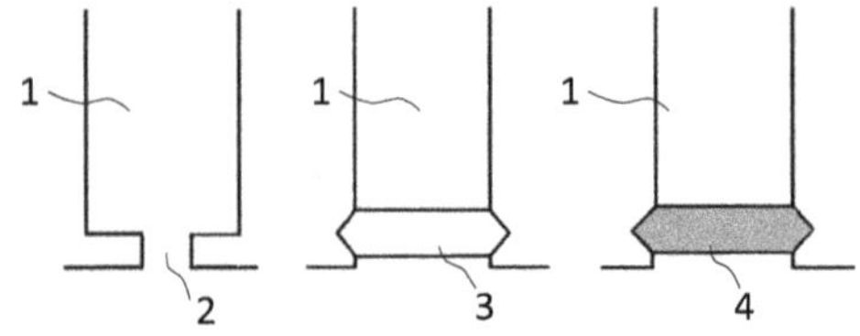

Fig. E.30 Slots (1): with narrow slot opening (2), with a regular wedge (3), with a magnetic wedge (4)

Example 20.4 In some variable frequency drives (VFDs), the built-in controller not only performs motor control, protection, and interface functions but can also be programmed to solve local automation tasks.

Example 20.5 The rotor, exciter, and cooling fan can be mounted on the same shaft of the machine.

Example 20.6 Magnetic slot wedges (Fig. E.30) both hold the winding in place within the slot and provide the required magnetic properties in the slot opening. The use of magnetic wedges improves the key electrical characteristics of electric motors by reducing the magnetic resistance of the air gap, decreasing winding and stator core losses, lowering temperatures, and increasing efficiency. A magnetic wedge can, for example, be made from a mixture of iron powder and epoxy compound.

See also Example 12.1.

21. Thermal Expansion

- *Use thermal expansion (or contraction) of materials. If thermal expansion is being used, use multiple materials with different coefficients of thermal expansion.*

Example 21.1 Joining components of electric machines (e.g., mounting bearings onto a shaft) can be achieved by methods such as heating the surrounding component to achieve an interference fit, cooling the enclosed component, or using a combination of both heating the outer component and cooling the inner component.

Example 21.2 Patent application US2014070634A1 describes a rotor structure with surface-mounted permanent magnets are secured by a retaining band. In this design, the thermal expansion coefficient ratio between the magnet material and the band material should not exceed 1.15.

Example 21.3 Special linear actuators can use shape memory alloys (SMAs). When current passes through the SMA, heat is generated, raising the SMA's temperature. Once the temperature surpasses a transition point, the atoms rearrange into a different crystalline structure, causing the material to contract upon heating and then re-elongate upon cooling, resulting in linear motion.

Example 21.4 In vertical shaft generators, differential thermal expansion between the stator core and stator housing often poses a problem. Patent CA1290011C offers a solution wherein sensors positioned on the stator housing and core monitor their temperatures. The temperatures are then compared, and if the housing temperature drops below that of the core, heaters around the housing are activated to maintain the housing temperature close to that of the core.

See also Example 1.2.

22. Spheroidality

- *Instead of using rectilinear parts, surfaces, or forms, use curvilinear ones; move from flat surfaces to spherical ones; from parts shaped as a cube (parallelepiped) to ball-shaped structures. Use rollers, balls, spirals, domes. Go from linear to rotary motion (or vice versa). Use centrifugal forces.*

Example 22.1 To reduce resistance to coolant flow, cooling channels can have rounded corners.

Example 22.2 To reduce localized electric stresses in electric machine insulation caused by high electric field concentrations around sharp corners of teeth, slots, and conductors—which can lead to insulation aging and breakdown—the corners of teeth, slots, and conductors are rounded.

Example 22.3 Cooling systems for electric machines are known to use pushing the cooling air radially by a fan mounted on the shaft.

Example 22.4 For actuators with a high number of degrees of freedom, spherical machines can be used—see Table 4.4 in Chap. 4.

Example 22.5 A linear actuator in patent application CN118030794A (Fig. E.31) is driven by a rotating electric motor. This rotational motion is converted to linear motion using a ball screw mechanism.

23. Inert Atmosphere

- *Replace a normal environment with an inert one. Add neutral parts, or inert additives to an object. Carry out the process in a vacuum.*

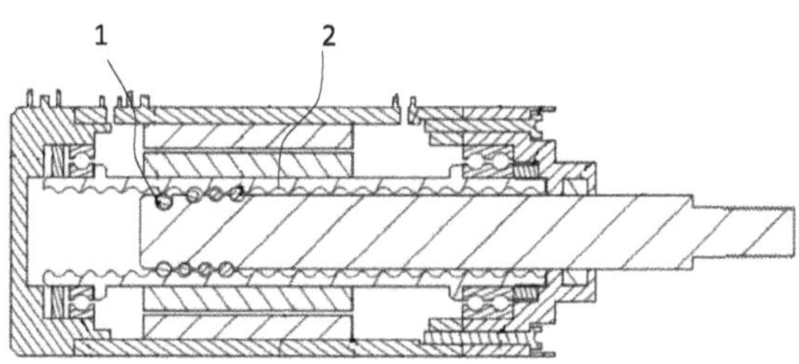

Fig. E.31 Actuator with ball-screw (1—balls, 2—screw)

Example 23.1 Explosion protection for electric machines can be achieved by purging the machine's housing with an inert gas before commissioning, or by filling the housing with an inert gas at a pressure higher than the surrounding atmosphere.

Example 23.2 A vacuum is used for impregnating windings—see Example 1.1.

Example 23.3 In electric machines with superconducting windings, liquid nitrogen is used to create ultra-low temperatures in the winding area.

Example 23.4 In some cases, electronics is manufactured in an inert environment (such as nitrogen, a pure and dry gas) to prevent oxidation.

24. Asymmetry

- *Change the shape or properties of an object from symmetrical to asymmetrical. Change the shape of an object to suit external asymmetries (e.g. ergonomic features). If an object is asymmetrical, increase its degree of asymmetry.*

Example 24.1 In a self-braking motor with a conical rotor (Fig. E.32), when the motor is turned off, a spring pushes the rotor out of the stator. This causes the rotor to shift left along its axis, and the brake pads press against the bearing shield due to the spring force, creating braking torque.

Example 24.2 Unlike machines with rotating rotors, machines with rolling rotors are asymmetric at any given moment of operation—the rotor is offset from the machine's center.

Example 24.3 The structure of some machine rotors can be asymmetric. Patent application US4074160A presents a rotor (Fig. E.33) with asymmetric flux barriers. Patent application WO2018188755A1 presents a rotor with an asymmetric arrangement of magnets (Fig. E.34).

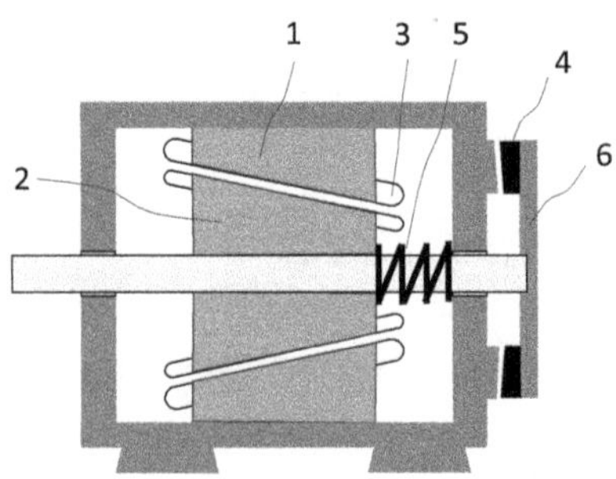

Fig. E.32 Self-braking motor with a conical rotor (1—stator, 2—rotor, 3—winding, 4—brake pad, 5—spring, 6—brake)

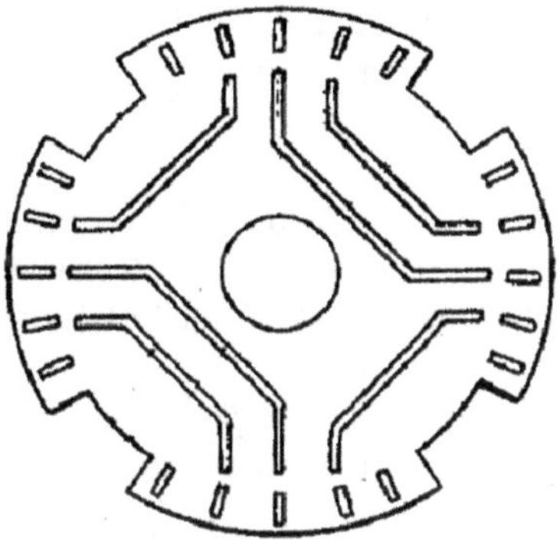

Fig. E.33 Rotor with asymmetric flux barriers

Fig. E.34 Rotor with an asymmetric arrangement of magnets

25. Flexible Shells and Thin Films

- *Use flexible shells and thin films instead of conventional three-dimensional structures. Isolate the object from the external environment using flexible shells and thin films.*

Example 25.1 In high-speed permanent magnet synchronous motors, retaining sleeves made of metal or carbon fiber are used to secure the permanent magnets on the rotor.

Example 25.2 To prevent bearing currents in drive systems with variable frequency drives, bearings with an electrically insulating coating on the outer surface are used. This coating, which can be based on oxide ceramics, is applied via thermal spraying.

Example 25.3 To prevent damage or deformation of power electronics PCBs, or to protect them from external influences, various coatings are used, such as polyurethane and acrylic coatings, which form a thin polymer film only fractions of a millimeter thick.

Example 25.4 For electrical insulation of a laminated core, an electrostatic epoxy coating can be used.

26. Phase Transitions

- *Use phenomena occurring during phase transitions (e.g., volume changes, release or absorption of heat, etc.).*

Example 26.1 Patent EP2637176B1 proposes a method to improve a transformer's ability to operate under peak loads. According to the invention, a phase-change material is embedded (encapsulated) in the casting resin of a cast-resin transformer. This phase-change material, distinct from the resin, has a phase transition temperature between 40 and 200 °C. When the material reaches this phase transition temperature under peak loads, it absorbs heat generated by losses without raising the winding temperature. Only when the entire quantity of the phase-change material has completed its phase transition does the temperature begin to rise further.

Example 26.2 In motors with frequent start-stop cycles, using a phase-change material (PCM) for cooling can reduce the average temperatures of the stator core and stator windings by several degrees. The effect of this method is greater with higher values of thermal conductivity and volumetric heat of fusion in the PCM.

Example 26.3 Patent application WO2014063982A2 proposes using combinations of several different PCMs with distinct critical temperatures to provide cooling in multiple stages (at several levels). Additionally, detecting the start of the phase transition can be used to monitor the machine's temperature.

Example 26.4 For cooling the electric machine in an electric vehicle, refrigerant from the air conditioning system can be used. As the refrigerant passes through the machine, it evaporates, providing significantly more effective cooling than conventional water cooling.

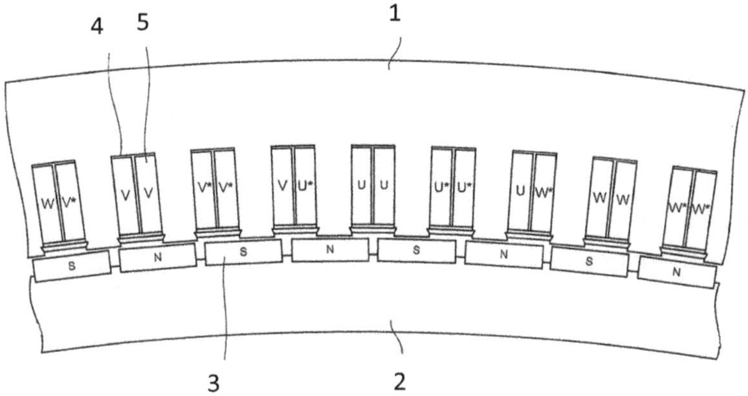

Fig. E.35 Active are of a PMSM with q <1 (1—stator yoke, 2—rotor yoke, 3—permanent magnet, 4—slot, 5—coil)

27. Blessing in Disguise

- *Use harmful factors (particularly, harmful effects of the environment or surroundings) to achieve a positive effect. Eliminate a harmful factor by combining it with another harmful factor. Amplify a harmful factor to such a degree that it is no longer harmful.*

Example 27.1 In conventional machines, harmonics higher than the fundamental (first) are generally considered undesirable. However, a permanent magnet synchronous motor with concentrated windings and a fractional slot-per-pole-per-phase number q (an example of the electromagnetic structure of such a machine taken from patent application US2015084342A1 shown in Fig. E.35) can use a higher harmonic, such as the 5th or 11th, for energy conversion instead of the fundamental harmonic, as is common in most machines.

Example 27.2 In submersible oil-filled machines, friction of rotating parts against the oil leads to losses. Oil movement can be used to move heat from the hottest parts of the machine to areas that are easier to cool. For this purpose, channels and structural elements can be created that bring the oil into organized motion along given circulation contours.

Example 27.3 Heat generated in the machine due to losses can be converted into electricity (via the thermoelectric or Seebeck effect) and used, for example, to power sensors.

Example 27.4 Failed permanent magnets (PMs) are often disposed of, which incurs certain costs. Recently, technologies have emerged for reusing old PMs, based on re-magnetization either after disintegration into powder or without disintegration.

Example 27.5 The vibration energy of a machine's housing can be harnessed to generate power, which can be used to supply vibration sensors, making them fully autonomous and wireless. Generators based on electromagnetic, electrostatic, piezoelectric, and triboelectric principles are known and can be used for such a purpose.

See also Example 2.1 (using winding resistance to preheat the machine with direct current).

28. Beforehand Cushioning

- *Prepare emergency means beforehand to compensate for the relatively low reliability of an object.*

Example 28.1 In a "master-follower" system, as described in Example 10.5, if the master variable frequency drive (VFD) fails, one of the "follower"-VFDs takes over control of the system.

Example 28.2 In an induction motor (IM) startup and control system, a bypass contactor is added to the VFD, enabling direct connection of the IM to the power grid in case of VFD failure.

Example 28.3 Self-healing insulation with integrated capsules containing "healing" material, as described in the example in Chap. 5.

Example 28.4 Two motors in tandem (see Example 3.3) form a redundant system. If one motor fails, the second one can continue to provide half of the power to the load.

Example 28.5 To ensure uninterrupted operation of the electromechanical system, an energy storage unit can be incorporated into the circuit.

See also Example 2.6.

29. Self-Service

- *Make an object serve or organize itself by performing auxiliary helpful functions. Use waste resources, energy, or substances.*

Example 29.1 The vibration energy of a motor housing can be recovered and used to power micro-sensors. For instance, patent application CN104104204A describes a vibration energy recovery device with self-adaptive control of ultrasonic piezoelectric actuation (see also Trick 38, "Mechanical Oscillation").

Example 29.2 Self-healing insulation with integrated capsules containing "healing" material, as described in the example in Chap. 5.

Example 29.3 In the cooling system of an electric machine shown in Fig. E.36, coolant is sprayed onto the end-windings under the influence of centrifugal forces.

Example 29.4 Self-aligning bearings can operate with significant angular misalignment between the shaft and housing (which occurs due to shaft bending or misalignment from structural flexibility or insufficient rigidity) thanks to the spherical rolling or sliding surface (Fig. E.37).

Example 29.5 The method for drying insulation after impregnation involves connecting the winding to an AC voltage source, causing current through the winding to heat steel parts via stray fluxes and conductors through copper losses. When connected to a direct current, heating occurs only through copper losses.

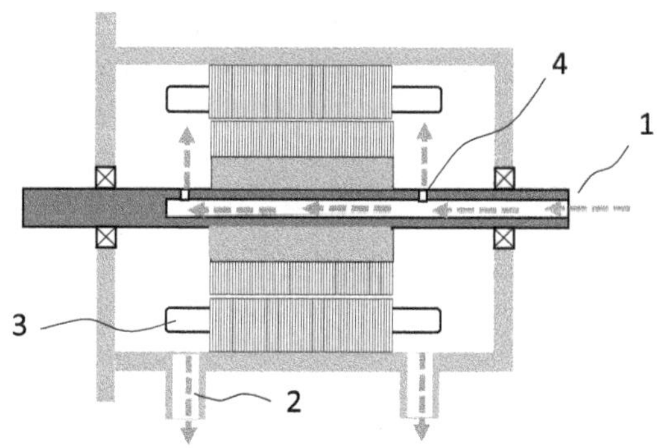

Fig. E.36 End-winding cooling system (1, 2—coolant inlet and outlet, 3—end-windings, 4—spray channel)

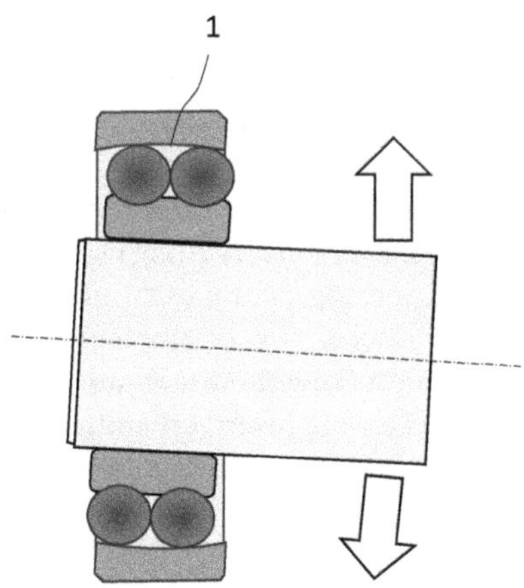

Fig. E.37 Self-aligning rolling bearing (1—spherical surface)

30. Strong Oxidants

- *Replace ambient air with oxygen-enriched air. Replace enriched air with pure oxygen. Expose air or oxygen to ionizing radiation. Use ionized oxygen. Replace ozonized (or ionized) oxygen with ozone.*

Example 30.1 In patent application EP0511409A1, a hard nitride layer is created on the shaft neck where the bearing is mounted to ensure durability.

Example 30.2 Films that form from interaction with a corrosive environment and inhibit further corrosion are called oxide (or oxidized) protective films. These films can be used to protect the machine housing and internal components.

Example 30.3 Patent US7806991B2 describes a technology for producing powder for permanent magnets with low eddy current losses, where specific concentrations and heat treatment of oxidizers like oxygen and fluorine are regulated.

Example 30.4 In patent application JPH0386036A, to prevent erosion of cooling channels, the concentration of dissolved oxygen in the cooling water is controlled by injecting additional volumes of oxygen or hydrogen.

31. Porous Materials

- *Make an object porous or add porous elements (inserts, coatings, etc.). If an object is already porous, use the pores to introduce a useful substance or function.*

Example 31.1 In USSR invention author's certificate Nr. 187,135 there is described an evaporative cooling system for electrical machines, characterized in that, to eliminate the need to supply a cooling agent to the machine, the active parts and certain structural components are made of porous materials, such as porous powdered steels impregnated with a liquid cooling agent. During the machine's operation, this agent evaporates, thereby providing temporary, intense, and uniform cooling.

Example 31.2 A gas-static bearing (Fig. E.38) used for supports with low loads and high rotational (sliding) speeds, consists of a chamber, porous inserts, a supply line, a housing, and a gas-tight sleeve. The bearing insert comprises a sleeve and porous inserts. Gas is supplied under pressure through the porous inserts at several points into the cylindrical gap between the shaft journal (not shown in Fig. E.38) and the bearing insert, creating a gas lubrication layer. In steady-state operation, the surfaces of the shaft and journal do not come into contact.

Example 31.3 Patent US5164256 and patent application EP0416120A1 present various solutions with plain bearings having a liner made of porous material, which is filled with lubricant, which gradually flows from the pores into the gap between the liner and the shaft, providing low friction sliding.

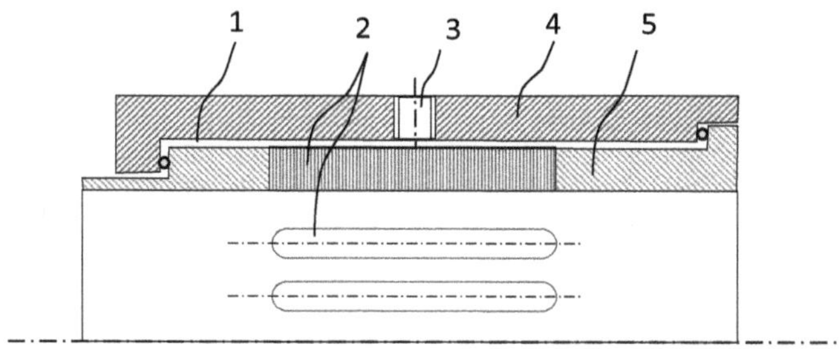

Fig. E.38 Gas-static bearing (1—chamber, 2—porous inserts, 3—supply line, 4—housing, 5—gas-tight sleeve)

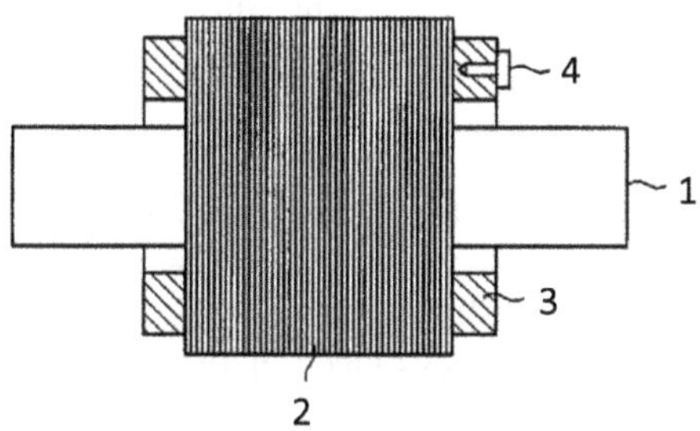

Fig. E.39 SCIM rotor from patent application DE102009008439A1 (1—shaft, 2—stack, 3—short-circuit winding ring, 4—balancing weight)

Example 31.4 The hollow shaft not only reduces the weight of the machine, but also opens up the possibility of using the channel passing through the center of the machine for various purposes.

Example 31.5 Rotor can have a "lightweight" supporting structure, with some kind of spokes supporting the rim.

Example 31.6 To increase the electrical strength, as well as the mechanical strength and flexibility of the insulation, it can be made multilayer, and some layers may contain pores with gas. A similar solution is presented, for example, in patent application US2020/0126694.

32. Anti-weight

- *Compensate for the weight of the object by connecting it to other objects that have lifting force. Compensate for the weight of the object by interaction with the environment (due to aero-, hydrodynamic, and other forces).*

Example 32.1 To prevent vibrations of the rotor, it must be balanced. To eliminate imbalance, a balancing weight, for example a bolt, is attached closer to the periphery of the rotor, for example, on the ring of the short-circuited winding in the SCIM (Fig. E.39).

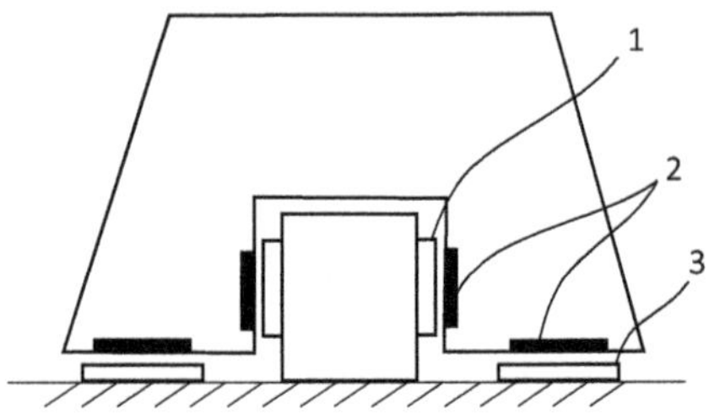

Fig. E.40 Maglev concept (1—propulsion coils, 2—superconducting magnets, 3—levitation coils)

Example 32.2 In Maglev transport systems (as shown in Fig. E.40), the weight of the train is compensated by a magnetic field, enabling magnetic levitation.

Example 32.3 Bearing supports for high-speed machines use active magnetic suspensions.

Example 32.4 In modern large vertical hydrogenerators, axial loads are very high. To absorb the axial load from the weight of the hydrogenerator's rotor and the axial pressure of the water and to transfer these loads through the supporting parts to the foundation, sliding thrust bearings are used. The sliding bearing is a structure of two rubbing surfaces placed in an oil bath.

33. Skipping

- *Conduct the process or its individual stages (e.g., harmful or dangerous) at high speed.*

Example 33.1 The machine's rotational speed is proportional to the supply frequency. When operating across a wide range of rotation speeds, undesirable resonant vibrations may occur in the machine-load system at certain frequencies. Modern frequency converters (FCs) incorporate a function to avoid resonance zones: If the setpoint falls within a "forbidden" frequency band, the rotational speed will only adjust when it reaches the upper or lower boundary of this band. The output frequency of the FC will then remain constant until the setpoint moves above or below the "forbidden" frequency range. This method of avoiding mechanical resonances is illustrated in Fig. E.41.

Example 33.2 In a high-speed motor system, in the event of a network power outage, the FC controller can be powered and remain functional, by regenerating rotational energy of the motor and the load.

Example 33.3 In Chap. 5, in the problem of expanding the ranges of operating torques and speeds in a hybrid or all-electric electric vehicle, a winding switching system is described. A similar system is implemented in Yaskawa's QMET-II product and is described in patent JP3596711B2.

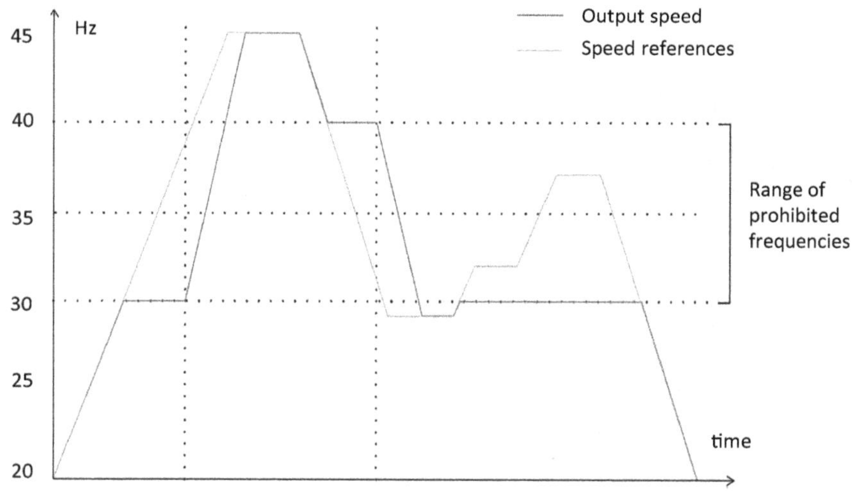

Fig. E.41 Illustration of the difference between the setpoint and the actual output speed value when passing through the prohibited frequency band

34. Use of Pneumatic and Hydraulic Structures

- *Instead of solid parts of the object, use gaseous and liquid ones: inflatable and hydrofillable, air cushion, hydrostatic, and hydroreactive.*

Example 34.1 Patent application US2011148229A1 describes a forced circulation liquid cooling system that includes a sump for collecting the liquid, a pump, and nozzles. The system provides direct cooling for both the stator and rotor from the air gap side as well as cooling for the end windings (Fig. E.42).

Example 34.2 In Example 38.1, gas springs are used to provide reciprocating oscillations of a runner inside the body of a linear machine.

Example 34.3 To ensure long-term operation of machine bearings at high rotation speeds, gas bearings are used. See also Example 31.2.

Example 34.4 Patent application US20080296909A1 describes using a pneumatic motor—a "turbo-pneumatic assistant"—to prevent high inrush currents when starting an induction motor (Fig. E.43). Current is supplied to the stator only after the rotor has accelerated.

35. Merging

- *Connect homogeneous objects or objects intended for related operations. Bring closer together (or merge) identical or similar objects or operations in space. Make objects or operations contiguous or parallel; bring them together in time.*

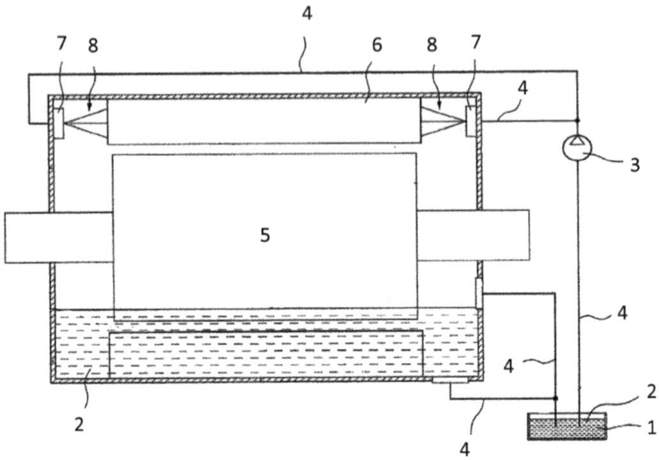

Fig. E.42 Cooling system for the electric machine: 1—reservoir with coolant, 2—coolant, 3—pump, 4—tubes, 5—rotor, 6—stator, 7—nozzle, 8—jets of sprayed liquid

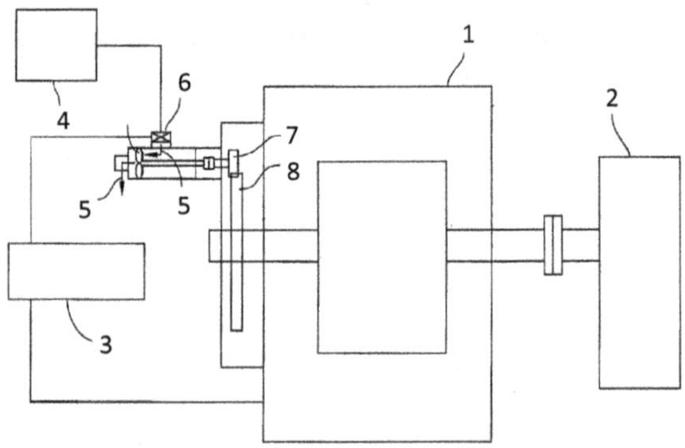

Fig. E.43 Turbo-pneumatic starting system for the electric motor: 1—electric motor, 2—load, 3—control system, 4—air supply, 5—air flow, 6—valve, 7, 8—gears

Example 35.1 In the PMSM rotor, a ferromagnetic yoke is used to bridge the flux between permanent magnets of different polarities. In the Halbach structure (Fig. E.44), the magnets themselves bridge the flux without creating leakage fluxes—the magnetic field goes only outward—toward the stator.

Example 35.2 Conventional high-voltage generators are designed for voltages up to 30 kV. To connect such generators to networks with voltages of several hundred kV, step-up transformers are required, which lead to additional losses, maintenance costs, and extra installation space. In the 1980s, generators (Fig. E.45) with cable windings insulated with polymer material were proposed, capable of withstanding voltages of about 500 kV. Such generators can be directly connected to the grid.

Example 35.3 In a pseudo-direct drive, an electric machine and a magnetic gearbox are combined into a single unit.

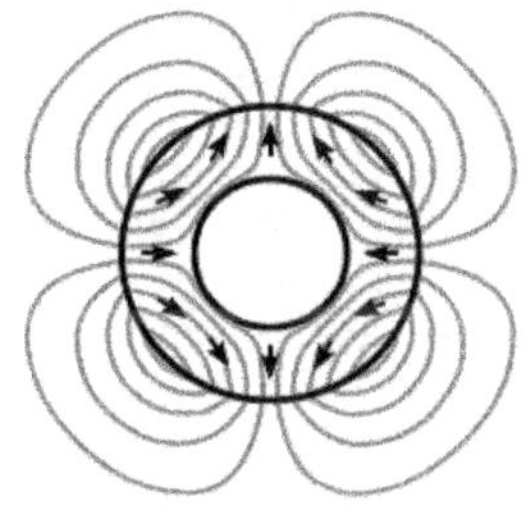

Fig. E.44 Ring of Halbach permanent magnets and the magnetic field (from patent application US2018219452A1)

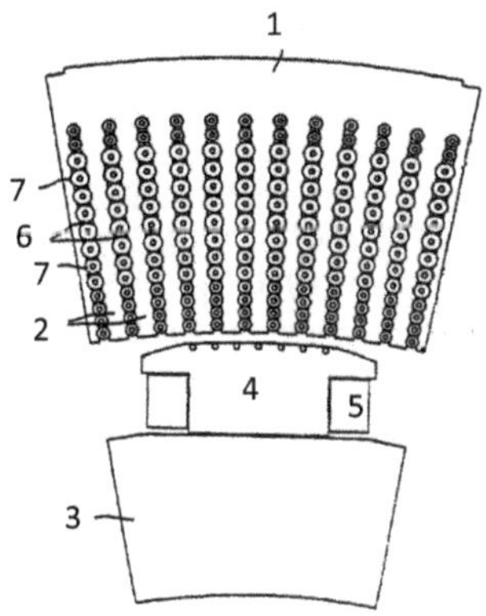

Fig. E.45 Structure of a generator segment with cable windings (1—stator yoke, 2—teeth, 3—rotor yoke, 4—rotor pole, 5—excitation winding, 6—slot, 7—cable winding, 8—air gap, 9—cooling channels)

Example 35.4 In a submersible oil-filled motor, the same oil can be used for filling the gaps and cavities around the rotor and for lubricating bearings, and the machine and the bearing cavities can be directly connected.

Example 35.5 Patent NO343796B1 describes an integrated thruster (POD) for ship propulsion, which is a combination of a gondola-based propulsor and a propulsor with a motor integrated into the nozzle (Fig. E.46). The propellers rotate in opposite directions.

36. Vice Versa

- *Instead of the action dictated by the conditions of the problem, perform the opposite action (e.g., instead of cooling the object, heat it). Make the moving part of the object (or the external environment) stationary, and make the stationary part move. Turn the object "upside down".*

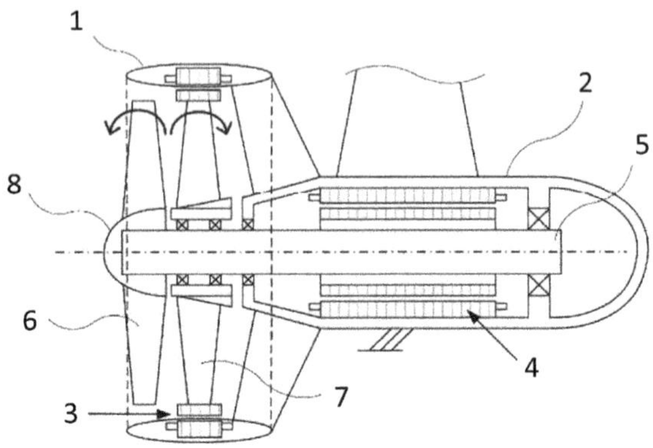

Fig. E.46 Integrated thruster (1—nozzle, 2—gondola, 3—rim drive electric motor integrated into the nozzle, 4—electric motor in the gondola, 5—shaft, 6—propeller nr 2, 7—propeller nr 1)

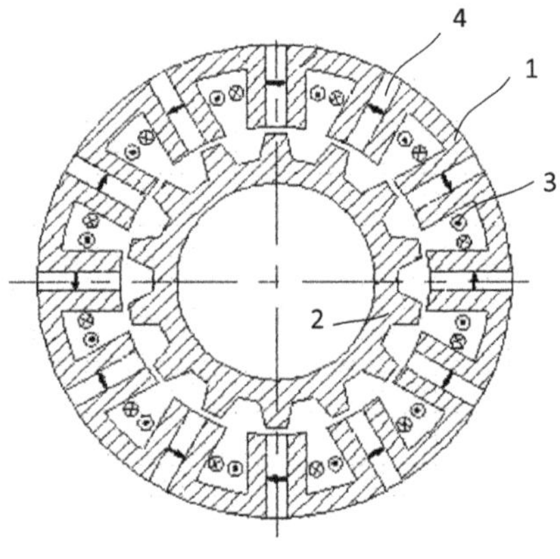

Fig. E.47 The structure of a flux-switching PMSM from patent CN103312104B (1—stator core, 2—rotor, 3—coils, 4—permanent magnets)

Example 36.1 Typically, in a PMSM, the magnets are located on the rotor. In a flux-switching PMSM, the magnets are located on the stator instead of the rotor (Fig. E.47).

Example 36.2 The stator is defined as a stationary component; however, in a birotative machine, the stator rotates (see also Example 7.1).

Example 36.3 In most electric machines, the rotor is located inside the stator, but in some cases (e.g., drone motors, wind turbines), using an external rotor topology can give certain advantages.

Example 36.4 Rotor mounting on the shaft is typically done by heating the rotor before assembly to create the necessary clearance. An alternative solution could be cooling the shaft before assembly. Cooling and shrinking of the shaft can be achieved, for example, by immersing the shaft in a liquid nitrogen bath. After assembly, the shaft heats up and, expanding, forms a tight fit with interference without deformation.

37. Color Change

- *Change the color of the object or external environment. Change the degree of transparency of the object or external environment. To observe poorly visible objects or processes, use coloring additives. If such additives are already applied, use marked atoms.*

Example 37.1 The type of surface material or its coating determines the cooling intensity of the surface. For example, white organic pigmented paint has both high reflectivity and high emissivity. Nanostructured coatings are being researched that can dissipate heat four times faster than was previously possible.

Example 37.2 In the utility model application CN202713047U, the motor has a transparent terminal box. The box is sealed and provides electrical insulation. The transparent walls allow visual monitoring of the condition of the electrical connections. In patent applications JP2000224796A and GB713608A, a "window" made of transparent material is used in the machine housing, opposite the inner parts that needs to be visually monitored. In patent application GB204499A, a plate made of material that changes color with temperature changes is used. This allows for a visual assessment of the machine's temperature.

Example 37.3 In patent application GB2450758A ("Device for reducing moisture in a rotating electrical machine"), it is proposed to use a desiccant agent inside the machine's housing that changes color depending on the humidity level inside the machine, along with a transparent window for visual monitoring of the desiccant's color.

38. Mechanical Oscillation

- *Bring an object into oscillatory motion. If such a movement is already taking place, increase its frequency (even up to ultrasonic). Use resonant frequency. Use piezo vibrators instead of mechanical ones. Use ultrasonic vibrations in combination with electromagnetic fields.*

Example 38.1 Patent RU2382477C2 describes a machine for generating significant pulsating axial forces (impulses). There are ring coils on the hollow ring housing (stator), and ring permanent magnets on the piston (Fig. E.48). The piston is a

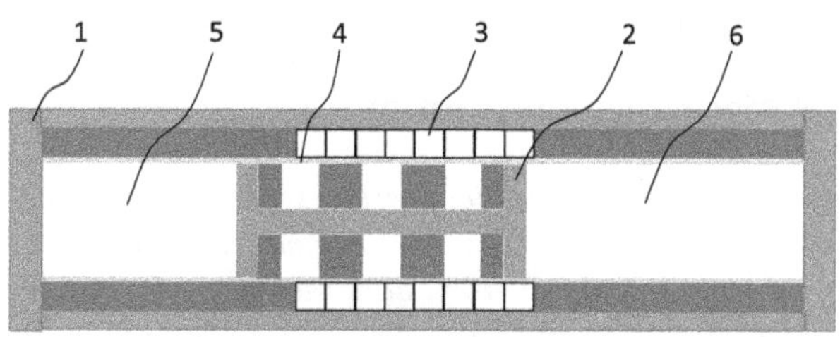

Fig. E.48 "Resonator" machine (1—housing, 2—piston, 3—annular coils, 4—ring magnets, 5, 6—gas chambers)

runner of a linear machine that performs reciprocating oscillations inside the housing. On both sides of the piston, sealed chambers filled with gas are formed, which act as "gas springs" when the gas is compressed. The piston and gas springs form a mechanical system with a specific resonant frequency. The linear machine's runner is driven at this frequency to achieve resonance and produce significantly larger axial impulses.

Example 38.2 Patent EP0294591B1 presents a solution for measuring the temperature in the windings of electric machines based on the temperature dependence of the speed of ultrasound propagation in certain media. A fiber optic cable is laid along the winding and connected to an ultrasonic source. Due to the temperature changes along the winding, the change in travel time or phase shift of the ultrasound can be measured electronically and interpreted as a temperature value.

Example 38.3 Patent application EP0100474A1 presents a method of impregnating or coating the windings of electrical machines with structurally viscous or thixotropic solvent-free reactive resins during the immersion process, characterized in that the reaction resin is exposed to ultrasound during immersion.

Example 38.4 Patent application CN101404442A presents a combination of a conventional electric motor and an ultrasonic motor, which utilizes the reverse piezoelectric effect and ultrasonic vibration of piezoelectric ceramics to convert the microscopic deformation of an elastic material into rotor movement through resonant amplification and frictional coupling.

Example 38.5 Patent FR 2645367B1 describes a method for manufacturing electrical machine cores from a magnetic material, including coating with a layer of particles by an electrostatic method, according to which the parts are degreased before applying a layer of insulating particles to them, characterized in that the parts are demagnetized before degreasing, and the demagnetized parts are subjected to mechanical vibrations (shaking) to remove sawdust particles that may settle on parts.

Example 38.6 In patent application DE3634421A1, to monitor the displacement of the rotor relative to the stator of an electric machine, the length of the air gap is measured using acoustic signals in the ultrasonic range. Continuous measurement can be performed both with the rotor stationary and with it rotating (Fig. E.49). Two measuring devices are installed at a certain angle along the circumference inside the machine housing to identify changes in the radial distance between the rotor and stator (s1, s2, s3), caused by rotational eccentricity or static eccentricity in any direction.

39. Mechanics Substitution

- *Replace the mechanical system with an optical, acoustic, or "scent-based" system. Use electric, magnetic, and electromagnetic fields to interact with the object. Transition from static fields to moving ones, from fixed fields to time-varying fields, and from unstructured fields to those with a defined structure. Use fields in combination with field-activated (e.g., ferromagnetic) particles.*

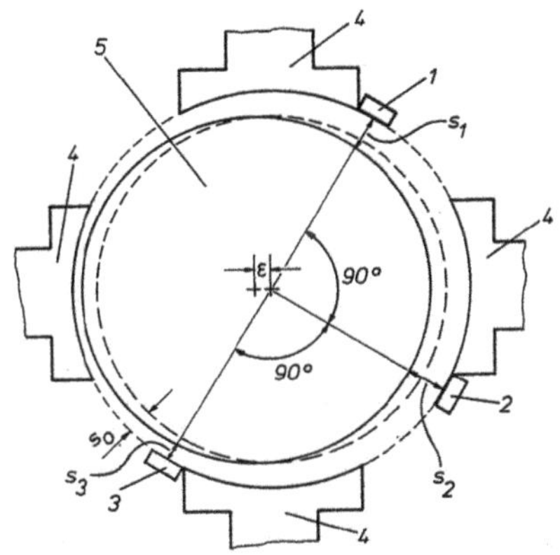

Fig. E.49 The principle of the measuring system (1, 2, 3—measuring devices, 4—stator poles, 5—rotor)

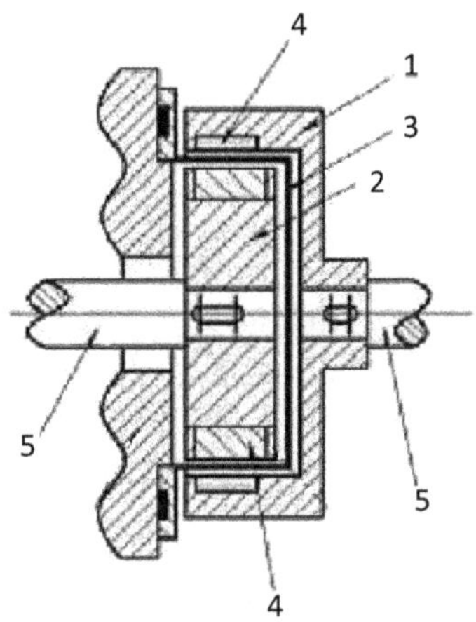

Fig. E.50 Magnetic coupling from patent application CN111030419A (1—outer coupling half, 2—inner coupling half, 3—sealing screen, 4—permanent magnets, 5—shafts)

Example 39.1 In a laser Doppler vibrometer (LDV), the Doppler effect is utilized—the change in frequency and, consequently, the wavelength of radiation perceived by the observer (receiver) due to the motion of the radiation source relative to the observer (receiver). LDVs can be used, for example, to detect faults in bearings by measuring vibrations. Contactless measurement of motor vibrations can also be used during end-of-line testing.

Example 39.2 Magnetic couplings (Fig. E.50) and magnetic gearboxes use a magnetic field instead of mechanical engagement of gear teeth.

Example 39.3 In passive magnetic bearings (Fig. E.51), support is provided by the repulsive forces of the magnets.

Example 39.4 In some applications, there is a transition from brushed DC machines with a stationary field to brushless AC machines with a rotating field.

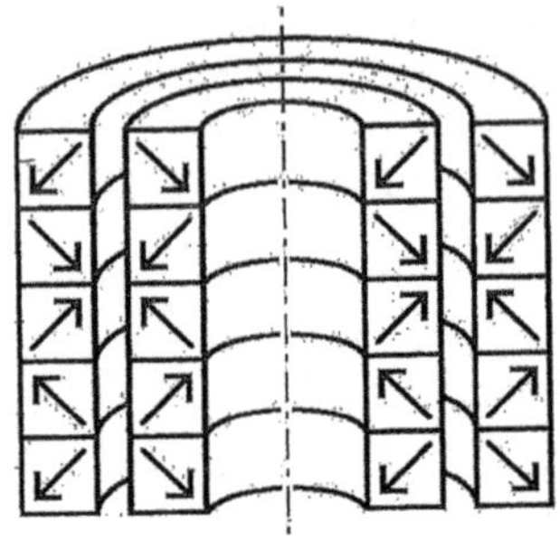

Fig. E.51 Magnetic bearing from patent application GB2564420A

Example 39.5 Electromagnetic powder clutches allow both rigid shaft connections and slipping between the half-couplings. The principle of powder clutches is based on the use of electromagnetic and mechanical forces acting within a gap filled with ferromagnetic powder. These forces alter the viscosity of the filler, thereby changing the connection between the driving and driven elements of the clutch. In the absence of a magnetic field, the ferromagnetic powder particles are not bonded together, and torque is not transmitted from one half-coupling to the other. When the winding is energized, a magnetic field is created in the gap, generating attractive forces between the powder particles and the half-couplings, which facilitates torque transmission. The stronger the magnetic field, the greater the transmitted torque.

40. Continuity of Useful Action

- *Carry on work continuously; make all parts of an object work at full load or optimum efficiency, all the time. Eliminate all idle or intermittent actions or work.*

Example 40.1 When operating electric drive systems over a wide range of rotational speeds, it may be practical to implement switching of the winding phase sections from a parallel to a series connection and back. In patent NO343095B1, a phase section switching system is implemented for two or more three-phase winding groups (Fig. E.52). The phase groups are switched alternately, thus ensuring continuous torque generation. See also Trick 7 "Dynamization".

Example 40.2 Patent US9553489B2 proposes a system for generating energy from ocean waves, in which two buoys provide reciprocating movement of the wires of two winch drums driving an electric generator. This solution ensures energy generation not only when the wave rises, but also when it goes down.

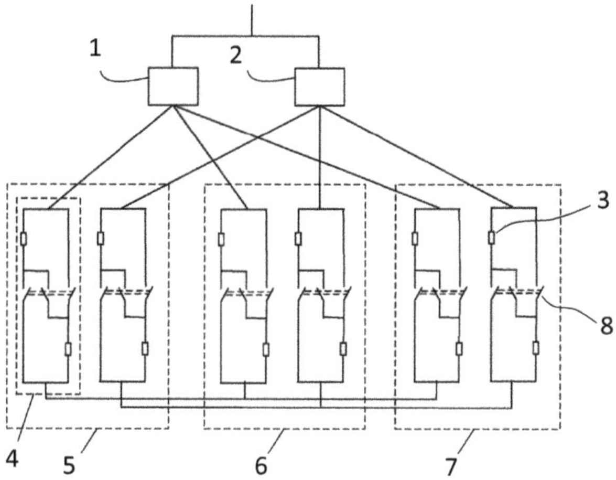

Fig. E.52 Winding switching circuit (1—first inverter, 2—second inverter, 3—part of the phase winding, 4—winding section, 5—phase A of the winding, 6—phase B of the winding, 7—phase C of the winding, 8—switch)

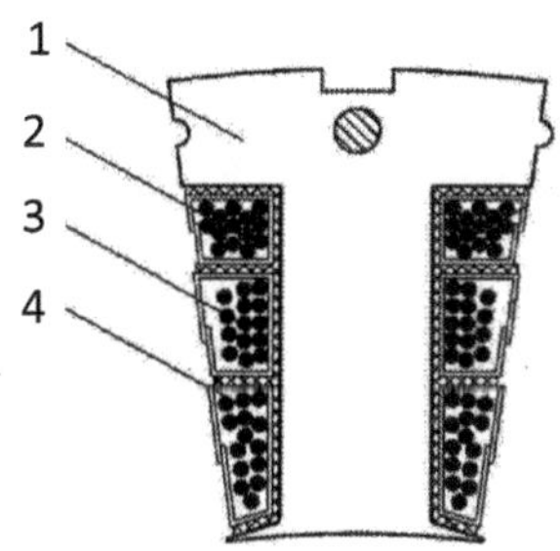

Fig. E.53 Winding made of three insulated parts (1—core, 2, 3, 4—winding parts)

Example 40.3 In the utility model CN207382083U (Fig. E.53), the winding on each of the teeth is divided into parts isolated from each other, so that if one of the parts of the winding fails, the machine can remain operational.

Example 40.4 Switched reluctance machines (SRM) are characterized by torque pulsations. Patent CN113285571B proposes a multi-packet SRM (Fig. E.54), in which, due to the displacement of the packages, a high torque is created almost continuously. This solution may also apply to the "Dynamization" and "Russian dolls" tricks.

Example 40.5 In a marine propulsion system, the propeller shaft can rotate at a constant frequency, while thrust is adjusted by changing the pitch angle of the propeller's blades.

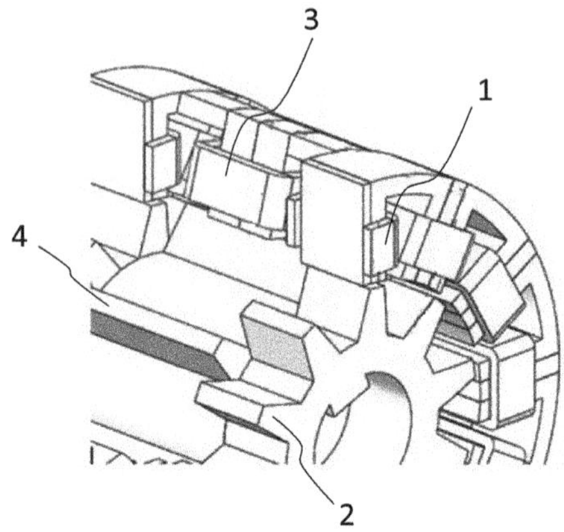

Fig. E.54 The structure of a multi-package switched reluctance machine (1—coils of the first package, 2—teeth of the first package, 3—coils of the second package, 4—teeth of the second package)

Example 40.6 In a hybrid vehicle or a hybrid aircraft, a gas turbine generator mostly operates at a constant load in its optimal mode. Load fluctuations are managed by energy storage devices such as batteries, capacitors, and flywheels.

Appendix F: The Correspondence Matrix

What gets worse when doing the change ---> / What needs to be changed:	1 Productivity	2 Adaptability or versatility	3 Use of energy by moving object	4 Reliability	5 Manufacturing precision	6 Illumination intensity	7 Device complexity	8 Difficulty of control and measurement	9 Ease of manufacture	10 Stress or pressure	11 Ease of repair	12 Loss of substance	13 Area of stationary object	14 Object-generated harmful factors	15 Volume of stationary object	16 Length of stationary object	17 Area of moving object	18 Harmful factors affecting the object	19 Volume of moving object	20 Length of moving object
01. Productivity		03, 01, 39, 21	01, 02, 30, 08	03, 01, 02, 30	37, 03, 38, 02	10, 19, 08, 03	09, 19, 39, 18	01, 38, 13, 05	01, 39, 05, 18	02, 21, 22	03, 37, 02, 29	39, 02, 01, 11	02, 01, 19, 14	01, 12, 38, 23	01, 21, 02, 05	25, 14, 22, 10	02, 10, 15, 31	27, 01, 36, 08	05, 20, 15, 02	38, 24, 39, 30
02. Adaptability or versatility	01, 39, 20, 21		08, 01, 34, 36	01, 36, 32, 18	-	20, 27, 10, 03	07, 34, 21, 39	03, 0	03, 36, 31	01, 16	03, 16, 14, 24	07, 02, 05, 36	07, 16	-	07, 31, 16, 12	03, 01, 16	01, 25, 34, 14	01, 28, 37, 31	07, 01, 34	01, 03, 34, 05
03. Use of energy by moving object	09, 39, 01	07, 19, 36, 16		08, 33, 28, 13	-	05, 07, 08	05, 34, 13, 39	01, 30	39, 10, 25	11, 22, 29	03, 07, 19, 39	01, 18, 38, 35	-	05, 01, 20	-	-	07, 08, 29	03, 01, 20, 13	01, 36, 38	09, 39
04. Reliability	03, 01, 34, 30	36, 01, 32, 18	33, 28, 13, 08		28, 37, 03	28, 37, 36	36, 01, 03	13, 17, 39	-	02, 18, 01, 08	03, 28,	02, 01, 34, 23	37, 01, 17, 24	01, 05, 17, 10	05, 01, 18	07, 34, 39, 28	19, 02, 22, 16	13, 01, 05, 17	12, 02, 22, 18	07, 04, 22, 24
05. Manufacturing precision	02, 38, 37, 23	-	37, 05	28, 37, 03		12, 37	10, 05, 38	-	-	12, 01	29, 02	01, 31, 02, 18	05, 34, 38, 26	24, 19, 15, 10	29, 02, 01	05, 37, 02	39, 06, 34, 37	10, 39, 02, 26	37, 39, 05	02, 39, 34, 21
06. Illumination intensity	05, 29, 16	07, 03, 08	37, 03, 08	-	12, 37		20, 37, 36	37, 07	08, 01, 39, 10	25, 01, 09	07, 19, 36, 16	36, 03	-	01, 08, 37, 23	-	-	08, 37, 10	07, 08	05, 36, 02	08, 37, 16
07. Device complexity	09, 19, 39	34, 07, 39, 21	13, 05, 34, 39	36, 01, 03	10, 18, 37	18, 19, 36		07, 02, 21, 39	13, 10, 03, 36	08, 03, 01	03, 36	01, 02, 39, 34	20, 26	08, 03	03, 16	10,	22, 03, 36, 16	27, 08, 34, 17	15, 10, 20	03, 08, 10, 18
08. Difficulty of control and measurement	01, 38	03, 07	01, 30	13, 17, 39, 32	-	05, 18, 10	07, 02, 21, 39		35, 39, 28, 34	01, 26, 21, 37	09, 10	03, 38, 02, 18	05, 23, 25, 16	05, 33	05, 38, 10, 31	10,	05, 36, 38, 19	27, 08, 34, 39	34, 03, 24, 16	16, 19, 10, 18
09. Ease of manufacture	01, 03, 02, 39	05, 36, 07	39, 10, 13, 03	04, 39, 13, 06	-	39, 18, 13, 03	13, 10, 03	20, 39, 28, 03		01, 08, 03, 21	01, 03, 28, 04	07, 15, 06	16, 17	-	01,	07, 19, 13	36, 03 10, 09	18, 05	36, 34, 03, 17	03, 34, 36, 19
10. Stress or pressure	02, 22, 01, 21	01,	22, 18, 02, 21	02, 36, 08, 01	12, 01	-	08, 03, 01	05, 26, 21	03, 01, 16		05,	02, 26, 12, 21	02, 07, 26, 21	05, 06, 13, 30	01, 18	01, 03, 22, 16	02, 07, 26, 39	27, 05, 21	20, 01, 02	01, 02, 26
11. Ease of repair	03, 37, 02	14, 03, 24, 16	07, 03, 39, 16	28, 02, 03, 16	29, 02	07, 03, 36	01, 03, 36, 28	-	03, 01, 28, 02	36,		05, 01, 15, 13	16, 29	-	03,	12, 38, 31	07, 36, 37	01, 02, 05, 16	29, 05, 01, 28	03, 39, 02, 29
12. Loss of substance	39, 01, 02, 11	07, 02, 05	01, 38, 18, 35	02, 34, 23, 01	01, 02, 18, 31	03, 20, 36	01, 02, 39, 18	01, 38, 02, 36	07, 15, 06	12, 26, 21, 02	05, 01, 15, 13		02, 38, 23, 31	02, 03, 15, 34	12, 23, 38, 31	02, 39, 18	01, 05, 02, 31	06, 27, 25, 17	03, 34, 25, 26	22, 34, 02, 23
13. Area of stationary object	02, 07, 19, 14	07, 16	-	37, 01, 17, 24	05, 34, 38, 26	03, 18, 01, 37	03, 38, 26	05, 01, 25, 38	17, 16	02, 07, 26, 21	16,	02, 22, 38, 23		27, 03, 17	-	10, 14, 04, 23	-	13, 05, 23, 01	-	-
14. Object-generated harmful factors	27, 01, 38, 23	-	05, 01, 20	18, 05, 17, 23	24, 19, 15, 10	08, 18, 23, 37	08, 03, 31	05, 33, 13, 03	-	05, 06, 13, 38	-	02, 03, 15	27, 03, 17		25, 38, 01, 24	-	19, 05, 38, 23	-	19, 05, 17	19, 07, 16, 27
15. Volume of stationary object	01, 21, 02, 05	-	-	05, 01, 16	01, 02, 29	01, 18, 39, 14	03, 31	05, 19, 10	01,	18, 01	03,	02, 23, 01, 15	-	25, 38, 01, 24		01, 32, 05, 22	-	15, 23, 08, 13	-	08, 22
16. Length of stationary object	25, 22, 14, 10	03, 01	-	07, 34, 39	05, 37, 02	12, 29	03, 10	10,	07, 19, 13	03, 22, 01	12,	02, 39, 18, 01	19, 14, 02, 17	-	01, 32, 05, 22		-	03, 38	25, 14, 07	-
17. Area of moving object	02, 10, 15, 05	07, 25	08, 37	34, 04	05, 37	07, 37, 08, 36	22, 03, 36	05, 26, 10, 38	36, 03, 10, 18	02, 07, 26, 39	07, 36, 02, 03	02, 01, 05, 23	-	19, 05, 38, 23	-	-		31, 06, 39, 03	14, 22, 19, 24	22, 07, 38, 24
18. Harmful factors affecting the object	27,01, 36, 18	01, 28, 27, 31	03, 18, 20, 13	13, 18, 05, 17	10, 39, 02, 38	03, 08, 37, 36	27, 08, 34, 17	27, 08, 34, 17	18, 01, 05	27, 05, 21	01, 02, 05	06, 27, 08, 17	13, 05, 23, 01	-	15, 23, 08, 13	03, 38	27, 03, 06, 39		27, 11, 21, 01	19, 03, 23, 24
19. Volume of moving object	02, 20, 05, 15	07, 34	01,	22, 03, 17, 28	29, 39, 05, 16	02, 36, 05	10, 03	34, 10, 24	34, 03, 17	20, 01, 26, 21	02,	26, 23, 15, 02	22, 14, 31	19, 05, 38, 23	-	14, 07, 24	03, 14, 24, 19	27, 33, 13, 01		03, 14, 01, 39
20. Length of moving object	22, 24, 39, 34	22, 07, 03, 16	32, 01, 18	02, 22, 34, 17	02, 39, 34, 21	37, 0	03, 08, 10, 18	01, 03, 10, 18	03, 34, 19	03, 32, 01	03, 39, 02	24, 34, 11, 02	-	19, 07	-	-	07, 19, 24	03, 07, 19, 18	14, 19, 39, 01	
21. Shape	19, 10, 15, 02	03, 07, 34	05, 20, 15, 22	02, 17, 16	37, 25, 17	36, 07, 37	16, 34, 03, 39	07, 36, 23	03, 37, 19, 39	15, 07, 02, 22	05, 36, 03	01, 34, 12, 35	-	01, 03	14, 05, 01	36, 22, 02, 14	35, 15, 24, 02	27, 03, 05, 01	22, 24, 07, 27	34, 15, 35, 24
22. Speed	-	07, 02, 10	32, 07, 01, 30	28, 01, 13, 39	02, 39, 37, 29	02, 36, 08	02, 39, 24, 15	12, 15, 13, 16	01, 36, 32, 03	20, 38, 30, 17	15, 05, 39, 13	02, 36, 39, 30	-	05, 18, 01, 33	-	-	34, 25, 15	03, 39, 01, 11	14, 34, 15	36, 22, 32
23. Moving object duration	01, 19, 22, 08	03, 01, 36	39, 20, 01, 38	28, 05, 36	12, 13, 16, 17	05, 08, 24, 01	02, 24, 34, 07	08, 34, 23, 01	13, 03, 24	08, 12, 13	34, 02, 13	39, 13, 12, 38	-	33, 23, 16, 27	-	-	12, 19, 08	27, 07, 06, 39	02, 05, 08, 25	05, 08, 04
24. Stationary object duration	40, 02, 16, 30	05,	-	15, 13, 20, 17	-	-	-	29, 15, 20, 01	01, 02	19, 24, 17	03,	13, 16, 38, 30	-	27,	01, 15, 30	03, 17, 01	-	19, 03, 17, 06	-	-
25. Loss of time	-	01, 39	01, 30, 08, 38	02, 25, 24	18, 10, 39, 38	03, 08, 10, 19	20, 34	38, 39, 37, 02	01, 39, 15, 24	21, 26, 24	37, 03, 02	01, 38, 02, 23	02, 01, 19, 24	01, 27, 38, 23	01, 16, 37, 38	25, 18, 22, 35	10, 24, 35, 16	01, 38, 15	05, 35, 15, 02	07, 05, 34
26. Quantity of substance	36, 34, 12, 13	07, 12, 34	15, 34, 16, 38	38, 12, 39, 17	06, 25	01, 39, 25, 31	12, 36, 13, 02	12, 13, 34, 38	34, 03, 01, 13	02, 26, 22, 12	05, 37, 02, 29	20, 12, 02, 18	05, 38, 17, 24	12, 01, 17, 23	-	-	07, 22, 34	01, 06, 34, 31	07, 40, 34	34, 22, 01, 38
27. Loss of information	36, 11, 07	-	-	02, 39, 11	-	08, 0	-	01, 06	37, 0	-	-	-	25, 16	02, 33, 27	5, 27	10, 0	25, 10	27, 02, 03	-	03, 10
28. Strength	34, 01, 02, 22	07, 12, 37	08, 01, 02	28, 12	12, 13	01, 08	05, 36, 39	13, 12, 07, 17	28, 12, 02, 37	02, 12, 38, 17	13, 28, 12	01, 39, 31, 17	04, 17, 39	07, 01, 27, 05	03, 22, 19, 07	07, 22, 39, 10	12, 15, 17, 34	38, 01, 21, 03	02, 07, 22, 14	03, 07, 32, 01
29. Stability of the object's composition	11, 01, 17, 12	01, 25, 15, 05	36, 08	-	38,	37, 12, 13, 07	05, 01, 27, 10	01, 27, 23, 11	01, 08	05, 01, 17	05, 01, 02, 16	05, 22, 25, 17	23,	01, 17, 13, 23	15, 39, 01, 17	21, 0	05, 28, 36	01, 18, 38, 25	39, 02, 08, 23	36, 07, 03, 39
30. Force	12, 39, 01, 21	07, 19, 38, 40	08, 19, 02	12, 01, 36, 33	39, 34, 21, 26	-	10, 01, 02, 38	26, 21, 02, 08	07, 21, 38, 03	38, 33, 28	07, 03, 28	32, 01, 17, 35	03, 38, 26, 21	36, 12, 26, 18	05, 26, 38, 21	39, 02	08, 02, 07	03, 01, 17, 38	07, 04, 09, 21	19, 08, 04, 26
31. Ease of operation	07, 03, 39	07, 15, 03, 16	03, 36, 18	19, 13, 32, 17	03, 37, 01, 11	36, 19, 03, 18	37, 29, 09, 19	-	05, 35, 09	05, 37, 09	09, 10, 03, 37	39, 37, 05, 18	38, 16, 07, 23	-	24, 38, 23, 31	03, 24	03, 19, 36, 16	05, 29, 39, 23	03, 16, 01, 07	03, 19, 36, 09
32. Weight of moving object	01, 12, 18, 21	34, 35, 07, 32	01, 09, 15, 31	12, 28, 03, 13	39, 01, 10, 38	08, 03, 37	10, 25, 26, 15	39, 34, 10, 37	31, 39, 03, 26	02, 26, 21, 17	05, 13, 39, 28	35, 01, 12, 31	-	31, 01, 31, 23	17, 05, 24, 14	-	34, 19, 30, 15	27, 33, 38, 13	34, 05, 17, 39	07, 32, 34, 15
33. Weight of stationary object	03, 39, 07, 01	08, 07, 34	-	02, 39, 32, 12	02, 03, 01, 19	01, 08, 37	03, 02, 10, 23	29, 39, 19, 07	39, 03, 37	36, 34, 02, 38	05, 13, 39, 28	35, 32, 36, 25	01, 25, 36, 05	01, 27, 03, 23	35, 01, 22, 05	02, 03, 34, 01	-	05, 08, 27, 21	-	-
34. Temperature	07, 39, 01	05, 38, 13	08, 07, 12, 19	08, 01, 12, 02	18	37, 25, 33, 16	05, 19, 16	12, 13, 01, 31	10, 13	01, 23, 08, 05	24, 02, 16	33, 26, 34, 31	01, 30	27, 01, 31, 23	01, 20, 24	07, 08, 04	12, 01, 23, 38	27, 06, 01, 05	15, 23, 17, 38	07, 08, 04
35. Measurement accuracy	02, 15, 39, 37	36, 01, 05	12, 20, 37	35, 28, 03, 11	-	20, 03, 37	13, 10, 02, 15	10, 18, 37, 39	20, 01, 29, 38	20, 39, 37	03, 37, 36, 28	02, 16, 31, 39	10, 39, 37, 12	12, 06, 23, 02	10, 18, 36, 31	37, 39, 12, 16	10, 39, 37, 12	39, 18, 27, 10	37, 36, 20	39, 10, 35, 16
36. Power	39, 01, 15	08, 19, 15	16, 20, 08, 21	08, 18, 10, 31	37, 05	16, 20, 08	40, 08, 25, 15	08, 01, 16	10, 02, 15	27, 02, 01	01, 05, 02, 15	39, 13, 38, 30	19, 37, 36, 30	05, 01, 38	25, 20, 29	-	08, 30	08, 27, 31, 05	01, 20, 30	03, 02, 01, 21
37 Extent of automation	35, 09, 01, 10	13, 24, 03, 01	05, 37, 36	28, 13, 37	39, 10, 38, 11	32, 37, 08	07, 18, 02	15, 13, 29	03, 10, 36	36, 01	03, 01, 36	01, 02, 38, 35	-	05,	10, 36, 18, 31	11,	19, 22, 36	05, 06	01, 36, 16	22, 36, 39, 19
38. Use of energy by stationary object	03, 20	-	-	02, 26, 11	-	08, 05, 01, 37	-	08, 01, 16, 29	03, 24	-	-	39, 13, 38, 31	-	08, 27, 38	-	-	-	02, 05, 27, 21	-	19, 24, 09
39. Loss of energy	39, 02, 34, 01	-	-	28, 02, 01	-	03, 36, 37, 07	14, 11	01, 12, 07, 11	-	-	05, 08	01, 13, 05, 21	19, 14, 25, 38	33, 01, 05, 27	14,	20, 30, 14	07, 10, 19, 25	33, 27, 01, 05	14, 38, 11	14, 05, 20, 36

What gets worse when doing the change ----> / What needs to be changed:	21 Shape	22 Speed	23 Moving object duration	24 Stationary object duration	25 Loss of time	26 Quantity of substance	27 Loss of information	28 Strength	29 Stability of the object's composition	30 Force	31 Ease of operation	32 Weight of moving object	33 Weight of stationary object	34 Temperature	35 Measurement accuracy	36 Power	37 Extent of automation	38 Use of energy by stationary object	39 Loss of energy
01. Productivity	22, 02, 15, 17	-	01, 02, 05, 38	40, 02, 16, 30	-	01, 30	36, 07, 11	01, 39, 02, 38	01, 12, 27,23	39, 07, 02, 26	03, 39, 14, 06	01, 10, 18, 21	39, 13, 07, 12	01, 33, 39, 02	03, 02, 15, 39	01, 40, 02	35,09, 01, 10	03,	39, 02, 34, 01
02. Adaptability or versatility	07, 21, 03, 32	01, 02, 22	36, 03, 01	05, 16	01, 39	12, 01, 07	-	01, 12, 37, 20	01, 25, 22	07, 19, 40	07, 15, 03, 16	03, 20, 07, 32	08, 07, 34, 16	13, 05, 12, 01	01, 35, 03, 02	08, 03, 34	13, 15, 01	-	38, 07, 03
03. Use of energy by moving object	09, 05, 34	32, 07, 01	39, 01, 20, 38	-	01, 30, 08, 38	15, 11, 16, 38	-	35, 08, 04, 01	08, 36, 19, 18	16, 10, 33, 05	08, 01	09, 38, 39, 31	-	08, 18, 12, 22	12, 03, 37	20, 08, 21, 38	37, 05	-	09, 27, 07, 18
04. Reliability	01, 03, 16, 28	33, 01, 28, 39	05, 01, 12, 29	15, 13, 20, 17	02, 25, 24	33, 39, 17, 12	02, 39	28, 39	-	32, 39, 02, 12	13, 19, 17	12, 32, 02, 17	12, 02, 32, 39	12, 01, 02	37, 12, 28, 11	33, 28, 10, 31	28, 36, 13	26, 11	02, 28, 01
05. Manufacturing precision	37, 25, 17	02, 39, 37	12, 13, 17	-	37, 10, 39, 38	37, 25	-	12, 13	25, 38	39, 08, 15, 26	03, 37, 01, 11	39, 37, 36, 38	10, 01, 13, 04	08, 10	-	37, 05	10, 39, 38, 11	-	36, 37, 05
06. Illumination intensity	37, 25	02, 36, 08	05, 08, 20	-	08, 03, 10, 19	03, 08	03, 20	01, 08	37, 12, 13	10, 08, 20	39, 10, 08	08, 03, 37	05, 01, 37	37, 01, 08	28, 07, 37	37, 0	05, 10, 02	37, 01, 03, 07	08, 16, 03, 20
07. Device complexity	34, 36, 39, 07	15, 02, 39	02, 24, 39, 07	-	20, 34	36, 12, 13, 02	-	05, 36, 39	05, 27, 19, 08	10, 16	13, 04, 10, 18	10, 25, 15, 26	05, 10, 01, 23	05, 19, 36	05, 10, 02, 15	40, 08, 25, 15	07, 03, 18	-	02, 01, 36, 05
08. Difficulty of control and measurement	13, 36, 03, 23	12, 24, 16, 01	08, 34, 29, 23	29, 15, 20, 01	38, 39, 37, 04	12, 13, 34, 38	01, 06, 13, 27	13, 12, 07, 39	28, 27, 23, 25	26, 39, 17, 08	05, 35	13, 10, 39, 36	20, 36, 39, 03	12, 13, 01, 16	10, 18, 37, 39	08, 03, 16, 02	15, 33	08, 01, 16	01, 12, 07, 08
09. Ease of manufacture	03, 39, 36, 13	01, 36, 32, 03	13, 03, 24	01, 16	01, 39, 15, 24	01, 11, 03, 18	37, 18, 38, 16	03, 12, 02, 37	28, 36, 03	01, 09	05, 35, 36, 16	39, 34, 07, 16	03, 13, 26, 36	13, 10, 38	03, 01, 09, 38	13, 03, 09, 18	32, 39, 03	03, 24	08, 01
10. Stress or pressure	01, 24, 07, 02	20, 01, 26	08, 12, 13	-	21, 26, 24	02, 22, 26	-	04, 38, 12, 17	01, 06, 05, 17	26, 01, 33	28,	02, 26, 21, 17	36, 34, 02, 38	01, 23, 08, 05	20, 39, 29	02, 01, 22	01, 18	-	05, 26, 29
11. Ease of repair	03, 36, 05, 24	15, 04	28, 34, 39, 13	03,	37, 03, 02, 29	05, 39, 02, 29	-	03, 28, 05, 04	05, 01	03, 28, 02	03, 09, 10, 07	05, 13, 01, 28	05, 13, 01, 28	24, 02	02, 05, 36	07, 02, 37, 05	15, 01, 14, 36	-	07, 03, 37, 08
12. Loss of substance	34, 01, 12, 35	02, 36, 39, 30	39, 13, 12, 38	13, 16, 38, 30	07, 38, 01, 02	20, 12, 02, 18	-	01, 39, 31, 17	05, 22, 25, 17	22, 07, 38, 17	37, 39, 05, 18	01, 20, 11, 17	01, 20, 27, 37	33, 26, 23, 31	16, 15, 31, 39	39, 13, 38, 30	01, 02, 38	39, 13, 09, 31	01, 13, 05, 31
13. Area of stationary object	-	-	-	05, 02, 08, 25	02, 01, 24, 38	05, 38, 17, 24	25, 16	17,	05, 30	03, 38, 01, 26	16, 24	-	25, 05, 22, 38	01, 23, 30	10, 39, 37, 12	19, 37	11	-	19, 14, 25
14. Object-generated harmful factors	01, 03	01, 39, 12, 11	07, 27, 06, 31	33, 23, 16, 27	03, 27	12, 18, 23, 03	02, 33, 34	07, 01, 27, 05	01, 17, 13, 23	01, 39, 03, 17	-	08, 27, 07, 23	01, 27, 03, 23	27, 01, 05, 18	12, 06, 10	05, 01, 38	05,	08, 27, 38	33, 01, 05, 27
15. Volume of stationary object	14, 05, 01	-	-	01, 15, 30	01, 16, 37, 38	01, 12	-	04, 22, 19, 07	18, 39, 01, 17	05, 38, 21	-	31, 25, 24, 08	01, 02, 08, 22	01, 20, 24	-	25, 20	-	-	-
16. Length of stationary object	36, 22, 07, 14	-	-	03, 17, 01	25, 34, 22	-	18, 10	07, 22, 39, 10	23, 21, 01	39, 02	05, 29	25, 31, 32, 17	01, 39, 17, 34	12, 01, 30, 38	37, 39, 12	09, 32	-	-	20, 39
17. Area of moving object	35, 15, 34, 24	34, 25, 24, 15	20, 12	-	10, 24	34, 25, 20, 36	25, 10	12, 07, 17, 22	28, 05, 36, 23	08, 25, 01, 05	07, 19, 36, 16	05, 19, 34, 39	12, 31, 05	05, 07, 16	10, 39, 37, 12	08, 02, 37, 38	22, 25, 39, 11	-	07, 19, 25, 10
18. Harmful factors affecting the object	27, 03, 12, 01	33, 27, 01, 39	27, 07, 06, 39	19, 03, 17, 06	01, 38, 15	01, 06, 34, 31	27, 02, 05	38, 01, 21, 03	01, 18, 25, 38	36, 01, 23, 38	05, 29, 39, 23	27, 33, 13, 23	05, 27, 36, 18	27, 06, 01, 05	39, 06, 11, 10	08, 27, 31, 05	06, 12, 15	02, 05, 27, 21	33, 27, 01, 05
19. Volume of moving object	03, 07, 34, 24	34, 24, 30, 15	20, 01, 24	-	05, 20, 15, 02	34, 25, 14	05, 27	04, 22, 07, 14	39, 02, 03, 23	07, 01, 26, 21	07, 36, 25, 09	05, 10, 34, 17	31, 17, 10	15, 23, 02, 38	29, 10, 39	01, 20, 36, 38	01, 15, 16, 18	-	14, 07, 36, 16
20. Length of moving object	03, 32, 02, 34	36, 24, 32	08,	-	07, 05, 34	34, 01	03, 18	32, 01, 34, 15	03, 32, 07, 15	19, 02, 24	07, 34, 01, 24	-	03, 19, 07, 24	02, 07, 08	39, 37, 24	03, 01	19, 18, 10, 16	-	14, 05, 01, 23
21. Shape		01, 07, 15, 38	22, 10, 04, 29	-	22, 02, 15, 19	26, 27	-	25, 22, 02, 17	06, 03, 38, 24	01, 02, 21, 17	37, 07, 10	32, 02, 34, 17	07, 02, 10, 12	27, 22, 10, 12	39, 37, 03	24, 20, 05	07, 03, 37	-	22,
22. Speed	01, 07, 38, 15		12, 08, 01, 35	-	-	02, 08, 34, 30	36, 10	32, 12, 10, 22	39, 06, 03, 38	36, 39, 07, 08	37, 39, 36, 09	05, 39, 36, 30	-	39, 25, 26, 05	39, 37, 03, 18	08, 01, 30, 05	02, 38	-	22, 40, 08, 01
23. Moving object duration	22, 10, 39, 29	12, 01, 35		-	40, 02, 39, 38	12, 01, 02, 17	02,	13, 12, 02	36, 12, 01	08, 05, 16	09, 13	08, 35, 15, 31	-	08, 01, 23	12	08, 02, 01, 30	20, 02	-	-
24. Stationary object duration	-	34, 24, 22, 36	-		39, 40, 02, 16	12, 01, 31	02,	-	23, 12, 01, 11	-	03,	31, 24, 07	20, 13, 08, 16	08, 38, 26, 17	02, 10, 18	16,	03,	-	-
25. Loss of time	24, 02, 15, 19	39, 36, 02, 24	40, 02, 39, 38	39, 40, 02, 16		01, 30, 38, 16	18, 10, 39, 37	34, 12, 39, 38	01, 12, 27, 35	02, 21, 26, 35	24, 39, 02, 15	02, 40, 21, 01	02, 40, 10, 35	01, 34, 33, 38	18, 15, 39, 37	01, 40, 02, 20	18, 39, 01, 25	03,	02, 35, 38, 37
26. Quantity of substance	01, 22	01, 34, 15, 39	12, 01, 02, 17	12, 01, 31	01, 30, 38, 16		18, 39, 01	22, 01, 15, 02	07, 05, 19, 17	01, 22, 12	01, 34, 02, 29	01, 20, 38, 31	13, 10, 38, 01	12, 19, 23	12, 05, 39	01, 0	32, 01	12, 01, 31	14, 38, 29
27. Loss of information	-	10, 37	02,	02,	18, 10, 39, 37	18, 39, 01		-	-	-	13, 27	02, 18, 01	02, 01, 35	-	-	02, 08	01,	-	08, 02
28. Strength	02, 25, 01, 17	32, 36, 10, 22	13, 12, 10	-	34, 12, 39, 02	34, 02, 13	-		36, 19, 01	02, 38, 12, 22	37, 17, 39, 05	03, 32, 17, 07	17, 10, 13, 03	25, 02, 17	12, 13, 16	02, 10, 01, 39	07, 0	01,	01,
29. Stability of the object's composition	27, 03, 38, 24	06, 07, 39, 38	36, 13, 02, 01	23, 12, 01, 11	01, 13	07, 37, 01	-	19, 04, 07		02, 01, 33, 16	37, 01, 25	33, 01, 05, 23	10, 23, 03, 17	01, 03, 37	36,	37, 01, 13, 31	03, 32, 01	13, 24, 34, 38	22, 05, 23, 20
30. Force	02, 01, 17, 15	36, 39, 07, 09	08, 05	-	02, 21, 26	22, 34, 38, 26	-	01, 02, 22, 13	01, 02, 33		03, 39, 12, 29	32, 03, 21, 38	38, 36, 03, 39	01, 02, 33	01, 02, 11, 18	08, 01, 38, 21	05, 01	03, 16, 26, 21	22, 07
31. Ease of operation	07, 15, 34, 39	38, 36, 15	34, 12, 32, 29	03, 16, 29	24, 39, 02, 15	09, 01	24, 02, 13, 27	37, 17, 12, 39	37, 01, 25	39, 36, 01		29, 05, 36, 07	20, 36, 03, 29	10, 13, 36	29, 36, 05, 15	01, 15, 05, 02	03, 15, 09, 12	-	05, 08, 36
32. Weight of moving object	02, 22, 01, 17	05, 32, 07, 30	35, 15, 31, 01	-	02, 01, 40, 39	12, 10, 38, 31	02, 18, 01	39, 13, 38, 17	03, 01, 08, 23	32, 02, 38, 21	01, 12, 05, 18		-	20, 34, 24, 30	39, 13, 01, 10	09, 26, 38, 31	10, 01, 38, 08	-	20, 05, 15, 08
33. Weight of stationary object	36, 02, 34, 22	01, 19, 25	-	05, 13, 08, 20	02, 40, 01, 10	08, 20, 38, 10	02, 07, 01	39, 05, 02, 13	10, 23, 03, 17	32, 02, 08, 01	20, 36, 03, 37	17, 31, 03		39, 08, 37, 27	38, 10, 39	07, 08, 38, 27	05, 10, 01	38, 08, 39, 03	38, 08, 39, 07
34. Temperature	22, 27, 08, 37	05, 39, 26, 25	08, 36, 23	08, 38, 26, 17	01, 39, 33, 38	12, 19, 25, 23	-	02, 25, 27, 17	03, 01, 37	01, 02, 12, 33	10, 13	26, 21, 20, 30	27, 01, 37		37, 08, 18	05, 22, 19, 29	10, 05, 08, 16	-	33, 19, 01, 30
35. Measurement accuracy	20, 39, 37	39, 36, 37, 18	39, 20, 37	02, 10, 18	18, 15, 39, 37	05, 20, 37	-	39, 20, 37	37, 01, 36	37, 05	03, 36, 19, 15	37, 01, 10, 39	39, 01, 29, 10	20, 08, 39, 18		12, 20, 37	39, 05, 02, 15	-	10, 37, 13
36. Power	34, 22, 05, 17	07, 01, 05	08, 01, 02, 30	16,	01, 40, 02, 20	24, 15, 08	02, 08	10, 02, 39	01, 37, 07, 31	10, 05, 26, 01	10, 01, 02	32, 26, 30, 31	08, 10, 19, 13	05, 22, 19, 29	37, 07, 05		39, 05, 19	-	02, 01, 30
37 Extent of automation	07, 37, 03, 36	39, 02	20, 04	-	18, 39, 01, 25	01, 36	01, 06	29, 36	38, 03	05, 01	03, 09, 15, 12	39, 10, 38, 01	39, 10, 01, 02	10, 05, 08	39, 10, 02, 15	39, 05, 13		-	11, 39
38. Use of energy by stationary object	-	-	-	-	02, 17, 14, 04	12, 01, 31	-	01,	13, 24, 34, 38	26, 21	-	-	08, 04, 20, 13	-	-	-	-		-
39. Loss of energy	-	16, 01, 30	-	-	02, 38, 37, 14	14, 38, 29	08, 02	10,	22, 05, 23, 20	26, 30	01, 37, 03	07, 20, 08, 39	08, 20, 38, 04	08, 30, 14	37,	12, 30	05,	-	

Appendix G: Physical and Chemical Effects Occurring in Electric Machines

Physical Effects

Ferromagnetism

Ferromagnetism is the appearance of spontaneous magnetization in some materials due to the ordering of magnetic moments, in which most of them are parallel to each other. This is the primary mechanism by which certain materials form permanent magnets or are attracted to magnets. Substances in which ferromagnetic ordering of magnetic moments occurs are called ferromagnets.

When the temperature rises above the Curie temperature, a ferromagnet loses its ferromagnetic properties.

Examples of use or manifestation:

- Permanent magnet: a product made from a ferromagnetic material capable of retaining residual magnetization after the external magnetic field is removed.
- Attraction of opposite poles of permanent magnets and electromagnets.
- Repulsion of like poles of permanent magnets and electromagnets.
- Attraction of ferromagnetic teeth of a rotor to the teeth of a stator with an excited winding, which acts as electromagnets, such as in switched reluctance motor.

In all the above examples, the force of attraction is beneficial when it contributes to creating useful torque or linear force, and harmful when, for example, it leads to vibration.

Force Acting on a Current-Carrying Conductor in a Magnetic Field

When a conductor carrying an electric current is placed in a magnetic field, each of the moving charges that constitute the current experiences the Lorentz force. Collectively, these forces create a macroscopic force acting on the conductor, known as the Laplace force. The magnitude of the Laplace force is determined by Ampère's law and depends linearly on both the current and the magnetic induction (flux density).

An example of the use of the effect is the creation of torque or linear force in many electric machines (EM).

Examples of negative manifestations:

- Electrodynamic deformation of busbars (conductors) in three-phase AC systems at substations under the influence of short-circuit currents.
- The spreading apart of conductors in railguns (linear EM) during a shot.

Electromagnetic Induction

Electromagnetic induction is the phenomenon where an electric current, electric field, or electric polarization arises when a magnetic field changes in a material medium over time or when this medium moves through a magnetic field.

When the magnetic field within a loop (through the surface bounded by this loop), created by a closed conductor, changes, an electromotive force (EMF) is generated in the conductor. The EMF is proportional to the rate of change of the magnetic flux through the surface bounded by the loop. This EMF induces a current in the conductor, known as the induced current. The induced current creates a field that opposes the change in magnetic flux that caused the induction.

A specific case of electromagnetic induction is mutual induction, the phenomenon where an EMF is induced in one loop when the current in a second loop changes, and vice versa.

Examples of application:

- Induction of currents in the squirrel-cage rotor by the rotating magnetic field of the stator in an induction motor.
- Energy transfer between the windings of a transformer.

Examples of negative manifestations:

- Eddy currents (see below).
- Skin effect (see below).
- Proximity effect (see below).

Eddy Currents (Foucault Currents)

Eddy currents are closed-loop electrical currents that arise in a solid conductor when the magnetic flux passing through it changes. These currents are induced by changing magnetic fields and are formed within a conducting material either due to a time-varying magnetic field surrounding the material or due to the movement of the material within a magnetic field, which alters the magnetic flux through the material or a part of it. The magnitude of Foucault currents increases with the rate of change of the magnetic flux. Unlike electric currents in wires, which follow well-defined paths, eddy currents form in the conductive mass itself, creating vortex-like loops. These current loops interact with the magnetic flux that generated them. According to Lenz's law, the magnetic field of the eddy currents is oriented to oppose the change in magnetic flux that induces these currents.

Example of application:

- Induction machines with a solid rotor.

Examples of negative manifestations:

- Losses in steel cores.
- Skin effect.

Skin Effect

An alternating current (I) in a conductor, as shown in Fig. G.1, generates an alternating vortex magnetic field, whose field lines (H) are perpendicular to the axis of the conductor. Due to electromagnetic induction, the alternating magnetic field induces a vortex electric field, which causes eddy currents (i) to flow. On the surface of the conductor, the eddy currents flow in the same direction as the main current in the conductor, while inside the conductor, they flow in the opposite direction. This phenomenon reduces the current density in the core of the conductor and increases the current density near the surface. The skin effect becomes more pronounced at higher frequencies.

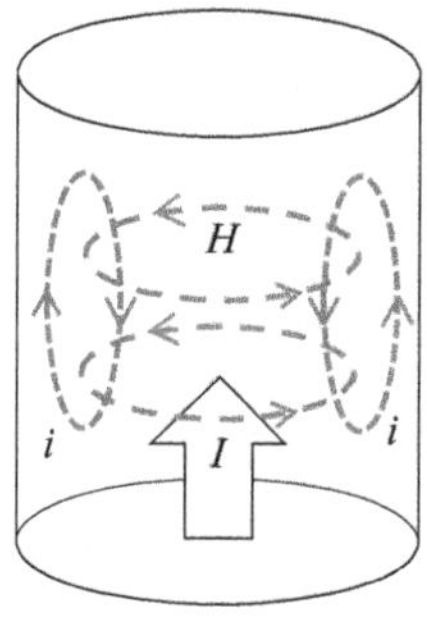

Fig. G.1 Skin effect

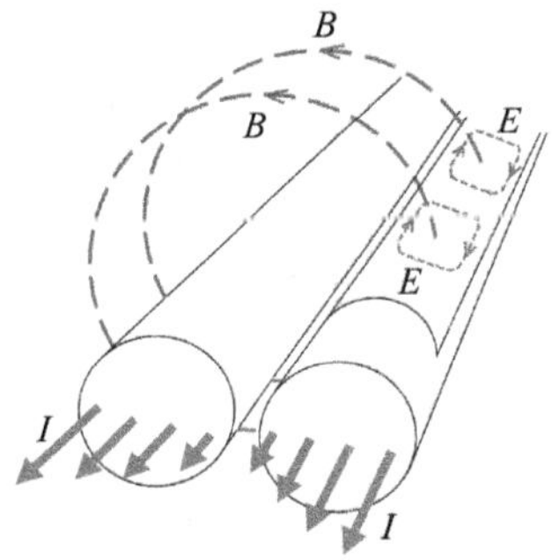

Fig. G.2 Proximity effect

Applications:

- At radio frequencies around hundreds of kilohertz, the skin effect in copper wires becomes so significant that using solid wires is impractical. Hollow thin-walled conductors (tubes) are used in such cases since having the area in the middle of the conductor does not provide any increase in conductivity.
- A thin silver coating on a high-frequency waveguide provides a greater increase in conductivity than coating a conductor carrying direct current.

Negative manifestation:

- Increased electrical resistance, leading to higher losses and heating of conductors.

Proximity Effect

In a bundle of conductors running parallel to each other, such as within a tightly wound coil, the current distribution in the conductors will be uneven. This concentration of current leads to an increase in the effective resistance of the circuit, which increases with frequency. The principle is illustrated in Fig. G.2, where two conductors with currents flowing in the same direction are shown. The first conductor, through its alternating magnetic field, induces an electromotive force (E) in the second conductor (a phenomenon of mutual induction) (B), which decreases the current in the region farther from the first conductor and increases it in the region closer to the first conductor.

Negative manifestation: Increased electrical resistance, leading to higher losses and heating of conductors.

Heat Transfer

Heat transfer is the physical process of transferring thermal energy from a hotter object to a cooler one, either directly (through contact) or through an intermediary material. There are three basic mechanisms of heat transfer:

- Conduction.
- Convection.
- Thermal radiation.

The primary types of heat transfer used in many electrical machines to cool active parts, which combine these basic mechanisms, include:

- Convective heat exchange between fluid or gas flows and the surface of a solid body, commonly referred to as "heat dissipation."
- Heat transfer from a hot medium (liquid, gas, or solid) to a cold one through a separating wall referred to as "heat transfer."
- Combined heat transfer through radiation and convection, known as "convective-radiative heat transfer."

Thermal Expansion and Contraction

Thermal expansion and contraction refer to the change in the dimensions of objects when exposed to thermal fields (heating and cooling). This can result in the development of significant stresses.

Useful application: Shrink fitting of the rotor onto the shaft.

Negative manifestation: Delamination (separation of layers) of, e.g., a stator and a housing.

Inertia

Inertia is the property of an object to remain at rest or in uniform motion in the absence of external forces, and to resist changes in its velocity when external forces are applied, due to its inertial mass.

Useful applications:

- Regeneration of kinetic energy stored in electrical machine during braking.
- Continued operation of a rotating machine during brief power interruptions due to inertia (stored kinetic energy).

Negative manifestations:

- Limited machine dynamics.
- Heavy startup.
- Prolonged coast-down time.

Heat Generation During Current Flow

When an electric current passes through a conductive medium, heat energy is generated. The amount of heat energy is inversely proportional to the conductivity of the material. The quantitative assessment of the thermal effect of electric current is given by the Joule-Lenz law.

Example of use: Motor winding heating systems for removing condensation.

Examples of negative manifestation: Losses in windings (copper, aluminum) and cores (steel).

Friction

Friction is a force that arises during the relative movement of two contacting bodies in the plane of their contact. There are different types of friction, including rolling friction (resistance to rolling), sliding friction, and static friction. Two main types of sliding friction are commonly distinguished: dry friction (or friction between unlubricated surfaces) and fluid friction (friction between lubricated surfaces).

Useful applications: Mutual holding of structural elements due to static friction force (interference fits, smooth connections of parts).

Examples of negative manifestation: Reduced torque on the shaft due to friction in bearings (rolling or sliding) or friction between the rotor of a high-speed machine and air or liquid surrounding it.

Gravity

Gravity is the force of interaction between masses at a distance.

Negative manifestations:

- Sagging and deformation of shafts in high-power machines.
- Impact of rotor mass and inertia on bearings.

Centrifugal Force

Centrifugal force is the force with which a moving material point acts on the object that restricts its freedom of movement and forces it to move in a curved path. The centrifugal force is directed along the principal normal to the trajectory, away from the center of curvature (from the center of the circle in the case of circular motion).

Examples of applications:

- Centrifugal switches, regulators.
- Retention of permanent magnets on a rotor located outside the stator in an electric machine with outer rotor.

Examples of negative manifestation:

- Increased forces on rotating parts of a machine as the rotational speed increases.
- The need to secure permanent magnets on the rotor, for example, by using a wrapping.

Superconductivity

Superconductivity refers to a significant increase in the conductivity of certain materials under specific conditions of temperature, magnetic field, and current density.

Example of use: Windings of machines made from superconducting materials.

Magnetostriction and the Villari Effect

Magnetostriction is the change in volume and linear dimensions of a body when its magnetization state changes.

Examples of use:

- Generation of mechanical oscillations with specified parameters.
- Small displacement actuators.

Negative manifestation: Vibrations and noise.

The Villari effect, or magnetoelastic effect, is the phenomenon opposite to magnetostriction, in which the magnetization of a magnetic material changes under mechanical stress.

Example of application: Magnetostrictive energy generators.

Peltier Effect and Seebeck Effect

The Peltier effect is the phenomenon of energy transfer when an electric current passes through the contact (junction) of two different conductors, where energy is transferred from one conductor to another. Depending on the direction of the current, in addition to Joule heating, additional heat is either generated or absorbed (Fig. G.3).

Example of use: Thermoelectric converter.

The Seebeck effect is the "reverse" of the Peltier effect.

Example of use: Thermoelectric generator, a device designed for direct conversion of thermal energy into electricity.

Electrostriction and Piezoelectric Effect

Electrostriction is the elastic, reversible increase in the size of a body (deformation) under an electric field, regardless of the field's polarity. Electrostriction is a property of all nonconductors or dielectrics. The deformation of a solid under the influence of an electric field depends on the field's polarity.

Examples of use:

- Electrostriction motors.
- Actuators.
- Displacement detectors.
- Generation of mechanical oscillations with specified parameters.

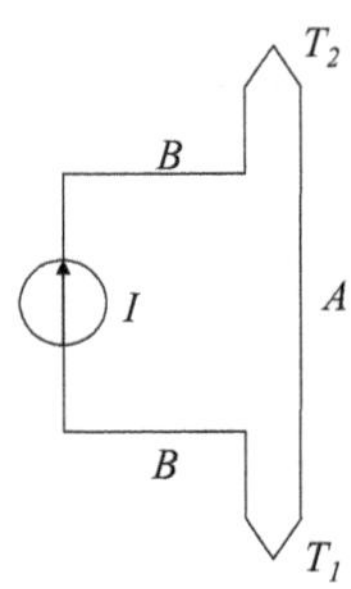

Fig. G.3 Peltier effect diagram (A, B - different materials)

Negative effects: Vibrations and noise.

The piezoelectric effect is the generation of charges on the surface of a solid under mechanical stress (elastic deformation) in the absence of an external electric field, resulting in an electrical potential between the deformed surfaces of the solid.

Examples of application:

- Electrostrictive energy generators.
- Stress and compressibility sensors.

Thomson Effect

The release or absorption of heat (in addition to Joule heating) within a conductor when a constant electric current passes through a nonuniformly heated homogeneous conductor or semiconductor (in the presence of a temperature gradient along the conductor).

Mechanical Oscillations

Mechanical oscillations are divided into free and forced oscillations. Free oscillations are the system's natural damped oscillations when it is displaced from its equilibrium position. Forced oscillations occur under the influence of a periodic force, typically external.

Vibrations and acoustic noise in electric machines are a combination of both types of oscillations.

Acoustic Oscillations

The propagation of sound waves through a medium. The effect depends on the frequency and intensity of the oscillations.

A negative manifestation is acoustic noise, resulting from machine parts vibrations.

Resonance

A sharp increase in oscillation amplitude when the forced and natural frequencies coincide.

The negative manifestation is increased mechanical vibrations of electric machine parts when the power supply frequency matches natural frequency of the machine rotor with an attached mechanical load.

Hall Effect

The occurrence of a potential difference (Hall voltage) in a conductor or semiconductor perpendicular to both the magnetic field and the current flow (Fig. G.4). The Hall voltage is proportional to the magnetic field and current.

Examples of use:

- Hall sensor—a measuring transducer used to detect the presence and strength of a magnetic field. The sensor operates based on the Hall effect, and its output voltage is directly proportional to the intensity of the magnetic field.

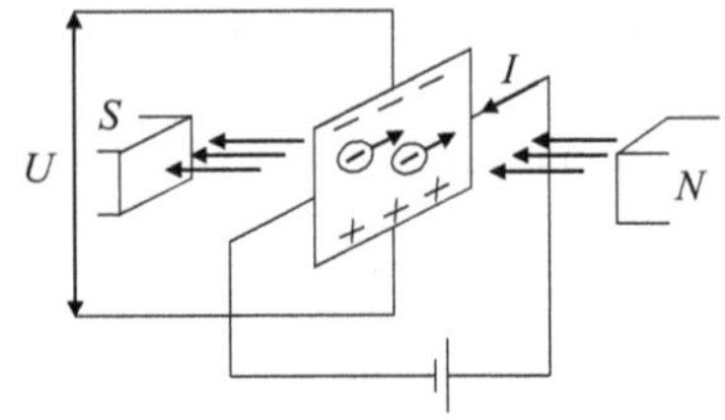

Fig. G.4 Hall effect scheme

- Current measurement.
- Measurement of power dissipated by a device.

Effect of Temperature on Electrical and Magnetic Properties

The electrical and magnetic properties of substances change significantly near certain temperatures (Curie points). Above the Curie point, a ferromagnet becomes a paramagnet.

Ferroelectrics have two Curie points, where magnetic or electrical anomalies occur. Antiferromagnets lose their properties at a temperature called the Néel point.

Examples of negative manifestation: Demagnetization of permanent magnets.

Magnetic Hysteresis

Magnetic hysteresis is the phenomenon where the magnetization vector M and the magnetic induction vector B in a material depend not only on the intensity H of the applied external field but also on the material's history. Magnetic hysteresis is typically observed in ferromagnetic materials such as Fe, Co, Ni, and their alloys.

Useful applications:

- Hysteresis motor.
- Magnetic hysteresis is the intrinsic feature of permanent magnets.

Negative manifestation: Hysteresis losses (see below).

Hysteresis Losses

The work done by the magnetizing force against the internal friction of the magnet's molecules produces heat. This energy lost as heat due to hysteresis is called hysteresis loss.

Laminar and Turbulent Flow

Laminar flow is the orderly movement of a viscous liquid (or gas) without interlayer mixing, with flow velocity decreasing from the center of the pipe to its walls. Turbulent flow is chaotic fluid (or gas) movement with random particle motion along complex paths and almost constant velocity across the cross-section.

Example of application: Cooling systems.

Pascal's Law

Pressure in liquids and gases is transmitted equally in all directions.

Examples of application:

- Cooling systems.
- Hydraulic protection of submersible motors using barrier liquids.

Gyroscopic Effect

Bodies rotating at high speeds can maintain the position of their rotation axis unchanged. Any external force applied to change the direction or angle of the rotation axis results in the gyroscope's precession, which is proportional to the applied force.

Example of application: The rotor of an electric machine can act as a gyroscopic element.

Example of negative impact: Complicating the system for maintaining the stable position of the rotor with electromagnetic bearings.

Capillarity

The spontaneous movement of liquid under the action of capillary forces in capillaries and partially open channels (microcracks and scratches).

Examples of application:

- Cooling of active parts of electric machines.
- Impregnation of windings.
- Capillary control in sliding bearings.

Example of negative manifestation: Penetration of moisture into insulation.

Liquids That Solidify in Magnetic Fields

Viscous liquids (oils) mixed with ferromagnetic particles solidify when placed in a magnetic field.

Example of application: Clutches for connecting the electric machine shaft to the load.

Electrical Discharges in Gases and Dielectric Breakdown

Dielectrics can exist in all states of matter: gaseous, liquid, and solid. For instance, air, carbon dioxide, and hydrogen, both in normal and compressed conditions, are used as gaseous dielectrics in practice. Electrical discharges (ED) can occur in any dielectric, being either partial or leading to a complete breakdown of the material.

ED in gas refers to the occurrence of electrical current in a gas due to its ionization and the influence of an electric field. The visible effects and characteristics of these discharges depend on controlling factors such as gas composition, pressure, space configuration, field frequency, and value of current.

A partial discharge (PD) is a localized electrical discharge that bridges only part of the insulation within an electrical insulation system. PD can occur in voids in solid insulation (e.g., paper or polyethylene), multilayer electrode/conductor systems with several layers of solid insulation, in gas bubbles (in the case of liquid insulation), or around electrodes in gas (corona discharge). They often involve air or oil gaps in the insulation.

Dielectric breakdown occurs when the dielectric's resistance collapses, and the material undergoes thermal destruction due to heating in a strong electric field.

Examples of negative impact:

- Breakdown of motor winding insulation due to high humidity or voltage surges.
- Breakdown of transformer oil in a transformer when impurities, such as water, enter the oil.

Static Electricity

The appearance of electric charges on the surface of substances, which can occur even without an external electric field (in pyroelectrics and ferroelectrics due to temperature changes). Exposing a material to a strong electric field, along with cooling or illumination, can create electrets that generate an electric field around them.

Example of negative impact: In motors operating in explosive environments, there's a risk of electrostatic discharge—paint on motor surfaces can attract charges suspended in the air, thus becoming statically charged.

First-Order Phase Transitions

First-order phase transitions involve a change in the physical state of a substance at a specific temperature, accompanied by the release or absorption of energy. These transitions occur both when a substance shifts from one state of matter to another and within the same state.

Examples of use:

- Cooling through evaporation (followed by condensation).
- Cooling with materials that undergo phase transitions.

Examples of negative impact:

- Freezing of lubricant in bearings.
- Melting of windings during short-circuit currents.

Second-Order Phase Transitions

These involve abrupt changes in properties such as heat capacity, thermal conductivity, magnetic properties, fluidity (superfluidity), plasticity (superplasticity), or electrical conductivity (superconductivity) at a certain temperature, without energy exchange.

Example of application: Superconducting windings (at ultra-low temperatures).

Example of a negative manifestation: Loss of magnetic properties when reaching a specific temperature (Curie point).

Magnetic Saturation of Ferromagnets

Magnetic saturation is the state in which a ferromagnet's magnetization reaches a maximum value, called saturation magnetization, which does not increase further even with an increase in the magnetizing field strength.

Example of application: Saturable current transformers used in differential current protection systems.

Example of a negative manifestation: Saturation of the steel core in magnetic circuits.

Shape Memory Alloys

Certain alloys (e.g., nickel-titanium) that have been deformed by mechanical forces can return to their original shape after heating, often producing significant force in the process.

Example of application: Nitinol motor—a motor based on the shape memory effect of the nickel-titanium alloy Nitinol, which restores its shape after being deformed when heated to a red-hot temperature.

Dielectric in an Electromagnetic Field

In a dielectric placed in an electromagnetic field, part of the energy is converted into heat. Heat is generated in the dielectric in both constant and alternating electric fields, but dielectrics can heat up particularly strongly in high-frequency fields. The magnitude of losses depends on the structure of the dielectric, the type of polarization, the presence of defects, impurities, and temperature.

A negative manifestation: Losses in the insulation of electrical machines.

Faraday Cage

A device invented by the English physicist and chemist Michael Faraday in 1836 to shield equipment from external electromagnetic fields. It is usually a cage made of a highly conductive material.

Examples of application: The body of a machine acts as a Faraday cage, shielding objects around the machine from electromagnetic fields generated by the machine's winding. This ensures, at least partially, electromagnetic compatibility.

Ferromagnetic Resonance

Selective (frequency-dependent) absorption of energy from an electromagnetic field. The frequency changes depending on the field intensity and temperature variation.

A negative manifestation: A voltage transformer and a power transmission line in an electrical system can form a resonant circuit, causing ferromagnetic resonance, which threatens the safety and stability of the electrical system.

Electroosmosis

The movement of liquids or gases through capillaries, solid porous diaphragms, and membranes, or through very fine particles under the influence of an external electric field.

Example of application: Electroosmotic drying of winding insulation.

Triboelectricity

The electrification of objects through friction. The magnitude and sign of the charge depend on the condition of the surfaces, their composition, density, and dielectric permittivity.

Example of application: Triboelectric nanogenerators for charging portable electronics using energy generated from everyday mechanical movements of the human body.

Other Effects That May Occur in Electric Machines.

Ettingshausen effect: the occurrence of a temperature difference in a direction perpendicular to the magnetic field and the current.

Nernst effect: the occurrence of an electric field during transverse magnetization of a conductor, perpendicular to the magnetic field and the temperature gradient.

Induced charges: the generation of charges on a conductor under the influence of an electric field.

Thomson effect (Magnetoresistive effect): a change in the conductivity of a ferromagnetic conductor in a strong magnetic field.

Barkhausen effect: a step-like behavior in the magnetization curve of a sample near the Curie point when temperature, elastic stresses, or external magnetic fields change.

Meissner effect: the expulsion of the magnetic field from a superconductor when it transitions to a superconducting state.

Contact potential difference (Volta's law): the occurrence of a potential difference when two different metals come into contact. The magnitude depends on the chemical composition of the materials and their temperature.

Piezomagnetism: the appearance of a magnetic moment under the influence of elastic stresses.

Hopkinson effect: an increase in magnetic susceptibility when approaching the Curie temperature.

Electrocaloric effect: an increase in the temperature of a substance when an electric field is applied to it and a corresponding decrease in temperature when the field is turned off under adiabatic conditions.

Some Chemical Effects

Polymerization

The process of forming a high-molecular-weight substance (polymer) by repeatedly attaching molecules of a low-molecular-weight substance (monomer, oligomer) to active centers in the growing polymer molecule.

Curing

A process in which liquid reactive oligomers and/or monomers irreversibly transform into solid, nonmelting, and insoluble cross-linked polymers. Example of polymerization and curing: Impregnation and casting of electric machine windings to achieve high mechanical and electrical strength, moisture resistance, low thermal expansion coefficients, and thermal conductivity.

Corrosion

The spontaneous destruction of metals and alloys as a result of chemical, electrochemical, or physicochemical interaction with the environment. Rust is typically the product of corrosion in iron and its alloys, such as steel, although many other metals also corrode.

Example of negative manifestation: Rusting of the electric machine housing.

Oxidation

A chemical reaction in which a substance combines with oxygen.

Example of negative manifestation: Oxidation of terminal connections.

Insulation Aging

An irreversible process involving changes in the structure and chemical composition of insulation under the influence of vibrations, friction, temperature, humidity, and other factors. Insulation aging is characterized by a decrease in insulation resistance, an increase in dielectric losses, and a reduction in electrical strength.

Lubrication Aging in Bearings

During friction and oxidation, the mechanical and chemical properties of the grease change. The type of oxidation depends on the operating conditions—physical aging dominates at lower temperatures and high speeds, while chemical aging prevails at higher temperatures. Physical aging leads to changes in rheological properties, resulting in leaks, worsening oil separation characteristics, and reduced ability to replenish grease in the contact zone. Chemical aging mainly results from oxidation. Antioxidants slow this process, but once depleted, oxidation leads to the loss of lubricating material due to conversion into volatile products and lacquer, which can no longer lubricate the bearing.

Some Physico-Chemical Effects

Combustion

The process of converting initial substances into combustion products through exothermic reactions, accompanied by the intense release of heat.

Example of a negative manifestation: Ignition of electrical machine as a result of insulation breakdown.

Explosion Phenomenon

The ignition of substances due to their instantaneous chemical decomposition, resulting in the formation of highly heated gases, accompanied by a loud sound, significant energy release (thermal, mechanical), and a bright flash of light.

Example of application: Explosive cladding (the application of coatings).

Electrolysis

A process in which the components of dissolved substances, or other substances resulting from secondary reactions at the electrodes, are deposited on the electrodes. This occurs when an electric current passes through a solution or molten electrolyte.

Example of application: Insulation monitoring during drying.

Appendix H: Dictionary of Some Terms Used

Technosphere: the part of the biosphere transformed by technical means for socio-economic purposes.

Electromechanics: a branch of electrical engineering that deals with the general principles of electromechanical energy conversion and their practical application in the design and operation of electric machines.

Engineering: a field of technical activity that includes a wide range of specialized areas and disciplines, aimed at the practical application of scientific, economic, social, and practical knowledge to utilize natural resources for the benefit of humanity.

Technology

1. A general term for complex devices, mechanisms, and systems. It can also refer to methods, processes, and ways of performing structured skillful activities.
2. A collection of methods and tools to achieve a desired outcome, broadly meaning the application of scientific knowledge to solve practical problems.

Paradigm: a set of concepts and thought patterns.

Concept

1. A speculative system expressing a certain way of presenting, understanding, and interpreting objects, phenomena, or processes; a set of interconnected views forming a unified system.
2. The unity of essential properties and relationships of objects reflected in thought.

Metaphysics: a branch of philosophy that studies the fundamental nature of reality, the world, and being as such.

Reality: the objectively manifested world; a fragment of the universe that constitutes the subject area of a particular science; objectively existing phenomena and facts.

Being: existence.

World (Universe): the material world (observable); the speculative (philosophical) world.

Truth: a thought corresponding to objective reality (independent of the observer). In science, truth has a dual characteristic. On one hand, it is traditionally the goal of scientific knowledge, while on the other, it represents an independent value ensuring the possibility of scientific knowledge coinciding with objective reality, or at least being a set of basic solutions to theoretical and practical problems.

Idea: a mental prototype of an action, object, phenomenon, or principle, highlighting its main and essential features.

Innovation: a newly introduced or significantly improved product (goods or services) or process.

References

1. A.V. Matveev, The new machine constant and estimation of weight and dimensions of electric machines. Elektrichestvo **11** (2019)
2. E. Arnold, *Die Gleichstrommaschine, Zweite Band* (Springer, Berlin, 1903)
3. A.V. Ivanov-Smolensky, *Electrical Machines* (Mir Publishers, Moscow, 1982)
4. D.C. Hanselmann, *Brushless Permanent-Magnet Motor Design*, 1st edn. (McGraw-Hill, New York, 1994)
5. J. Pyrhönen, T. Jokinen, V. Hrabovová, *Design of Rotating Electrical Machines* (John Wiley & Sons, New York, 2008)
6. J.R. Hendershot, T.J.E. Miller, *Design of Brushless Permanent-Magnet Motors* (Magna Physics Pub, Madison, 1994)

A. Matveev, *Shaping the Future of Electric Machines*,
https://doi.org/10.1007/978-3-031-95248-7

Index

A. Matveev, *Shaping the Future of Electric Machines*,
https://doi.org/10.1007/978-3-031-95248-7

www.ingramcontent.com/pod-product-compliance
Lightning Source LLC
Chambersburg PA
CBHW050435190825
31302CB00003B/30
9783031952470